Location in Space

Location in Space

Theoretical Perspectives in Economic Geography

THIRD EDITION

Peter Dicken
University of Manchester

Peter E. Lloyd
University of Liverpool

HarperCollins*Publishers*

Project Editor: Susan Goldfarb
Art Direction/Cover Coordinator: Heather A. Ziegler
Text Art: Textmark
Cover Design: Wanda Lubelska Design
Cover Photo: H. Armstrong Roberts
Production Manager: Jeanie Berke
Production Assistant: Paula Roppolo

LOCATION IN SPACE: Theoretical Perspectives in Economic Geography,
Third Edition

Library of Congress Cataloging-in-Publication Data

Lloyd, Peter E.
    Location in space : theoretical perspectives in economic geography
    / Peter E. Lloyd, Peter Dicken.—3rd ed.
        p.   cm.
    Includes bibliographical references.
    ISBN 0-06-041677-7
    1. Geography, Economic.   2. Space in economics.   3. Industry—
Location.   I. Dicken, Peter.   II. Title.
    HF1025.L584   1990
    330.9—dc20   89-28665
                                                              CIP

90 91 92 93 9 8 7 6 5 4 3 2

Contents

Part Two / Location in Space: Alternative Perspectives   253

Preface

It is now 12 years since the second edition of *Location in Space* was published and fully 17 years since the book's original appearance. Judged by its continued use in many courses, the book—especially its particular structure and approach—continues to meet a need. But it was showing its age in a world that had emphatically not stood still since the last revision. Indeed, many observers would argue that the 1980s witnessed particularly rapid change in the geography of economic activity at all scales of analysis. At the same time, the concerns and the approaches adopted by economic geographers have changed substantially themselves. In the second edition, we tried to incorporate some of the early changes in chapters that dealt with the vagaries of locational decision making and the spatial behavior of multiplant enterprises. The aim then was to incorporate additional variation and complexity into the classical and neoclassical model-based framework that continued to form the core of the book.

During the 1970s, the value of the neoclassical model-based approach in human geography came under increasing attack, initially from the behaviorists and corporatists but more sweepingly from Marxist and Marxist-influenced writers. Such critiques reflected, in part, the immense economic changes—often catastrophic in their effects on local communities—that gathered momentum from around 1973 on. The critiques certainly had more impact under such external conditions. While economic geography students were grappling with flat plains populated by uniformly distributed populations and occupied by small factories operating under conditions of perfect competition, the world around those students came to appear increasingly remote from such abstract notions. To some writers, the response was to reject that approach entirely—geographers have always had a strong tendency to respond to change by throwing away old ideas and concepts only to find that they had perhaps gone too far (to "throw Weber out with the bath water," as one writer has observed).

There was a real need to adopt an approach that was more explicitly consistent with a world made up of massive multilocational and often multinational firms, a world of globally expanding production systems made possible in part by globally shrinking technology (though still driven by the same imperative of profit). Most of all, there was a pressing need to place geographic space, especially the location decision, in a more meaningful context. It has now become accepted—more explicitly than before—that the act of locational choice in a modern capitalist economic system is essentially an *investment* choice. As such, it has to be seen from the much wider perspective of why as well as how such choices are made. Geographers had tended to focus their attention on the specifically spatial act itself, often forgetting the prior questions for all would-be business owners: whether to invest and what to invest in, both of which have to be addressed before the question of where to invest emerges. Geographers, then, were rather naive about the realities of the production process, so bound up had they become with space and place. Equally, while recognizing the realities of competition and exchange in the simplified neoclassical model (with its uncomplicated rules), they had failed to translate these into the complicated realities of the modern world.

But in trying to get a better understanding of the complex and volatile world in which we live, it is a mistake to assume that the simplistic approach of the neoclassical model is worthless. It is merely one way of viewing and understanding this complex world. Indeed, this approach originally gave economic geography its first great step forward from mere description into theory. It has considerable merit in the rigor of its analytic approach. It enables us to "creep up on complexity" one variable at a time. As such, the model has particular value for the geographer, who has not only to try to understand the complex mechanics of economy and society but also to try to understand the working of such systems in space and time. One way of learning about a complex spatial world, then, is to adopt a modeling approach whereby we attempt to understand the basic parameters of the economic problem in capitalist societies, adding complexity in a controlled way to analyze the impact of each additional factor. In this third edition of *Location in Space*, therefore, we retain the essence of this model-based approach in a new Part One, which consists of revised and slightly modified versions of what were Chapters 2 through 7 plus Chapter 10 in the second edition. The focus of Part One is explicitly spatial; it is concerned with locational factors and locational choices in a simple world. In Part One, therefore, we try to retain the best of the still valuable insights of neoclassical location theory.

Part Two of the third edition is entirely new. In these four newly written chapters, we move from the simplistic world of Part One to a more realistic look at the organization of production and the process of competition in the contemporary capitalist world, pulling together insights that have been emerging in the literature of economic geography. Some of that literature has consisted of Marxist critiques of approaches to understanding; some has come from the literature of business. Our purpose in Part Two is to use insights from a variety of sources to rework our understanding of location in space from the specific viewpoint of the organization of production under modern competitive conditions.

Our aim is the same as it was in the second edition: to introduce students easily and in a readable way to major concepts and principles. No previous knowledge of either geography or economics is required; concepts and terminology are defined and spelled out in nontechnical language. A large number of graphic illustrations form a key part of the presentation. Above all, our aim is to keep what was generally regarded as good and useful in the second edition and to introduce a number of new ideas that address more directly the world with which modern students of economic geography are concerned. We believe that the revised version of the book will be highly relevant both to geography majors and to the many students of business who take some geography courses. In particular, the new material in Part Two should be attractive to such students.

PETER DICKEN
PETER E. LLOYD

Acknowledgments

The first and second editions of *Location in Space* represented bursts of frantic activity on top of two busy academic lives. To have attempted a third edition was perhaps asking too much of all those who have stood by us and supported our efforts while we have become even busier in day-to-day academic affairs. It could not have been contemplated without the willing and always professional support of two members of the Manchester University Department of Geography. Jean Woodward word-processed the manuscript draft with all her usual skill and complete dedication. Nick Scarle, who lived through the second edition, was with us again to produce artwork for the third. We are immensely grateful to both of them.

Between editions both authors had their separate professional interests and groups of professional colleagues and graduate students to provide the impetus for new directions. Peter Lloyd wishes specifically to acknowledge the help of research assistants and associated graduate students in the North West Industry Research Unit, among whom John Shutt, Jamie Peck, Graham Haughton, and Richard McArthur were driving intellectual forces during ten exciting years. Graduate students were always a further stimulus—pushing and inquiring, forcing a sharpening of ideas and concepts. Peter Dicken wishes to thank Adam Tickell for helping to nudge him along particular academic routes, as well as all those colleagues in North America and Europe who have made suggestions for the third edition and have been a continuing source of intellectual stimulus. As always, of course, none of those named or unnamed are responsible for any deficiencies in the book. Every effort has been made to contact owners of copyrighted materials, and due acknowledgment is given in the text to the original sources.

We would also like to thank the following reviewers of the manuscript for this third edition for their comments and suggestions:

Dr. Edward Taafe, Ohio State University
Dr. Luc Anselin, University of California, Santa Barbara

Professor Howard Botts, University of Wisconsin—Whitewater
Professor Ruth York, University of Nebraska at Lincoln
Professor Richard Preston, University of Waterloo (Ontario)

Above all, we both have to acknowledge the debt to our families. Valerie and Ros kept us going, and our debts to them are too large to calculate. Our now adult children seem to have survived relatively unscathed by the pressures of ever-too-busy, always-writing fathers. Michael and Christopher, Jennifer and Murray have all launched their own lives and careers, and we take this opportunity to thank them and to wish them well in the future.

PETER DICKEN
PETER E. LLOYD

Location in Space

Introduction: The Perspective of Economic Geography

Most people today have at least a general idea of what an economist does, although they might be hard-pressed to provide a precise definition. But the same cannot be said of many people's perception of what an *economic geographer* does. All of us who ply our trade as economic geographers are often asked, "What exactly does an economic geographer do?" or "How does economic geography differ from economics?" Providing an answer to such questions is important not merely as an exercise in self-justification or to keep economic geographers out of the unemployment statistics; it is important because it should help to demonstrate that economic geography has a distinctive and significant role to play in comprehending the complex world in which we live and, indeed, in helping to make it a better place. The particular perspective of the economic geographer has long made a valuable contribution to many areas of business and government, and it is important that it should continue to do so. One of the aims of this introductory chapter is to answer the questions posed above. The other aim is to outline the kind of approach taken in this book—to map the route we plan to follow.

Both economic geographers and economists share a common interest in certain kinds of subjects. Both are interested in the *production, distribution, and consumption of wealth* (primarily material wealth) in society. But al-

1

though both disciplines share a common interest in these processes, they are distinguished from each other by distinctive perspectives and viewpoints. However, before discussing the specific perspective of the economic geographer, we need to know a little about economic systems in general.

WHAT IS AN ECONOMIC SYSTEM?

The world economy is made up of a number of economic systems. Each system represents a different way of trying to solve a fundamental problem that all societies face, whatever their political complexion or level of development. The fundamental—indeed, universal—problem is how to provide for the *material needs and wants* of a society's population. Different types of economic systems produce different answers to this question, and in so doing, they produce *different geographies*.

Of course, the most basic and fundamental material needs of any human population are those necessary to ensure *survival*: food, shelter, clothing, and the like. But beyond these basic needs is a whole spectrum of *wants*—items whose consumption is more discretionary than obligatory. The problem is that above a certain basic level, such as the minimum calorific requirements to sustain life, the distinction between needs and wants becomes blurred. What is regarded as basic in one society or social group may be an unattainable luxury in another. Above a very fundamental level, therefore, the distinction between needs and wants becomes *relative* rather than absolute. As the economist Lester Thurow observed, "Wants become necessities whenever most of the people in a society believe they are, in fact, necessities." For example, in a society where no public transportation is available, private transportation (such as an automobile or a bicycle) becomes a necessity.

At the base of all economic systems lie two fundamental considerations:

1. Very few of an individual's basic needs and wants can be satisfied by individuals acting alone; they require some form of *social collaboration* with others.
2. Beyond a very simple level, such joint efforts need to be *organized, coordinated, controlled,* and *regulated*.

An economic system, therefore, is a particular form of *social organization* whose purpose is to provide for the *material* needs of society.

BASIC DECISIONS IN ALL ECONOMIC SYSTEMS

Certain basic decisions have to be made in all economic systems, regardless of their specific nature. What distinguishes economic systems from each other is how such decisions are made. The key questions are these:

1. What kinds of goods and services should be produced, in what quantity and quality.

2. How are such goods to be produced, given the availability of resources and factors of production? In other words, how much of each resource or factor should be applied to the production of particular goods?
3. At what relative values should goods be exchanged? In money-based economies, the exchange economy is expressed in terms of the relative prices of goods, hence the question becomes, At what price shall various goods be sold?
4. How should the total product be shared (distributed) among the population? In other words, who gets what?
5. How should the geographic (spatial) pattern of production be organized?

TYPES OF ECONOMIC SYSTEMS

At first sight it might seem that the "economic problem" could be solved in myriad ways. In fact, the solutions that have been adopted during human history can be boiled down to three major strategies (used in various combinations). Each strategy has the necessary ingredients for solving the economic problem: first, a means of mobilizing effort; second, techniques for allocating effort; and third, a means of distributing output in such a way as to permit further production to take place. The three solutions can be summarized briefly as follows in terms of their controlling or coordinating mechanism.

1. Tradition The classical solution to the economic problem grew out of the slow evolution of methods of coping with nature's scarcity and humanity's increasing aspirations. Tried and tested practices from the past were retained and built into the fabric of society by becoming intertwined with laws, customs, and beliefs. Deviations from accepted practice were suppressed by heavy sanctions and taboos.

The production problem—that is, the allocation of resources to various uses—was solved by recourse to tradition, to what had been done before. In the case of work allocations, the traditional solution simply assigns children to the jobs of their parents, thus ensuring a continuation of the necessary social fabric over time. Similarly, kinship and family links are applied to the solution of the distribution problem with the fruits of production allocated by rank and status in the social hierarchy. This is no less an economic system than the ones with which the modern Westerner is familiar.

2. Command This strategy for solving the economic problem is widely applied in the present day, particularly in the Communist countries of the Soviet Union, Eastern Europe, and China. It is a quasi-military solution under which allocations and assignments are made by an economic "commander in chief" or by some designated group in society. Decisions are made in accordance with some more or less conscious objectives set out by those in authority. Those in positions to command may be either elected or self-appointed. Their programs for solving the economic problem may be based on social justice or self-interest or some particular combination of both.

Regardless of the moral or philosophical basis of the command system, it provides yet another way of organizing human activity to produce a society's material needs. The production problem is solved by direct allocation of people and resources to particular tasks and the distribution problem by some conscious assignment of the fruits of labor by some precepts determined by the group in authority.

3. The Market This is the kind of economic system with which you are probably most familiar. Yet its very familiarity may mean that its basic characteristics are taken for granted rather than understood. Since this kind of economic system forms the focus of this book, we need to devote close attention to how such a system operates. Robert Heilbroner (1972) provides a particularly illuminating "parable":

> Suppose, for instance, that we were called on to act as consultants to one of the new nations emerging on the continent of Africa.
>
> We could imagine the leaders of such a nation saying, "We have always experienced a highly tradition-bound way of life. Our men hunt and cultivate the fields and perform their tasks as they are brought up to do by the force of example and the instruction of their elders. We know, too, something of what can be done by economic command. We are prepared, if necessary, to sign an edict making it compulsory for many of our men to work on community projects for our national development. Tell us, is there any other way we can organize our society so that it will function successfully—or better yet, more successfully?"
>
> Suppose we answered, "Yes, there is another way. Organize your society along the lines of a market economy."
>
> "Very well," say the leaders. "What do we then tell people to do? How do we assign them to their various tasks?"
>
> "That's the very point," we would answer. "In a market economy, no one is assigned to any task. In fact, the main idea of a market society is that each person is allowed to decide for himself what to do."
>
> There is consternation among the leaders. "You mean there is no assignment of some men to mining and others to cattle raising? No manner of designating some for transportation and others for weaving? You leave this to people to decide for themselves? But what happens if they do not decide correctly? What happens if no one volunteers to go into the mines, or if no one offers himself as a railway engineer?"
>
> "You may rest assured," we tell the leaders, "none of that will happen. In a market society, all the jobs will be filled because it will be to people's advantage to fill them."
>
> Our respondents accept this with uncertain expressions. "Now look," one of them finally says, "let us suppose that we take your advice and allow our people to do as they please. Let's talk about something specific, like cloth production. Just how do we fix the right level of cloth output in this 'market society' of yours?"
>
> "But you don't," we reply.
>
> "We don't! Then how do we know there will be enough cloth produced?"
>
> "There will be," we tell him. "The market will see to that."

"Then how do we know there won't be *too much* cloth produced?" he asks triumphantly.

"Ah, but the market will see to that too!"

"But what is this market that will do these wonderful things? Who runs it?"

"Oh, nobody runs the market," we answer. "It runs itself. In fact, there really isn't any such *thing* as 'the market.' It's just a word we use to describe the way people behave."

"But I thought people behaved the way they wanted to!"

"And so they do," we say. "But never fear. They will want to behave the way you want them to behave."

"I am afraid," says the chief of the delegation, "that we are wasting our time. We thought you had in mind a serious proposal. What you suggest is inconceivable. Good day, sir." (pp. 25–27)

In fact, this is precisely how the "pure" market system works. Unlike the command system, in this system there is no direct control of either production or distribution. Economic life is governed by what has been called the "invisible hand" of the market. Decisions to produce or consume are dispersed, not centralized. The key to the operation of the market system is the *price* set by the market, which is determined by the interaction between *supply* and *demand*. What is produced and in what quantity depends, theoretically, on what consumers are prepared to buy. This in turn depends on the price asked by the producer and on what the consumer can afford. The end result of all these uncoordinated decisions is—again in theory—a balance, or *equilibrium*, between supply and demand.

Figure I.1 shows how this process works. The price of a particular good is plotted on the vertical axis of the graph; the quantity produced of the same good is plotted on the horizontal axis. From a producer's point of view, the higher the price customers are prepared to pay, the more the producer will be prepared to produce; conversely, the lower the price, the less the producer will be willing to produce. This relationship is shown as the upward-sloping *sup-*

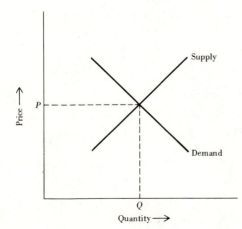

Figure I.1 The central feature of a market economy: the equilibrium relationship of demand, supply, and price.

ply curve in Figure I.1. The reverse relationship between price and quantity applies to the consumer. When the price is high, less of the good will be demanded; when the price is low, more will be demanded. This relationship is shown as the downward-sloping *demand curve* in Figure I.1. If we assume that the supply curve represents the preferences of all suppliers of a good and that the demand curve represents the preferences of all customers of the good, then the *market price* will be the point on the price axis P where the supply and demand curves intersect. But this will also fix the quantity of production, Q on the horizontal axis. The system is in balance or equilibrium. Whether or not such an equilibrium is either attained or maintained is less important for our argument at this stage than the fact that the pure market system operates as if it is.

A particular type of market system is the *capitalist system*. A capitalist system is an economic system in which not only is the allocation of resources determined by the price system (as in Figure I.1) but also in which the *means of production are privately owned*. The "great engine" of the capitalist system is the *profit motive*, whereby individual producers strive to sell their products for more than it costs to produce them. Profit, therefore, is the difference between the revenue (R) that a producer receives from selling its products and the cost (C) of manufacturing and distributing them to its customers:

$$P = R - C$$

The capitalist market economic system prevails throughout most of the industrialized world outside the command economies and the few traditional or subsistence economies that still survive. But it does not operate in the pure, unfettered manner we have described. Most of the industrialized (and industrializing) economies are "impure" capitalist market systems in which, for various reasons, governments intervene. Most commonly, such intervention occurs for "social" reasons—to offset or reduce the detrimental effects of the operation of a pure market system. While it is certainly true that the capitalist market system has shown itself to be highly efficient in *creating* wealth (though with an inbuilt tendency toward cyclic fluctuations—booms and slumps), it is less successful in *distributing* wealth. It produces highly uneven economic development both vertically (between different social groups) and horizontally or spatially (between different places).

THE PRODUCTION SYSTEM

At its simplest, the production process consists of transforming inputs (such as raw material) into outputs (finished products), as Figure I.2 shows. The production system, however, consists of much more than simply converting materials into products. As Figure I.3 shows, the production system consists of five major components or processes:

1. The production process itself: the transformation of materials into finished and semifinished (intermediate) products

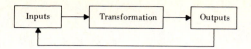

Figure I.2 Basic elements of the production process.

2. The circulation process: the intermediary activities that connect the various parts of the system—for example, transportation and communications systems, financial systems, and other producer-service systems (such as advertising)
3. The distribution process: the activities that make goods and services available to the final consumer—for example, retailing
4. The regulation process: the various ways in which the system is controlled—for example, laws, regulations in business activity
5. Final demand or consumption

These five components are tied together in complex ways such that any major change in one component will have repercussions on the others. We may choose to study the system as a whole, or we may choose to focus on one element. But we must always keep in mind the overall interconnectedness of the component parts. That is what is meant by the term *system*.

THE ECONOMIC GEOGRAPHER'S PERSPECTIVE

Everything we have said so far would be regarded as familiar—even proprietary—territory by economists. What, then, is the economic geographer's role in the understanding of economic systems? Fundamentally, the economic geographer is concerned with the *spatial organization* of economic systems: with *where* the various elements of the system are located, *how* they are connected together in space, and the *spatial impact* of economic processes. In contrast, most economists are relatively unconcerned with spatial questions. Indeed, as one (maverick) economist has observed, most economists have

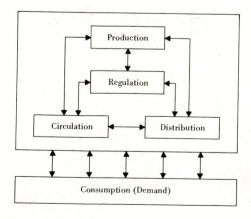

Figure I.3 Organization of the production system. *Source:* After Bailly, Maillat, and Rey (1987), fig. 1.

inhabited a "spaceless wonderland" and behaved as if all human activity takes place on the head of a pin!

Thus the distinctive perspective of the economic geographer is spatial. But this should not mean that the role of the economic geographer is to separate out space and treat it as some unique and distinct object of study in its own right. On the contrary: space cannot and should not be treated in this way. One of the criticisms of much of the economic (and other human) geography of the 1960s was that it tended to do just that. As Doreen Massey (1985) points out:

> Geography set itself up as "the science of the spatial." There were spatial laws, spatial relationships, spatial processes. There was a notion that there were certain principles of spatial interaction which could be studied devoid of their social content. . . . There was an obsession with the identification of spatial regularities and an urge to explain them by spatial factors. The explanation of geographical patterns, it was argued, lay within the spatial. There was no need to look further. . . . [However,] this is an untenable position. . . . There are no such things as purely spatial processes; there are only particular social processes operating over space. (p. 11)

Unfortunately, in making these kinds of criticisms of the geography of the 1960s, many critics went too far in the other direction. In denying the special position of space as an object of study, they tended to throw space out altogether and to infer that space—and place—did not matter at all. Fortunately, that extreme position is rapidly disappearing, and we are beginning to see the relationships between the spatial and the social (including the economic) more clearly.

In adopting a spatial perspective, then, modern economic geographers pose three basic, interconnected questions:

1. In what ways are economic activities organized spatially on the earth's surface, and how do such spatial forms or patterns change over time?
2. Why are economic activities organized spatially in particular ways; that is, what are the underlying processes at work?
3. How does the spatial organization of economic activities itself influence economic and other social processes?

These three questions are summarized diagrammatically in Figure I.4. The diagram emphasizes the fact that the spatial organization of economic ac-

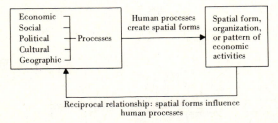

Figure I.4 The spatial perspective as interplay between process and form.

tivities—the spatial patterns we observe on maps and satellite photographs—is the result not merely of "geographic" processes such as distance and direction but also of processes operating in the economic, social, political, and cultural spheres. In other words, although economic geographers ask predominantly spatial questions, they need to seek answers to such questions in both the aspatial as well as the spatial domain. But Figure I.4 also indicates that the particular spatial organization of economic activities itself exerts an influence—often a very important one—on economic and other human processes. In other words, while it is true that economic geographers need to be familiar with economic processes, it is also true that economists need to be aware of the influence of geographic space and spatial organization. Hence the approaches of economic geographers are complementary but distinctive.

ALTERNATIVE EXPLANATIONS

We have established that the modern economic geographer is concerned with explaining the spatial organization of economic activities. That is the common ground. But such explanation may be approached in a variety of ways. In this book, our focus of attention is the capitalist market economic system. But we do not attempt to provide a single explanation of the spatial organization of such an economic system. Rather, we provide several alternative explanatory perspectives, each of which throws a rather different light on the same basic phenomenon.

Why do we choose to do this rather than opt for a single explanatory approach? One reason is that we believe that we can actually learn from different viewpoints or perspectives. It is too easy to become trapped into interpreting the facts we observe about the world through one particular set of theoretical spectacles. Once a particular perceptual position is fixed in the mind, it becomes difficult to appreciate alternative approaches. Figure I.5 illustrates this point in a rather simplistic way. The drawing is part of a sketch by Henri de Toulouse-Lautrec that the geographer William Kirk used to develop his view of the behavioral environment. Kirk's (1963) description of the sketch is appropriate for our point as well:

> To most of us, for reasons which need not detain us here, this is a sketch of the head and shoulders of a young Parisienne, complete with neck ribbon. But if the neck-band is seen as a mouth, the pattern changes and reconstitutes itself as the head of a hook-nosed old crone. The patterns are mutually exclusive, but can be changed at will to give the impression of movement. Yet once established, each picture resists change and has a certain momentum. (p. 366)

The same can be said of the adoption of alternative theoretical positions: adoption of one viewpoint often precludes appreciation of an alternative. Over the years, geographers have been especially prone to throwing out concepts that are no longer fashionable. Sometimes this is no doubt justified, but sometimes we can often continue to learn from the knowledge of a variety of

Figure I.5 Alternative perceptual interpretations of the same basic factual information.

theoretical perspectives. Certainly, students of economic geography should be aware of different approaches.*

In this book, we explore three broad approaches to the explanation of the spatial organization of economic activities in capitalist market systems. We begin with the approach of *neoclassical location theory,* which originated in the work of location theorists working in the nineteenth and early twentieth centuries but whose full development was a product of the 1950s and 1960s. As David Smith (1987) has pointed out, neoclassical location theory is not merely a branch of neoclassical economics, although it shares some common features, notably its normative approach. Normative theory is concerned with what should happen, given certain assumptions, rather than with what may actually happen in reality. To that extent, it is an "artificial" approach based on a set of rigid assumptions. In Chapter 1, the neoclassical approach is used in a specifically logical manner whereby all variables except one (distance) are initially held constant to create a simplified model of the spatial organization of economic activities. Then, in the succeeding chapters of Part One, each of the restrictive assumptions is relaxed to move the model closer to reality.

The neoclassical model has been heavily criticized in recent years, but it is wrong to dismiss it out of hand. Indeed, the simplistic approach of the neoclassical model is one way of understanding our complex world. It enables us to observe in some depth, and without the interference of too much additional complexity, the operation of very important economic processes. Indeed, this

---

* R. J. Johnston (1983a, 1983b) provides brief and very readable introductory surveys of the major alternative approaches and philosophies in human geography as a whole.

approach originally gave economic geography its first great step forward from mere description into theory. It has considerable merit in the rigor of its analytic approach. It enables us to "creep up on complexity" one variable at a time. As such, the model has particular value for the geographer who has not only to try to understand the complex mechanics of economy and society but also to try to understand the working of such systems in space and time. One way of learning about a complex spatial world is, then, to adopt a modeling approach in which we attempt to understand the basic parameters of the economic problem in capitalist societies, adding complexity in a controlled way to analyze the impact of each additional factor.

We would argue, therefore, that the neoclassical location theory approach remains very valuable but clearly has limitations.One alternative viewpoint, developed in the first two chapters of Part Two, is a *behavioral-organizational approach* to understanding the spatial organization of capitalist market systems. This explanatory perspective came into being in the second half of the 1960s largely as a reaction to what were seen to be unrealistic behavioral assumptions in neoclassical location theory. Initially, behavioral approaches focused on the nature of individual decision-making behavior, particularly locational decision making. As such, it tended to run into a blind alley. More recently, largely through the efforts of geographers such as Michael Taylor, it has taken on a new lease of life. As Chapters 7 and 8 show, this perspective now emphasizes the key importance of corporate organizations and their interrelationships, which reflect different degrees of power. In Chapters 7 and 8, we stress the increasingly dynamic and volatile nature of the external environment in which today's business organizations exist and the competitive strategies that firms adopt in their struggle to survive. This approach crosses the disciplinary boundaries with business studies and management sciences. But even though these chapters differ in approach from the neoclassical chapters, they do make use of some of the concepts developed within the neoclassical framework.

This same basic notion applies to the third explanatory approach adopted in this work, the *structural approach*, based on Marxist and neo-Marxist writing, the topic of Chapters 9 and 10. At the very least, an understanding of the Marxist-structuralist perspective is greatly enhanced by a prior understanding of the alternative perspectives. The structuralist critique of neoclassical and of behavioral locational analysis emerged in the late 1960s and early 1970s and gained impetus from the severe economic dislocations that swept the industrialized countries from the mid-1970s on. The argument was that economic change in general—and, therefore, spatial change—could only be understood by analyzing not the decisions of individual firms but the underlying structure of capitalist society. This involves a recognition of the social relations of production (the contradictory relationship between capitalists, who own the means of production, and labor). Most important, the emphasis of leading "structuralist" writers in economic geography, such as Doreen Massey, Richard Walker, and Michael Storper, was on the need to understand the production process itself rather than to examine individual location factors.

Parts One and Two of this book, therefore, present three alternative explanatory perspectives toward understanding the spatial organization of capitalist market economies. Some experts would argue that the differences between these alternatives are so great as to be totally incompatible. Others would insist that we can, indeed, learn something from each, that each throws a different light on what is essentially the same system. Some of these issues of complementarity versus contradiction are explored very briefly in the Postscript that brings the book to a close. Note, finally, that even though our emphasis in this book is essentially theoretical rather than descriptive, the theories and concepts themselves are applied to the specific, concrete situation of modern capitalist market economies. They should not be regarded as universals, independent of a specific context of time and space.

FURTHER READING

Heilbroner, R. L. (1972). *The Economic Problem*, 3d ed. Englewood Cliffs, N.J.: Prentice-Hall.

Johnston, R. J. (1983). *Geography and Geographers: Anglo-American Human Geography Since 1945*, 2d ed. London: Edward Arnold.

Johnston, R. J. (1983). *Philosophy and Human Geography: An Introduction to Contemporary Approaches*. London: Edward Arnold.

Kuhn, A. (1966). *The Study of Society*. London: Associated Book Publishers, chap. 1.

Massey, D. (1984). "Introduction: Geography Matters." In D. Massey and J. Allen (eds.), *Geography Matters: A Reader*. Cambridge: Cambridge University Press, chap. 1.

Massey, D. (1985). "New Directions in Space." In D. Gregory and J. Urry (eds.), *Social Relations and Spatial Structures*. London: Macmillan, chap. 2.

Smith, D. M. (1987). "Neoclassical Location Theory." In W. F. Lever (ed.), *Industrial Change in the United Kindgom*. Harlow: Longman, chap. 2.

PART ONE

Location in Space: A Model-based Approach

A major problem facing any economic geographer attempting to unravel the complexity of the world is deciding where to begin. There is no unique point of entry, no best starting point, simply because so many of the variable factors that interact to produce the economic landscape contribute toward both causes and effects. In everyday language, we face the classic "chicken and egg" dilemma. But as Kenneth Boulding has so aptly pointed out, the egg theory of chickens is every bit as good as the chicken theory of eggs! Thus wherever we break into the circle of complexity, we do so in the knowledge that other points of entry are possible.

This does not mean, however, that our starting point is purely arbitrary, because one clue to an appropriate starting point can be found in the nature of geography itself. As we emphasized in the Introduction, geography is, first and foremost, a spatial discipline. Intrinsically, therefore, one of its central concerns is distance; indeed, one very much used expression is the "friction of distance." This simply means the impediment to movement that occurs because places, objects, or people are spatially separate. Movement involves a cost, whether it is actual payment of money to travel by bus, train, or jet or to transport materials or goods; wear and tear on the soles of the feet; or perhaps the time involved in moving between places. Distance—and especially the cost involved in overcoming it—is clearly a logical starting point for geographic analysis and is thus the one we use here. It is the approach adopted in a particular theoretical perspective, neoclassical location theory.

In Chapter 1, we begin by asking ourselves the question, What kind of spatial pattern of economic activities would we expect to find if the only factor affecting the pattern were the friction of distance? To answer this question, we have to try to isolate distance, separate it from all the many other variables—economic, social, cultural, and even psychological—that together produce the geographic pattern of economic activities that we see around us. Obviously, we can't just go out and "stop the world." But we can use our imagination and assume that for the time being, all these other variables are not variable at all but constants (that is, they are the same everywhere). If we do this, we can deduce the kind of spatial organization of economic activities that would occur if the friction of distance were the only influence.

The model of the economic landscape created in Chapter 1 is orderly and regular. It contains a little of the real world, but it is a long way from the world we actually observe. Many other variables enter the determination of production costs in space and of spatial variations in demand. Having artificially held these constant in Chapter 1, we then progressively relax them one at a time. Factors of production, such as raw materials, capital, and labor, are not available everywhere in uniform quantity and quality, nor are they all infinitely and equally mobile. Transportation costs are also highly variable in space and differ with such variables as the type of transportation media, the nature of the land surface, and the length of journey. Hence, Chapters 2, 3, and 4 present a progressively more realistic analysis of the spatial organization of economic activities under the impact of these important cost variables, both individually and in combination.

Even when the complexities of real cost variations in space are taken into account, however, the locational problem is only partly solved. The minimization of costs is not the sole criterion governing the location of economic activity. Producers may seek to maximize their profit, and to do this, they may attempt to keep their costs as low as possible or to increase their returns. Better still, they may attempt to achieve both. Thus the spatial distribution of demand is a key variable. When demand for a product is high, it becomes possible for a producer to manufacture a larger volume of output. As such volume increases, the cost of producing each item tends to fall; economies of scale are achieved. In Chapter 5, therefore, demand and scale of production are treated together.

The approach adopted in Chapters 2 through 5 is relatively static. In each chapter, a little dynamism is introduced, but no attempt is made to deal more comprehensively with change over time. In Chapter 6, we do precisely that by examining the process of uneven spatial development through such concepts as cumulative causation, economic multipliers, and spatial growth poles.

Chapter 1

Spatial Organization of Economic Activities: A Simplified Model

The essence of any simplified model, as the name suggests, is that it abstracts from the complexity of the real world. In building a simplified model of the spatial organization of economic activities, we make certain assumptions with the aim of concentrating on one variable, geographic distance. The simplifying assumptions made in this chapter are, in effect, the rules of the game, and as with rules for all games, we have to adhere to them for the duration of the game. We can divide our rules or assumptions into two categories: those that relate to the land surface and its characteristics and those that relate to the population living on that land surface. Let us look at each of these in turn.

SIMPLIFYING ASSUMPTIONS

Simplifying Assumptions About the Land Surface

Our basic assumption about the land surface is that it is completely flat and homogeneous in every respect. It is, in technical terms, an *isotropic* plain. *Isotropic* is a Greek word that means "having equal physical properties in all directions." Thus on our isotropic plain are none of the features so dear to the hearts of physical geographers: no mountains or U-shaped valleys, no hills or rivers. Soil fertility is the same everywhere, as is climate. All the raw materials needed by economic activities are also available everywhere and at the same cost.

On such a plain there are no barriers to movement. In fact, we can assume initially that movement can occur in all directions with equal ease and that there is only one type of transportation. Transportation costs—the costs of overcoming the friction of distance—are exactly proportional to distance. As Figure 1.1 shows, this means that it costs exactly twice as much to travel (or to move materials and goods) 100 miles as it does to go 50 miles. We also assume that the plain is limitless or unbounded, so that we do not have to deal with the many complexities that tend to occur at boundaries.

Simplifying Assumptions About the Population

The most important initial assumption about the population is that it is spread perfectly evenly over the plain: that is, the density of population is everywhere the same (see Figure 1.2). We also assume that every one of these people is identical in the sense that each possesses the same financial resources and spends money in the same way. In other words, their demands and tastes are the same. These people are also given quite remarkable mental qualities. We assume that each one is fully aware of all possible behavioral alternatives (that is, they have perfect knowledge) and that they act in a totally rational way. Because we are dealing with economic matters, by such rationality we mean that every producer of goods and services aims to make the largest possible profit and that every consumer aims to satisfy his economic needs as cheaply as possible. All the actions of producers and consumers are therefore guided by the desire to *optimize:* to get the best possible outcome for their efforts. In economic terms, the world is a *perfectly competitive* market in which there are large numbers of producers, none of whom can individually influence the price at which goods are sold.

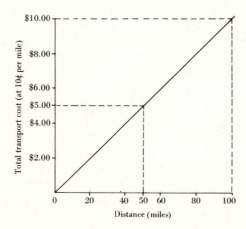

Figure 1.1 Transportation costs increase in direct proportion to distance.

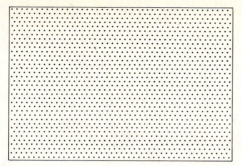

Figure 1.2 An even spatial distribution of population.

In summary, then, the assumptions on which we will begin to build our simplified model are as follows:

1. The land surface is an unbounded plain that is homogeneous in all respects:
   a. The surface is perfectly flat, with no barriers to movement. Movement is therefore possible in all directions.
   b. Transportation costs are proportional to distance, and there is a single uniform transportation system.
   c. Physical resources are evenly distributed; that is, soils are of equal fertility throughout, and raw materials are ubiquitous and of equal cost.
2. The population living on the plain has the following characteristics:
   a. Residents are distributed evenly.
   b. They have identical incomes, demands, and tastes.
   c. Both producers and consumers have perfect knowledge and act perfectly rationally with respect to this knowledge. They are able, therefore, to behave in an optimal fashion. As producers, for example, they are assumed to seek a single goal—the maximization of profits. As consumers, they seek to minimize their outlays in meeting their consumption needs.

It is vital that these assumptions be borne in mind throughout this chapter. They form the basis of a particular approach to understanding our complex world. The merits of such an approach were well argued by one of the founders of location theory, J. H. von Thünen. Writing well over a century ago (1842), he prefaced his analysis of the location of agricultural production in this way:

> I hope that the reader who is willing to spend some time and attention on my work will not take exception to the imaginary assumptions I make at the beginning because they do not correspond to conditions in reality, and that he will not reject these assumptions as arbitrary and pointless. They are a necessary part of my argument, allowing me to establish the operation of a certain factor, a factor whose operation we see but dimly in reality, where it is in incessant conflict with others of its kind.
>
> This method of analysis has illuminated—and solved—so many problems

in my life, and appears to me to be capable of such widespread application, that I regard it as the most important matter contained in all my work.

The "certain factor whose operation we see but dimly in reality" is the friction of distance. By stripping away all the other complicating factors (that is, by using the "other things being equal" formula), we can focus explicitly on the role of distance in the spatial organization of economic activities.

SPATIAL ORGANIZATION OF THE PRODUCTION OF ONE GOOD

One Producer

Suppose that one of the inhabitants of our plain decides to produce a good—say, sausages—in quantities greater than he must for his own needs. Because all the materials required are available everywhere at the same cost, production can be established at home. However, production is small-scale, the price charged for sausages if people come to buy them from the point of production will be the "going rate," or market price, for sausages (see Box 1.1). In the spaceless world on which many economists have tended to base their analyses, the market for a good is a point. In fact, we are concerned with a *market area*, and this has very important implications for the spatial organization of the production of sausages (and all other goods and services as well, of course). The quantity of goods any individual is prepared to buy depends on the *real price* that has to be paid, and this is made up of two distinct elements:

1. The market price, that is, the price at the point of sale
2. The cost of traveling to and from that point

In our case, it is the second of these elements that helps to determine both our sausage producer's chances of selling enough sausages to stay in business and how extensive the market area will be.

Suppose that the market price for sausages is $1.50 per pound. Any customer living immediately next to the sausage producer pays that price for sausages. If $6.00 per week is set aside for sausages, 4 pounds can be bought at $1.50 per pound. Because we have assumed that every individual has the same demands, every inhabitant of our plain is willing to spend $6.00 per week on sausages. But how many sausages each one actually gets depends on the two elements just identified: the market price plus the cost of travel. The farther away a person is from the point of sale, the fewer sausages can be bought for $6.00 simply because a larger proportion of that $6.00 has to be spent overcoming the friction of distance. Figure 1.3 (p. 22) illustrates this relationship between distance and quantity demanded on the assumption that transport costs are 10 cents per mile for a return trip (i.e., 5 cents per mile each way). A person living 15 miles from the point of sale has to spend $1.50 on traveling. This leaves $4.50 of "sausage money" to buy 3 pounds of sausages. In other words, for this person, the real price per pound is $2.00, compared with $1.50 for the person living next door to the producer. As distance increases, so too does this real price. At 30 miles, the cost of travel ($3.00) leaves only enough money to buy 2 pounds of sausages, at the equivalent of $3.00 per pound. At 45 miles, transportation costs leave exactly $1.50, just enough to buy 1 pound of sausages. But to the person living at a 45-mile distance, these are really expensive sausages, working out at the equivalent of $6.00 per pound! If the producer is not prepared to sell quantities of less than 1 pound, it is clear that anybody living beyond 45 miles from the point of sale remains sausageless.

If we tilt Figure 1 in Box 1.1 over to the left through 90 degrees, we get a clearer view of the "ideal" market area for sausages. With an evenly distributed population, the area served is a circle with a radius of 45 miles. This distance, in fact, represents the *range of the good;* beyond 45 miles, the real price of sausages is greater than people are prepared to pay. Since we know that the population is evenly distributed over our plain and we also know how many pounds of sausages each person is prepared to buy, depending on the distance from the point of sale, we can calculate the total demand per week for sausages for the market area. Figure 1.4 (p. 22) shows three circles, each of which represents both a specific distance and the per capita demand for sausages at that distance. By adding the demand per head for each customer, we find that the total demand is 96 pounds of sausages within the 45-mile range.

Before we leave our first producer of sausages and consider the locations of other sausage producers, we need to answer another basic question. If the producer is to set up in business to produce and sell sausages and to remain in business, it must be possible to sell a minimum quantity of sausages to cover basic costs and to give some profit. This fundamental requirement also has a spatial form. Suppose that our producer needs a weekly return of $105 to cover costs. This represents the equivalent of 70 customers buying 1 pound of sausages per head. But as we have seen, those living near the point of sale can afford to buy more than 1 pound per head. In Figure 1.4, the minimum demand needed is contained within a radius of 30 miles. Such a minimum level of demand needed to ensure the survival of a producer is generally known as the *threshold* value. These two important concepts, threshold and range, are very closely related. One represents the minimum demand necessary to support a business; the other represents the maximum distance over which the business can sell its goods. The threshold must lie within the range or at least equal it; otherwise, as Figure 1.5 (p. 23) shows, the business must fail.

More than One Producer of the Same Good

A glance back at Figure 1.4 shows that there are large numbers of people who cannot buy sausages simply because they live outside the range of sales from point *P*. Clearly, there is room for other producers. Assuming that conditions facing them are the same as for our first producer (that is, a market price of $1.50 per pound, transportation costs of 10 cents per mile, and a maximum expenditure of $6.00 per consumer), the threshold and range will be identical for each sausage producer. But one very important complicating factor arises as soon as we consider second, third, or *n*th producers of the same good: each has to take

Box 1-1   # Demand, Supply, and Price

The price of a good can be established in a variety of ways, depending on the kind of economic system being considered. Here we assume that we are dealing with a market system in which no individual can influence the prices of goods. Instead, price reflects the balance between demand and supply. *Demand* and *supply* are terms that refer to the willingness of people in general to purchase or sell goods at different prices. Suppose that the demand and supply schedules for our hypothetical sausages are as follows:

Demand schedule for sausages		Supply schedule for sausages	
Price per pound ($)	Pounds of sausages demanded per week	Price per pound ($)	Pounds of sausages demanded per week
3.00	0	3.00	132
2.50	18	2.50	106
2.00	36	2.00	80
1.50	52	1.50	52
1.00	70	1.00	26
0.50	88	0.50	0
0.00	106		

Clearly, the higher the price of sausages, the less quantity demanded, and the lower the price, the greater the demand (this is plotted graphically in Figure 1). On the other hand, producers are more willing to supply greater quantities of sausages if prices are high, and their enthusiasm fades as prices drop lower (Figure 2). Theoretically, there is one point in common between the two schedules. As shown in Figure 3, this is where the demand curve and the supply curve cross. The crossover point—the point where demand and supply are in equilibrium—is at a price of $1.50 per pound. At that price, suppliers are prepared to sell 52 pounds of sausages per week, and this is exactly the amount consumers are willing to buy at that price. In this case, therefore, the market price is $1.50 per pound.

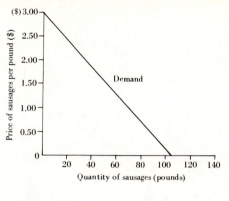

Figure 1 A hypothetical demand curve for sausages.

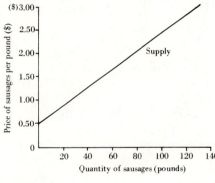

Figure 2 A hypothetical supply curve for sausages.

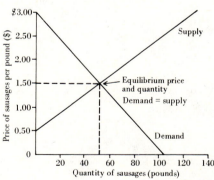

Figure 3 The market price of sausages depends on the relationship between supply and demand.

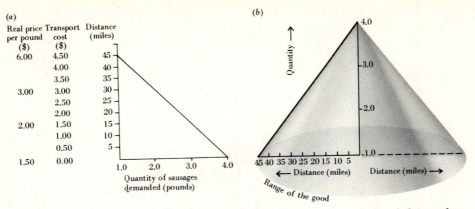

Figure 1.3 Distance and the "real price" of sausages: (*a*) Hypothetical demand curve incorporating distance costs; (*b*) Hypothetical demand cone showing the range of the good.

into account the existing location and market area of already established producers. As an example, consider our first producer with a market area of 45 miles' radius. It would be irrational for a second producer of sausages to locate less than 90 miles from the first one, simply because the two producers would be competing for some of the same customers (Figure 1.6).

The same principles apply to all sausage producers, each of whom will sell to a circular market area of equal size. However, as the number of producers

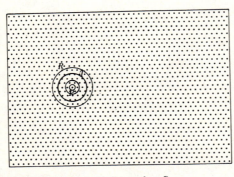

Variations in demand with distance from P

Distance from P	Number of customers	Per capita demand (pounds)	Aggregate demand (pounds)	Value of sales ($)
0	1	4	4	6.00
>15 miles	6	3	18	27.00
16–30 miles	24	2	48	72.00
31–45 miles	26	1	26	39.00

P = Point of production
R = Range of the good
T = Threshold

Figure 1.4 Variations in total demand for sausages with distance from the point of production.

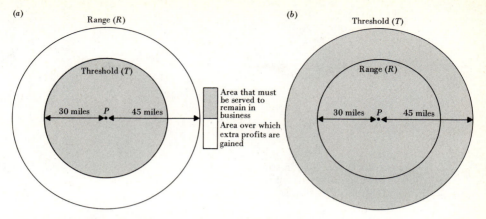

Figure 1.5 Relationship between threshold and range: (*a*) Threshold is within range, business survives; (*b*) Threshold is outside range, business fails because people living between *R* and *T* cannot afford to purchase from *P*.

increases and covers more and more of the plain, other kinds of problems arise. Figure 1.7*a* shows that if producers are located so that the edges of their market areas just touch, there is no competition between producers but, at the same time, there are potential consumers who live outside the range of any producer who are left unserved. On the other hand, if the circles overlap, although they give complete coverage from the consumer viewpoint, this is unsatisfactory for the producer because it generates competition, which may lower profits (Figure 1.7*b*). One suitable solution, for both producers and consumers, is shown in Figure 1.7*c*: the circles are transformed into hexagons (in effect by bisecting the zones of overlap so that overlap is removed). This is a perfectly logical solution to the problem because we have already assumed that consumers will purchase their goods from the cheapest—that is, nearest—producer. Thus a network of equal-sized hexagonal market areas is the most likely spatial structure for producers of the same good, under the simplified conditions of our discussion.

We can, however, take the argument just one stage further before we introduce the complication of different types of goods. Recall that the threshold for the sausage producer is located within the range *R* (Figure 1.5*a*). This means, in effect, that the area between the two generates sales over and above

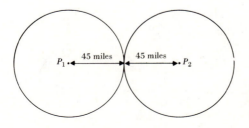

Figure 1.6 The location of a second producer is constrained by the location of the first producer.

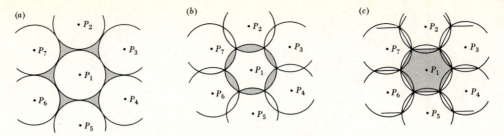

Figure 1.7 Theoretical spatial arrangement of the market areas of competing producers of the same good: (*a*) Population of interstitial areas is unserved; (*b*) Overlap of circular market areas results in competition between producers; (*c*) One suitable solution: hexagonal market areas.

the level necessary to cover the producer's basic costs plus some profit. In other words, the producer gains "excess profits," though these are unlikely to persist for very long. The existence of excess profits is likely to attract more producers because we assume that there is *free entry* into the industry. The result is that individual market areas become progressively smaller until they approach the threshold size (Figure 1.8). Any further shrinkage would drive producers out of business. We assume that the best solution for both producers and consumers is one in which all producers make the same profit and all consumers are served at the lowest cost consistent with this level of profit. The result is a compact, uniform lattice of production centers, each serving hexagonal market areas of identical size, equivalent to the threshold size for that good.

SPATIAL ORGANIZATION OF PRODUCTION OF OTHER GOODS

Even the most avid hot dog enthusiast would probably agree that man cannot live by sausages alone (if nothing else, he needs the buns to go with the sausages). Other goods and services will be demanded by the inhabitants of

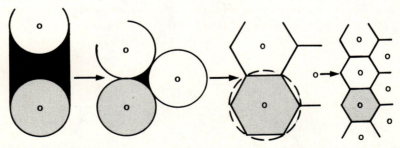

Figure 1.8 Theoretical development of market areas from initial circular form to final hexagon of threshold size. *Source:* A. Lösch (1954), *Die räumliche Ordnung der Wirtschaft* [The Economics of Location] (Stuttgart: Gustav Fischer), fig. 23.

our isotropic plain. How would the production of these other goods and services be arranged in space, and what would their market areas be like? Fundamentally, the same basic principles would apply whatever the good or service in question. Each will have a specific threshold and a specific range that will apply to every producer of the same good. But thresholds and ranges differ from one good to another, according to such variables as their basic price (including their relative transportation costs) and the frequency with which they are demanded. Some goods, such as bread or milk, are required very frequently and are relatively cheap. Thus a producer of such goods need not serve a very wide geographic area. By contrast, most people that we know do not buy a new Rolls-Royce or Cadillac every week; a dealer selling such luxury cars must therefore be able to reach a very large number of customers. If, as we have assumed, the population for both bread and Cadillacs is evenly distributed, the spatial threshold and range will be very small for goods such as bread and very large for such rarely purchased and expensive goods as Cadillacs. Thus we can envisage a *continuum* of goods and services from those of *low order* (small threhold, small range) to those of *high order* (large threshold, large range). All goods, whatever their order, will be offered in hexagonal market areas, but the size of these areas varies directly according to the order of the good. The network of market areas of low-order goods is made up of small hexagons; the network of market areas of high-order goods is made up of large hexagons.

SPATIAL ORGANIZATION OF PRODUCTION OF "BUNDLES" OF GOODS: THE ARRANGEMENT OF CENTRAL PLACES

The existence of a threshold requirement will influence both the number and the relative location of producers. High-order goods will be available at only a few locations, while low-order goods will be provided at a large number of locations. For reasons that will be explored further in Chapter 5, producers will tend to *cluster* together in certain locations. Christaller called such clusters *central places*, whose primary function is to provide its surrounding population with goods and services.* Christaller argued that central places would be arranged in a very precise manner both vertically (in terms of their relative importance or "pecking order") and horizontally (in terms of their relative location).

The relative importance of each central place depends on both the number

---

*Central place theory originated with the pioneering work of the German geographer Walter Christaller. In his major published work, *Central Places in Southern Germany* (1966; originally published 1933), Christaller posed the basic question, "How can we discover the laws?" He proceeded to seek an answer by deductive reasoning subsequently tested by a detailed empirical study of settlement patterns in southern Germany. Preston (1985) has recently evaluated Christaller's contribution to the study of the evolution of central places.

and the order of goods and services it provides. Christaller suggested that the central place system takes the form of a regular hierarchy of central places. The precise form of the hierarchy is based on the principle that a place on a particular level in the hierarchy provides not only goods and services that are specific to its level but also all other goods and services of lower order. To explain this more clearly, we need to rank all the goods and services that are demanded by our evenly dispersed population in order of their threshold values. For simplicity, assume that ten goods of different order are demanded by the population of our plain. Table 1.1 shows how the organization of these ten goods, and of the centers providing them, can be fitted together to form a regular hierarchy of centers. Good 1 is the highest-order good, which, by definition, must be located so that it can serve the largest possible market. Good 1 will therefore be produced only in the highest-order central place, here called an A center. How many A centers there will be depends on the total demand for the highest-order good and the size of its threshold value. Suppose that the total population of our plain is 1 million and that good 1 has a threshold value equivalent to the demand of 200,000 people: then no more than five A centers can exist. These will be spaced at an equal distance from each other in exactly the same way as our sausage producers and for the same basic reasons.

The good of next lowest threshold, good 2, will also be located in an A center, though, of course, it will serve a slightly smaller area than good 1. Similarly, good 3 will also be located in the A centers. However, the extent of good 3's market area, based on each of the A centers in which it is located, is

Table 1.1 RELATIONSHIP BETWEEN THE ORDER OF A GOOD AND THE CENTRAL PLACE HIERARCHY

Threshold	Goods and services ranked by descending threshold value	High — Level in the hierarchy — → Low		
		A centers	B centers	C centers
High	1	✓		
	2	✓		
	3	✓	✓	
	4	✓	✓	
	5	✓	✓	
	6	✓	✓	
	7	✓	✓	✓
	8	✓	✓	✓
	9	✓	✓	✓
Low	10	✓	✓	✓

✓ denotes that the good is provided in the central place.

sufficiently small as to leave a considerable number of people out of reach of good 3. If, and only if, the unserved demand is equal to good 3's threshold value, a new producer of the good can come into existence. As Table 1.1 shows, it will be located at the central place of the next lowest order (which we can call a B center). Figure 1.9 shows us where the B centers will be located: they will be located at the midpoint between three A centers. Reference to Figure 1.7 should make it clear why this is so. The market area for a good from a production point should not overlap with market areas for the same good from other production points. Also, under our assumptions, a good can be produced only if the appropriate threshold demand exists. By locating at the midpoint between three A centers in Figure 1.9, therefore, the B center can serve a market area that is just large enough to meet the threshold requirement of the good (good 3 in Table 1.1 in this case) without overlapping with the market areas for the same good from centers of next highest order. In Figure 1.9, the market areas for the B centers are shown by the solid lines. Good 3 can be termed a *hierarchical marginal good* because it is a good that "defines" a new level in the central place hierarchy; it is the highest-order good provided by a central place B.

We can continue this process to produce as many levels in the hierarchy as there are hierarchical marginal goods (that is, goods whose market areas based on particular central places leave sufficient unfulfilled demand for a new center to be established). Table 1.1 and Figure 1.9 show just one more level in the central place hierarchy—C-level centers. These are defined by good 7 in Table 1.1 and are located in exactly the same way as B centers: at the midpoint between three centers of the next highest order. Clearly, then, each C center nests within three B centers and each B center nests within three A centers.

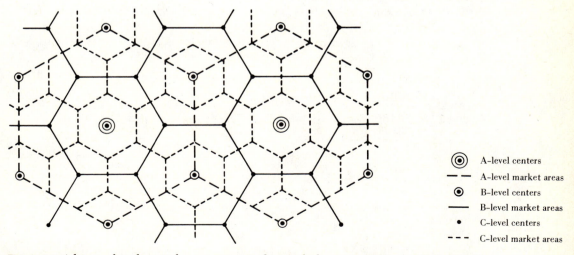

◉	A–level centers
– – –	A–level market areas
⦿	B–level centers
——	B–level market areas
•	C–level centers
- - -	C–level market areas

Figure 1.9 A hierarchical spatial arrangement of central places according to Christaller's $k = 3$ principle.

Thus Table 1.1 shows that each level of the hierarchy is characterized by a specific mixture of goods. A centers, the highest-order centers, provide all goods with highest to lowest threshold values. B centers provide goods 3 to 10 but not the higher-order goods 1 and 2. Similarly, C centers provide goods 7 to 10 but not goods 1 to 6, which are of a higher order. Each level in the hierarchy is also characterized by equal distances between centers on that hierarchical level. Thus all A centers are the same distance from each other, each B center is an equal distance from another B center, and B-level distances are less than A-level distances, and so on down the central place hierarchy.

Christaller's model, then, implies a fixed relationship between each level in the hierarchy. This relationship is known as a k value (k meaning a constant) and indicates that each center dominates a discrete number of lower-order centers and market areas in addition to its own. Figure 1.9 shows a $k = 3$ system in which the hierarchical arrangement is according to the "rule of threes." The diagram is interpreted in the following way. Recall from the discussion of hierarchical marginal goods that each new center was located midway between three centers of the next highest order. In effect, therefore, each new center (and its market area) is shared among the three higher-order centers. Within the hinterland of every A center is the equivalent of two B centers (one-third of each of the six B centers surrounding an A center) and six C centers. Christaller used the term *marketing principle* to describe the organization of a $k = 3$ system because it was based on the principle of supplying the maximum number of evenly distributed consumers from the minimum number of central places. Table 1.2 shows the pattern for a five-level hierarchy in which a metropolis would supply highest-order goods for the equivalent of two cities (plus itself), six towns, 18 villages, and so on.

Under these conditions, a system of central places evolves in accordance with two basic principles. The first is that all parts of the plain are supplied with all conceivable goods from a given number of centers; the second is that a central place of given rank provides the goods and services appropriate to its own rank and all goods and services of lower order. Modification of one or both of these principles would present alternative hierarchical arrangements.

Christaller in fact recognized that a $k = 3$ system of central places was not

Table 1.2   A CENTRAL PLACE SYSTEM ORGANIZED ACCORDING TO CHRISTALLER'S MARKETING PRINCIPLE ($k = 3$)

Level of hierarchy	Equivalent number of central places dominated by the highest-order center	Equivalent number of market areas dominated by the highest-order center
1. Metropolis	1	1
2. City	2	3
3. Town	6	9
4. Village	18	27
5. Hamlet	54	81

the only conceivable form for the hierarchy to take. He suggested two other organizing principles in which the size and orientation of the hexagonal market areas is changed, thus altering the functional relationships between centers on different levels of the hierarchy (Figure 1.10).

Christaller's Traffic Principle: $k = 4$

Even on our isotropic plain, where movement is possible in all directions, some movement paths are more likely to be followed than others. Christaller (1966) pointed out that the marketing principle ($k = 3$) is an awkward arrangement in terms of connecting different levels of the hierarchy:

> In a system of central places developed according to the marketing principle, the great long-distance lines necessarily by-pass places of considerable importance, and the secondary lines built for short-distance traffic can reach the great places of long-distance traffic only in a roundabout way—often even in remarkably zig-zag routes. (p. 74)

As an alternative arrangement, Christaller suggested that central places could be organized according to what he called the traffic principle:

> The traffic principle states that the distribution of central places is most favorable when as many important places as possible lie on one traffic route between two important towns, the route being established as straightly and as cheaply as possible. The more unimportant places may be left aside. According to the traffic principle, the central places would thus be lined up on straight traffic routes which fan out from the central point. (p. 74)

Where central places are arranged according to the traffic principle, therefore, lower-order centers are located at the midpoint of each *side* of the hexagon rather than at the corners as in a $k = 3$ system. Thus the traffic principle produces a hierarchy organized in a $k = 4$ arrangement in which central places are nested according to the "rule of fours." Figure 1.10b shows how this can be interpreted. Take the central places numbered 1 to 6 around the higher-order center, A. In a $k = 4$ system, central place 1 is "shared" equally between the

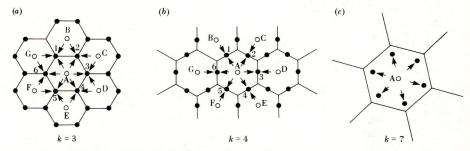

Figure 1.10 Christaller's three alternative spatial arrangements of central places: (a) Marketing principle ($k = 3$); (b) Traffic principle ($k = 4$); (c) Administrative principle ($k = 7$).

two higher-order centers, A and B. Place 2 is shared equally between A and C, and so on for each of the six lower-order centers. Center A, therefore, dominates or serves the equivalent of four market areas of next lower order (six times one-half of each of the market areas 1 to 6 together with its own market area). In a $k = 3$ system, the market area relationship is one to three. Table 1.3 shows the hierarchical relationship for a five-level system and should be compared with Table 1.2 to clarify the difference between a $k = 3$ and $k = 4$ system.

Christaller's Administrative Principle: $k = 7$

Christaller's other suggested organizing principle was based on the realization that from a political or administrative viewpoint, centers that are shared pose problems. Any pattern of control that cuts through functional units is potentially problematic (an extreme example would be the divided city of Berlin). Consequently, Christaller suggested that an arrangement whereby lower-order centers were entirely within the hexagon of the higher-order center would obviate such problems. Such a pattern is shown in Figure 1.10c: all six lower-order centers are fully subordinate to center A which, therefore, dominates the equivalent of seven market areas at the next lowest level (market areas 1 to 6 plus its own).

Whichever of these three hierarchical arrangements ($k = 3$, 4, 7) is adopted, all three share one basic characteristic. Once one of the k values is established in an area, according to Christaller, this value remains constant so that the hierarchical ordering of central places and of market area sizes follows this strict progression throughout. The result is a very regular, rigid, and discrete-level hierarchy of centers in which the relationship between levels is identical throughout.

Lösch's Approach to the Hierarchical Arrangement of Centers

August Lösch, writing just a few years after Christaller published his study of central places, took a far less rigid view of the hierarchical arrangement of

Table 1.3   A CENTRAL PLACE SYSTEM ORGANIZED ACCORDING TO CHRISTALLER'S TRAFFIC PRINCIPLE ($k = 4$)

Level of hierarchy	Equivalent number of central places dominated by the highest-order center	Equivalent number of market areas dominated by the highest-order center
1. Metropolis	1	1
2. City	3	4
3. Town	12	16
4. Village	48	64
5. Hamlet	192	256

centers.* Using an argument that is often difficult to follow, he shows how this more flexible hierarchy can be derived. He pointed out that the $k = 3, 4,$ and 7 arrangements were but the three smallest of a very large number of possible market area structures. Figure 1.11 shows the ten smallest areas, though Lösch in fact identified 150 such areas altogether. Recall that each good in our simple landscape possesses a hexagonal mesh of market areas, the fineness or coarseness of the mesh being determined by the good's threshold value. In Lösch's view, there is no reason why these market areas and their associated production centers should, or could, be arranged in the rigid manner Christaller proposed.

The question Lösch posed was, How might this confusion of hexagonal nets of various sizes be combined to form a spatial structure that would be efficient for both producers and consumers? Begin, says Lösch, by arbitrarily choosing just one production center from the entire set of production points established on the plain. (Each production point itself is established on the basis of the principles discussed earlier in this chapter.) Then arrange the nets so that this one center is common for all of them. Figure 1.11 shows this for ten goods and their hexagonal nets. This center, at which every good would be available, could be regarded as the metropolis, the highest-order center of all. The nets are then arranged so that alternating 30-degree sectors radiate out from the metropolis.† There will be 12 such sectors (see Figure 1.12), arranged alternately so that six sectors have very many economic activities and six have

---

* August Lösch was, without doubt, one of the most original, stimulating, and at the same time most obscure of the modern location theorists. His major work, *The Economics of Location*, was originally published in German in 1939 but was not available in an English translation until 1954, some nine years after his death at the age of only 39. Lösch's view of the world was extremely idealistic: "The real duty of the economist," he wrote, "is not to explain our sorry reality, but to improve it. The question of the best location is far more dignified than determination of the actual one" (p. 4). Lösch's work shows a breadth and depth of thought unequaled by most other location theorists. We have already employed some of his ideas as building blocks in this chapter. For example, it was Lösch who first demonstrated how hexagonal market areas for different goods could be derived using spatial adaptations of the demand curve.

† This particular section of Lösch's work has been the subject of much controversy. Lösch seems to suggest that having focused all the nets on one center, the nets are then rotated around that central point until we get "six sectors with many and six with only a few production sites" (p. 124) with a consequent maximum coincidence of production sites. This conjures up a picture of Lösch sitting at his desk with 150 sheets of transparent paper with a pin through the middle spinning each of them round until he reached the sectoral pattern he wanted. This may have been so, though Tarrant (1973), and Beavon and Mabin (1975) suggest a rather different interpretation. The mathematical detail of their argument need not concern us here, but their conclusions are particularly illuminating. According to both studies, the production of "city-rich" and "city-poor" sectors is not the *result* of rotation, as many have believed, but a *constraint* on it. In other words, if a sectoral pattern is to be achieved, there is a very limited number of ways in which the hexagonal nets can be arranged. Once certain ones are oriented in a particular way, the positions of the others are fixed. It is perhaps worth reemphasizing the idealistic basis of Lösch's view. He was concerned with finding the "best" spatial solution to the arrangement of producers, and he regarded a sectoral pattern as part of this best arrangement.

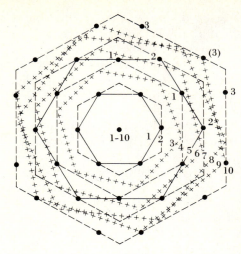

Figure 1.11 Lösch's ten smallest market areas focused on the "metropolis." *Source:* A. Lösch (1954), *Die räumliche Ordnung der Wirtschaft* [The Economics of Location] (Stuttgart: Gustav Fischer), fig. 27.

relatively few. Lösch called these "city-rich" and "city-poor" sectors, but these terms are misleading. Each sector has the same number of production points (central places), but the number of activities at each center varies between the two types of sector. In other words, there are more higher-order centers (in Lösch's scheme these are centers with a larger number of economic activities) in the activity-rich sector than in the activity-poor sector. Figure 1.13 shows this more clearly for just part of the total system. The left-hand sector of the diagram has far more centers with a larger number of activities at each. For example, the activity-rich sector has 34 centers with more than five economic activities at each and 39 centers with either three or four activities. The activity-poor sector, by comparison, has only 23 centers with more than five activities and 24 with three or four. Conversely, the activity-poor sector has far more production centers with only one activity (52 percent of the total) than the activity-rich sector (39 percent of the total).

Despite the cloudiness of some of Lösch's statements, there is no doubt that he produces a spatial organization of production centers whose form is very appealing. Lösch himself saw this spatial pattern of economic activities as offering several advantages. In particular, he claimed that the total distances between production points are minimized, and therefore, both the volume of shipment and the length of transport routes needed to satisfy the demands of the systems are reduced. At the same time, because the largest number of production locations coincide, the maximum amount of purchases can be made locally. Thus Lösch suggested that this spatial arrangement of urban centers was consistent with what he saw to be a basic element in human organization: the *principle of least effort*. Lösch's economic landscape both maximizes the number of firms operating within the market and minimizes aggregate transportation costs.*

---

* For a further discussion of Lösch's equilibrium model, see Mulligan (1981) and Mulligan and Reeves (1983).

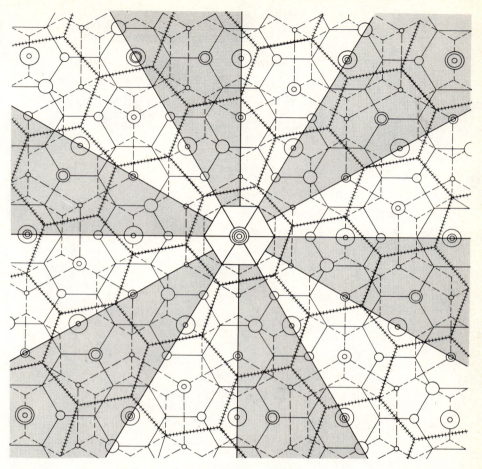

Figure 1.12 Theoretical spatial arrangement of market areas and production centers according to Lösch. *Source:* A. Lösch (1954), *Die räumliche Ordnung der Wirtschaft* [The Economics of Location] (Stuttgart: Gustav Fischer), fig. 28.

Using similar assumptions but dissimilar approaches, Christaller and Lösch generated rather different urban hierarchies. In Christaller's scheme, the hierarchy is composed of a series of discrete levels in which a center produces exactly the same mix of goods as every other center on the same hierarchical level. Lösch's hierarchy, by contrast, is far less rigid. The less regular coincidence of centers producing different orders of goods means that centers of the same size (in terms of the number of economic activities) may produce quite different combinations of goods. In Lösch's scheme, functional mixture and hierarchical position are not synonymous. Figure 1.14 illustrates some of the flexibility of Lösch's structure (the lines represent delivered prices so that the production point of each good is where the appropriately numbered line is at its lowest point or points). Place A may be regarded as the metropolis in which all goods (1 to 7) are produced, but there is no fixed relationship

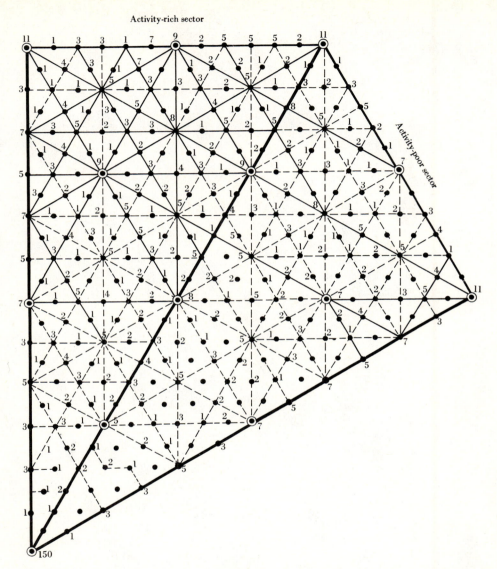

Figure 1.13 Transport lines and clusters of economic activities in Lösch's "activity-rich" and "activity-poor" sectors. *Source:* A. Lösch (1954), *Die räumliche Ordnung der Wirtschaft* [The Economics of Location] (Stuttgart: Gustav Fischer), fig. 32.

between the type of goods supplied and the other centers as postulated by Christaller. In the first place, it is possible in Lösch's scheme to have different combinations of functions in centers of the same size or rank. For example, centers E, I, and K each provide two functions. E and I both produce goods 2 and 4, but K produces goods 2 and 5. Second, functions of the same order may be produced in centers of different size. Thus good 7 is produced not only in the metropolis but also in what is clearly a lower-order center, H. Christaller's

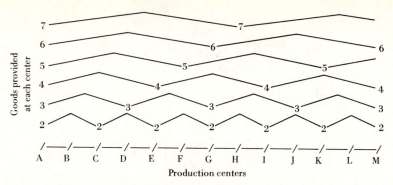

Figure 1.14 Hypothetical distribution of goods produced by centers of different order. *Source:* W. Stolper (1955), "Spatial Order and the Economic Growth of Cities," *Economic Development and Cultural Change* 3: 140. Reproduced with permission of the University of Chicago Press, © 1955 by the University of Chicago.

condition that a center on a given hierarchical level will provide the goods of all lower-order centers is not fulfilled with the single exception of the metropolis. Thus Lösch's scheme allows the existence of *specialized* production centers while Christaller's does not, except insofar as each level in his hierarchy is distinguished by a specific hierarchical marginal good.

Despite the obscurity of certain parts of the Löschian model, it is more flexible and more comprehensive than Christaller's. Indeed, Lösch claims that it is unnecessary to envisage three separate and conflicting principles as Christaller does, on the grounds that his complete regional system of centers and hexagonal market areas subsumes all of these principles at the same time. Two important consequences follow from the kind of economic landscape Lösch envisaged (neither of which he identified himself). One relates to the implications of the sectoral arrangement on movement and the other to population distribution. Let us examine each of these in turn.

A major advantage of Lösch's sectoral patterning of production centers, as he saw it, was that movement was more efficient when channeled along discrete routes (Christaller's $k = 4$ principle was based on a similar view). As Figure 1.13 shows, the transport network is denser in the activity-rich sectors, though for some reason Lösch positioned his major transport routes, of which there were 12 radiating outward from the metropolis, along the edges of each sector, whereas they should logically run *through* the activity-rich sector. However derived, the fact remains that the introduction of linear transportation routes greatly modifies the spatial arrangement of economic activities by the frictional effects of distance. Movement will likely be easier, quicker, and possibly cheaper along established routes than "across country" (indeed, movement may well be possible only along such routes). The result will be the distortion of what would otherwise be regularly shaped hexagonal market areas and the circular land-use areas. Figure 1.15 shows how a hexagonal market area might become elongated along a route; clearly, the impact of routes on a full network of hexagonal areas would greatly alter the overall spatial pattern of market areas.

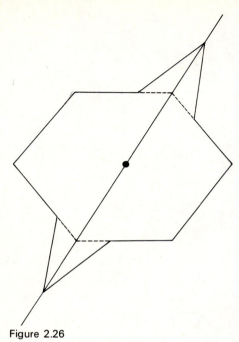

Figure 2.26

Figure 1.15 Distortion of a hexagonal market area by a transportation route.

The second implication of Lösch's kind of economic landscape is that it inevitably destroys our initial assumption of an evenly distributed population. Not only will population densities (the number of people per square mile or square kilometer) be greater in the individual urban centers, but also differential population densities occur on a larger scale because of the sectoral clustering of activities Lösch envisaged. Inevitably, such differences in density will cause variations in the size of market areas. Figure 1.16 shows this effect diagrammatically. Recall that in a fully developed system of market areas for a particular good, the size of each hexagon will be equivalent to that good's threshold value. Suppose that a producer needs the equivalent of 100 customers to keep the business going. The geographic extent of the threshold market area will be determined by the population density. If the density is 100 people per square mile, the market area will have that geographic dimension.

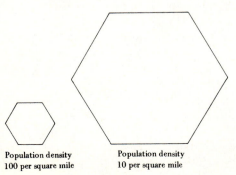

Population density
100 per square mile

Population density
10 per square mile

Figure 1.16 Market area size varies with population density.

If people are less densely settled—say, 10 per square mile—the market area will be ten times larger, though serving exactly the same number of people. Thus in areas of high population density, especially nearer the metropolis, a high level of sales can be achieved in a small market area. At a greater distance from the metropolis, however, where population densities are lower, this same level of sales would require a much larger market area. Isard (1956) attempted to modify Lösch's system diagrammatically by allowing hexagon size to vary, although, as Figure 1.17 shows, such variation makes it very difficult to preserve a true hexagonal structure.

However, the introduction of such modifications as linear transportation

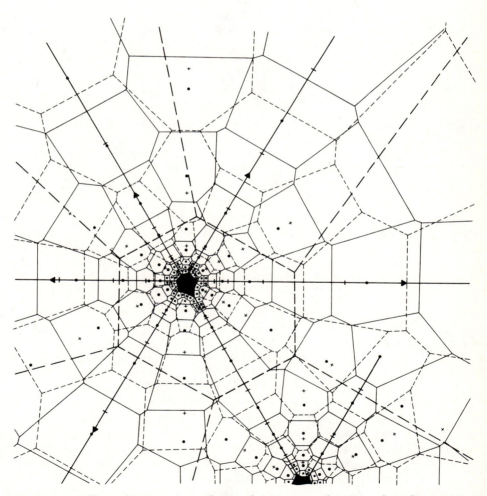

Figure 1.17 Effect of uneven population densities on the size of market areas in a Löschian landscape. *Source:* W. Isard (1956), *Location and Space Economy* (Cambridge, Mass.: MIT Press), fig. 52. Copyright © by the Massachusetts Institute of Technology. Reproduced by permission.

routes and varying population densities does not destroy the underlying logic of the argument. The spatial organization of economic activities on our simplified landscape is still ordered fundamentally by the frictional effect of distance. Movement and its costs are still the basic influence.

THE CENTRAL PLACE SYSTEM IN THEORY: A SUMMARY VIEW

Having begun by considering the logic underlying the decision of an individual to offer a good for sale to his neighbor, we have been able to extend that same logic to make some interesting discoveries about the spatial organization of central places in our simplified landscape. On the basis of the distance variable alone, it has been possible to generate on the isotropic plain a complex network of hierarchically ordered centers with predictable functional and locational characteristics. This network takes the form of a spatially organized system. The objects of the system, the central places and the households of the evenly distributed consumers, are linked together by the flows of goods and cash returns as supplies and demands are matched by exchange. The dynamic force that gives the system its structure is the cyclic exchange process in which *inputs* of money (demand) from a dispersed population are transformed into *outputs* of goods and services by the individual production subsystems (butcher, bakers, printers, etc.) making up the central places. This exchange process, on which both the individual production subsystems and the central places (as aggregates of these subsystems) depend for survival, has a spatial form because of the differences in location between points of demand and points of supply. In moving to a central place to exchange income for goods and services, a consumer must use up scarce resources (money, time, physical energy) to overcome the friction of distance. At a certain distance from the supply point, this expenditure is so great that when it is added to the price that must be paid for the goods and services, it reduces demand for them to zero. At this distance the exchange process stops because the demand input is absent (this is the range of the good).

The spatial extent of the process varies from one good to another, being most extensive for high-order goods (generally those of high value that are purchased relatively infrequently) and least extensive for low-order goods. The entire form and structure of the central place system, therefore, is dependent on these systematic variations in the spatial amplitude of the exchange process. The perfect hierarchical structure developed in our discussion in fact represents the *steady state* of the central place system—a condition of dynamic equilibrium in which the nature of the functional organization achieves a balance between inputs and outputs. In other words, the demands generated by the population are supplied most efficiently. Thus individual central functions, the central places themselves, and the network of central places in the hierarchy are all in perfect dynamic equilibrium with their respective environments. Can we find evidence of this structure in the real world?

THE CENTRAL PLACE SYSTEM IN PRACTICE: EMPIRICAL EVIDENCE

Hierarchies of Central Places

In empirical investigations, there has been a strong tendency for workers to follow the framework laid down by Christaller rather than the more flexible hierarchy devised by Lösch. Much of the pioneering work was carried out in the late 1950s by the geographers Garrison and Berry, work that stimulated an enormous number of studies of central place systems in many parts of the world during the 1960s.

The consensus of most central place studies is that a definable hierarchy of central places does exist and can be identified at a variety of scales and in quite varied geographic conditions. Table 1.4 summarizes the results of seven such studies carried out in a diversity of areas, from the northwest coast of the United States through the almost homogeneous landscape of southwestern Iowa to the Niagara peninsula and southern Ontario; from the heavily industrialized South Wales coalfield to the undulating agricultural region of Baden-Württemberg in Germany. In each of the seven areas, a distinct grouping of centers according to their functional significance is apparent, though the boundaries between hierarchical levels vary from one area to another. The number of levels also differs, ranging from three in Snohomish County, Washington, to six in southwestern Iowa.

The data in Table 1.4 are arranged into groups on the basis of distinct breaks in the overall distribution of central places. Figure 1.18 (p. 42) illustrates this point for the Niagara peninsula, one of the seven examples summarized in Table 1.4. In Figure 1.18, the 37 central places are arranged in order in terms of the number of central functions in each place (central functions are goods and services such as retail shops, professional services, and leisure facilities but excluding manufacturing activities). This rank distribution seems to fall into five broad categories, shown by the dotted lines in the diagram. These lines represent the divisions between hierarchical levels. Quite clearly, there is considerable variation within each hierarchical level in the number of functions per central place. Certainly the pattern is some way removed from the very rigid Christaller pattern and closer to the more continuous distribution proposed by Lösch.

Table 1.4 shows that each level in the hierarchies of the seven areas is also associated with a particular population size or range of population sizes. Much work on central place systems has in fact been concerned with establishing the existence of a functional relationship between such variables as the number of central functions or the number of establishments of each functional type, on the one hand, and the population sizes of central places, on the other. The underlying logic is that settlements with large populations will be able to support more central functions and more units of each function than settlements with small populations. For example, Figure 1.19 (p. 42) is a graph showing the number of functional units in each of the Niagara peninsula settlements on the vertical axis and the population of each settlement on the

Table 1.4   SELECTED EXAMPLES OF CENTRAL PLACE HIERARCHIES

Level of hierarchy	No. of centers	No. of central functions	Population
1. Snohomish County, Washington			
Low			
1	20	1–16	15–2,586
2	9	22–42	600–2,996
3	4	56–64	1,684–3,494
2. Southwestern Iowa			
Low			
1	29	<10	<150
2	32	10–25	150–400
3	15	28–50	500–1,500
4	9	>55	2,000–7,000
3. Southwestern Ontario			
Low			
1	10	1–12	25–1,702
2	2	19–22	408–486
3	2	28–32	673–676
4	1	78	3,507
5	1	99	22,224
6	1	150	77,190
4. Niagara peninsula			
Low			
1	20	1–22	49–2,064
2	9	27–43	1,900–5,400
3	4	55–59	6,900–9,800
4	2	69–73	18,000–44,000
5	2	89–98	57,000–101,000

[a] Functional index (see Davies, 1967). Briefly, this weights each function in a central place by its frequency of occurrence in the general study area, summing them to the index.

[b] Not available.

Sources: Data from Berry and Garrison (1958a); Berry, Barnum, and Tennant (1962); Murdie (1965); Marshall (1969); Davies (1967); Barnum (1966).

horizontal axis. The graph gives the impression that there is a straight-line (linear) relationship between the population of settlements and the number of functional units they possess. Obviously, however, the relationship is not perfect; if it were, we would be able to predict precisely the number of functional units in a town by calculating its population.

Although it is clear that the rigid hierarchical organization Christaller proposed is not found in more than a few instances, it is equally clear that there *are* distinct groupings of central places according to systematic variations in

Level of hierarchy	No. of centers	No. of central functions	Population
5. Owen Sound area, Ontario			
Low			
1	10	7.42–15.76[a]	65–213
2	12	34.52–179.67	73–1,501
3	6	257.55–655.09	1,090–3,450
4	1	3,013.88	17,421
6. South Wales			
Low			
1	25	2.33–11.74[a]	N.A.[b]
2	18	11.69–58.67	N.A.
3	4	83.99–176.05	N.A.
4	3	249.27–450.60	N.A.
5	1	2,077.96	N.A.
7. Baden-Württemberg, West Germany			
Low			
1	162	2–24	57–780
2	134	25–50	781–2,150
3	44	51–97	2,200–5,501
4	19	108–207	6,200–24,000
5	2	248–292	73,500–90,000

their functional significance. Not surprisingly, the hierarchy is at its clearest in circumstances in which conditions are closest to the assumptions of the original model, that is, towns acting predominantly as centers serving a surrounding area with central place goods and services in fairly homogeneous physical conditions. (Iowa is as close as we are likely to get to a flat plain.) The pattern also holds good where a center's manufacturing functions use local or ubiquitous (universally available) materials and serve a local market. These, in fact, were the kinds of manufacturing function Lösch envisaged when devising his

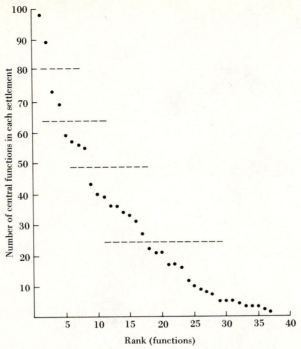

Figure 1.18 A functional hierarchy of central places in the Niagara peninsula.

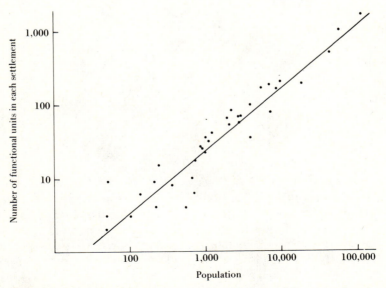

Figure 1.19 Relationship between the number of functional units and population of central places in the Niagara peninsula.

more flexible hierarchical structure. In other words, manufacturing activities that are most similar to central place activities do nothing to distort the hierarchical organization of centers.

A geographer who has more recently done much to explore the theoretical and empirical nature of central place systems—John Parr—has suggested that we can best think of the hierarchical models of central places as forming a spectrum with the rigid Christaller model at one end of the spectrum and the Lösch "family" of models at the other. Parr suggests that his own general hierarchical model falls between these two. It allows for a variation in the *k* value even within the same hierarchical system and appears to fit empirical observations of the real world quite well.

Rank-Size Relationships

Quite apart from studies based on central place hierarchies, other kinds of empirical regularities have been recognized in the system of urban centers. By far the best known of these is what has become known as the *rank-size rule*, a phenomenon that has a very long history but was brought to general attention by G. K. Zipf in 1949. If we take all the urban centers in a particular country and arrange them in order from largest population to smallest, in many cases the size relationship between the towns of each rank is extremely regular. If the rank-size rule applies exactly, the population of, say, the third-ranking center would be one-third of the population of the first-ranking center, the population of the twentieth-ranking center would be one-twentieth of the first-ranking center, and so on. Expressed in a formal way, the rank-size rule is

$$P_r = \frac{P_1}{r^q}$$

where r = rank of a city
P_r = population of a city of rank r
P_1 = population of the largest city
q = exponent generally with a value close to 1

If we plot the population size against rank for every urban center in a country, the relationship (on a logarithmic scale) appears as a downward-sloping line with a slope defined by q. This simply means that the size of a center is inversely proportional to its population rank.

Figure 1.20 shows the extent to which the overall rank-size relationship holds for U.S. and Canadian cities in the early 1980s.* The relationship is by no

---

* A vast number of empirical studies of city size distributions in general and of the rank-size relationship in particular have been conducted. Carroll (1982) provides details of almost 60 such studies across a whole range of countries. International studies tend to reveal one of three general patterns of city size distribution. In addition to the rank-size pattern, a number of countries exhibit a *primate* distribution in which one or perhaps two very large cities dominate and medium-sized

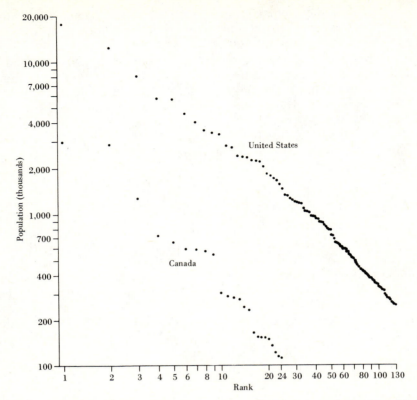

Figure 1.20 Rank-size relationships in the urban systems of the United States (1984) and Canada (1981). *Sources: Statistical Abstract of the United States, 1985; Canada Yearbook, 1985.*

means perfect in either case. Graphically, the relationship is closer in the case of the United States than it is in Canada. Earlier work by Madden (1956) showed that the general rank-size relationship remained remarkably stable in the United States for more than 150 years, from 1790 to 1950, although the rank of *particular* cities changed substantially.

---

cities are absent. Yet other countries seem to have a mixed or intermediate pattern of city size distribution; Canada would seem to fall into this category. Some writers have tried to relate these patterns to a country's level of economic development. But although some underdeveloped nations do have primate patterns, others do not. Numerous factors probably account for these differences: level of economic development, changes in political status, and length of time urbanized among them. Berry (1961) suggests that as the economic, political, and social life of a country becomes more complex, its urban size distribution will tend to develop toward a rank-size pattern that represents the steady state of an urban system. But the explanation is probably far more complex than this. As Sheppard (1982) has pointed out, "Comparisons of city size distributions can be misleading, since the same pattern may be a symptom of very different situations" (p. 149).

We have evidence, therefore, of both *hierarchical* distributions of centers, based on functional complexity, and of *continuous* (rank-size) distributions of centers, based on population size. Not surprisingly, there has been a good deal of controversy as to whether these two distributions are compatible or conflicting. The problem is most acute in the context of Christaller's central place hierarchy. Stewart (1958) and Vining (1955), for example, argued that the central place hierarchy based on function and the rank-size distribution based on population are quite incompatible. Beckmann (1958) claimed that the two may well be compatible on the grounds that centers of similar functional status will not have exactly the same population size but a range of such sizes. As a result, the steps of a strict hierarchy tend to be smoothed out, resulting in a more continuous distribution of population sizes. The problem is less acute if we take either Lösch's form of the hierarchy or Parr's general hierarchical model, both of which produce a more continuous functional hierarchy.

THE HIERARCHY AND THE MOVEMENT OF CONSUMERS TO CENTRAL PLACES

In looking at empirical evidence of an urban hierarchy so far we have concentrated on the "objects" of the system—the urban centers themselves—in terms of their functional structure. But how far does the actual movement of consumers to these centers to satisfy their demands for goods and services correspond to the predictions of central place theory? The theory tells us that consumers will travel to the nearest center providing the desired good, so movement for low-order goods, which are available at a large number of centers, should tend to be short-distance, while movement for high-order goods should be characterized by longer distances. In other words, distance traveled to acquire central place goods should be directly related to the order of the good. Let us take just one example each of low- and high-order goods and see how far the spatial pattern of consumer purchases corresponds to expectations based on central place theory.

This aspect of central place theory is generally tested by questioning a large number of people about their shopping habits for different goods. For each good, a map can be drawn connecting the location of each consumer with that of the center in which the good is purchased with a straight line known as a *desire line*. Patterns of consumer behavior for low-order goods would therefore show many very short desire lines focusing on a large number of small places. Patterns for high-order goods would show much longer desire lines focused on a smaller number of larger centers. Let us take the example of food purchases as a typical low-order good. Figure 1.21 shows the spatial pattern of food purchases in a part of eastern Ontario, the area between Ottawa and Cornwall on the St. Lawrence River. It is clear that most of the dispersed population of the area shopped for food in local centers; indeed, the pattern of desire lines not only indicates predominantly short-distance movement but also gives some idea of the extent of each center's general market area for food.

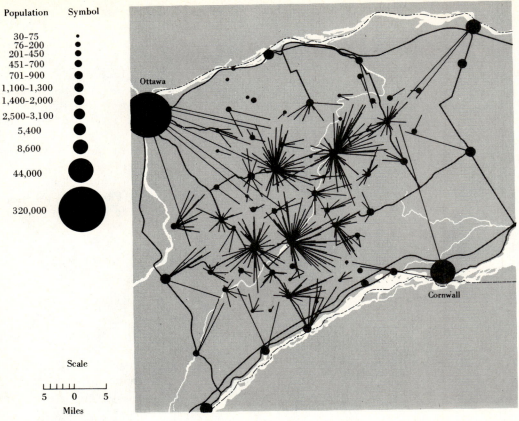

Figure 1.21 Spatial pattern of food purchases in a part of eastern Ontario. *Source:* D. M. Ray (1967), "Cultural Differences in Consumer Travel Behavior in Eastern Ontario," *Canadian Geographer* 11: 143–156, fig. 3. Reproduced by permission.

At the other end of the scale, Figure 1.22 shows the spatial pattern of the purchase of a higher-order good, optical services. The movement for optical services in eastern Ontario consisted of longer-distance travel, predominantly to the two major urban centers, Ottawa and Cornwall. Desire lines are much longer than in the case of food, reflecting the lack of provision of optical services in most of the lower-order centers.

These examples, which could be replicated many times over from the very large number of central place studies in different areas, clearly suggest some correspondence with theory. On the whole, the length of a shopping trip is directly related to the order of the good or service, and it does seem that to a certain extent, consumers attempt to minimize their movement costs by visiting nearby urban centers. However, a classic study of consumer behavior in Iowa (Golledge, Rushton, and Clark, 1966) revealed that this relationship

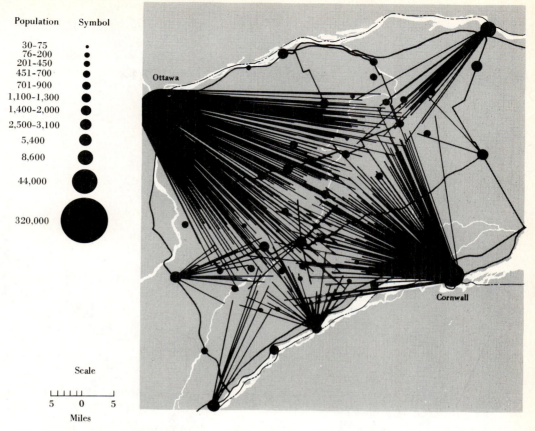

Population   Symbol

30–75
76–200
201–450
451–700
701–900
1,100–1,300
1,400–2,000
2,500–3,100

5,400

8,600

44,000

320,000

Ottawa

Cornwall

Scale

5   0   5
Miles

Figure 1.22 Spatial pattern of movement for optical services in a part of eastern Ontario. *Source:* D. M. Ray (1967), "Cultural Differences in Consumer Travel Behavior in Eastern Ontario," *Canadian Geographer* 11: 143–156, fig. 8. Reproduced by permission.

holds to a greater extent for low-order "convenience" goods than for high-order "shopping" goods. Table 1.5 shows the distances traveled by a sample of the Iowa farm population for various types of good. Column 1 shows the mean distance traveled to the "maximum purchase town" (calculated on the basis of dollars spent on a good at different places in a single year), while column 2 indicates the mean distance traveled to the "nearest purchase town" (the nearest town in which an actual purchase was made).

The distances in Table 1.5 reveal two interesting features. First, they confirm that on the whole, distance traveled does vary with the order of the good as already suggested (for example, compare distances traveled for food with those traveled for women's clothing). Second, there is considerable variation between "maximum purchase" distances and "nearest purchase" dis-

Table 1.5 CONSUMER TRAVEL IN IOWA

Good or service	Mean distance to maximum purchase town (miles)	Mean distance to nearest purchase town (miles)
Beauty/barber	7.4	6.1
Food/groceries	7.8	5.2
Dry cleaning	10.5	10.3
Major appliances	14.5	13.8
Men's clothing	15.6	8.2
Boys' clothing	15.6	11.4
Furniture	18.7	17.6
Car purchases	19.7	18.8
Girls' clothing	29.3	13.4
Women's clothing	30.3	14.1

Source: After R. G. Golledge, G. Rushton, and W. A. V. Clark (1966). "Some Spatial Characteristics of Iowa's Dispersed Farm Population," *Economic Geography* 42: 261–272, table 1. Reproduced with permission.

tances, implying that consumers do not always purchase goods at the nearest center. This, too, is related to the order of the good. Differences tend to be least for low-order goods; in the case of high-order "shopping goods," consumers appear to travel a great deal farther than the nearest center providing the good. For example, there is a difference of only 2.6 miles between the mean maximum purchase town and the nearest purchase town for food and groceries. In comparison, there is a 16.2 mile difference for women's clothing. Such differences may be related to a desire on the part of the consumer to "shop around" for higher-order goods, to compare competing products at different outlets, and so on. The relation with order is not total, however; for example, there is very little difference between the two distances for furniture or car purchases.

One final point should be made before leaving this very brief discussion of shopping distances. Reference to the desire line map of low-order purchases (Figure 1.21) reveals that although most trips are short, there are a few very long desire lines to the high-order centers. This is probably because when making a trip to such centers for high-order goods, consumers may at the same time purchase low-order goods. In effect, this reduces the total travel expenditure: a single trip to a higher-order center replaces the need to make separate trips to different centers. Where only low-order "convenience" goods are being purchased, of course, the nearest center is likely to be chosen. One implication of such multipurpose trips is that the spatial range of low-order goods provided by high-order centers is likely to be greater than the spatial range of the same goods provided from low-order centers.

SPACING OF URBAN CENTERS

In our discussion of the urban hierarchy, we found that if urban centers are considered in terms of their central place functions, a hierarchy is evident, though its structure is less rigid than central place theory would suggest. But if urban centers of all types are considered in terms of their population size, the result is a more or less continuous (rank-size) distribution, although, as we have suggested, the two may well be compatible. When we turn to consider the spatial distribution of urban centers, we are faced with a similar problem. We would expect to find some spatial regularity between central places. But if urban centers of all types are considered, we would now hardly expect to find complete spatial uniformity (such as a hexagonal distribution). We would, however, expect evidence of some regularity.

One means of objectively measuring the spatial pattern of urban centers is the technique of *nearest neighbor analysis*.* King (1962) used the nearest neighbor technique to investigate the spatial pattern of urban places in 20 sample areas in the United States, and his results are summarized in Table 1.6. There is clearly considerable variability in the spacing of settlements due, presumably, to the differential influence of physical, economic, cultural, and other forces. In fact, 12 of the 20 areas (including, of course, Iowa) had patterns that could be described as "approaching uniform," though in no case was the approach especially close to the predicted value of 2.15. The remainder had predominantly random patterns or, in the cases of Utah and Washington, patterns tending toward the clustered.

But if we concentrate on urban centers, which are predominantly central places in a functional sense, then, presumably, the correspondence with theory should be closer. Central place theory tells us that there should be a characteristic distance separating centers on a given hierarchical level and that high-order centers should be widely spaced, with the distances separating centers on each level decreasing as we move down the hierarchy.

Numerous writers have claimed to identify "typical" distances separating central places on different hierarchical levels. Lösch claimed to find a close correspondence in Iowa between theoretical distances and actual distances among central places in what appeared to be a $k = 4$ hierarchy. Brush's classic (1953) study of central places in southwestern Wisconsin also identified mean distances between central places that appeared to vary in accordance with

---

* Nearest neighbor analysis is a method for determining how far a distribution of points (and at an appropriate scale we can regard urban centers as points) differs from a random distribution. The method consists simply of measuring the straight-line distance between every urban center and its nearest neighbor in the area in question. The resulting mean nearest neighbor distance value can then be compared with the values expected if the pattern were either completely clustered (a nearest neighbor value of 0), random (a value of 1), or completely uniform (a value of 2.15). A perfectly hexagonal arrangement of centers would have a value of 2.15.

Table 1.6   SPATIAL PATTERN OF SETTLEMENTS IN 20 SELECTED AREAS OF THE
UNITED STATES

Clustered (ideal value = 0.00)		Random (ideal value = 1.00)		Approaching uniform (ideal value = 2.15)	
Area	Nearest neighbor value	Area	Nearest neighbor value	Area	Nearest neighbor value
Utah	0.70	California	1.08	Georgia	1.32
Washington	0.71	Florida	0.94	Iowa	1.35
		Louisiana	1.08	Kansas	1.33
		New Mexico	1.10	Minnesota	1.38
		North Dakota	1.11	Mississippi	1.28
		Oregon	1.02	Missouri	1.38
				Ohio	1.27
				Pennsylvania	1.22
				Texas (NW)	1.23
				Texas (SE)	1.16
				Virginia	1.22
				Wisconsin	1.24

Source: After L. J. King (1962), ''A Quantitative Expression of the Pattern of Urban Settlements in Selected Areas of the U.S.,'' *Tijdschrift voor Ekonomische en Sociale Geografie* 53: 1–7, table 1.

theory. Brush recognized a three-tiered hierarchy in the area of hamlets, villages, and towns; their relative distances are shown in Table 1.7, and they appear to correspond closely to theoretical expectations. However, Dacey (1962) rejected this conclusion after carrying out a nearest neighbor analysis of the same centers. He found, in fact, that the central place system in southwestern Wisconsin more closely approximated a random pattern than a uniform pattern. As Dacey pointed out, this did not necessarily deny the existence of a central place hierarchy in southwestern Wisconsin but rather suggested that the hierarchy as identified by Brush did not conform to the uniform spatial pattern he had claimed. Thus there appear to be two possible explanations in this case. Either the levels of the hierarchy were wrongly identified or the central places were not uniformly spaced because of other

Table 1.7   SPACING OF CENTRAL PLACES IN SOUTHWESTERN WISCONSIN

Between-center distances	142 hamlets	73 villages	19 towns
Theoretical distance	5.6 miles	10.0 miles	19.8 miles
Mean measured distance	5.5 miles	9.9 miles	21.2 miles
Range of variation	1.0–12.0 miles	3.5–18.5 miles	7.0–38.0 miles

Source: J. E. Brush (1953), ''The Hierarchy of Central Places in southwestern Wisconsin,'' *Geographical Review* 43: 380–402, table 2. Reprinted by permission. Copyrighted by the American Geographical Society of New York.

factors. A third possibility, however, is that the measure of distance itself is inappropriate. Indeed, it is extremely doubtful whether straight-line physical distance is a truly valid measure of the functional spacing of central places.

A more realistic measure might well be one that takes into account variations in population density because population density has a marked effect on the size of market areas. Where population densities are high, threshold market areas tend to be smaller and, presumably, centers will be located closer together, and vice versa.

Figure 1.23 illustrates the general relationship between population density and the size of market area for samples of central places in four quite disparate geographic regions of the United States. The graph relates three variables: the area served (in square miles), the total population served, and the population density. Note how the symbols for the four areas are, on the whole, clearly distinguishable from each other and that the relationship is

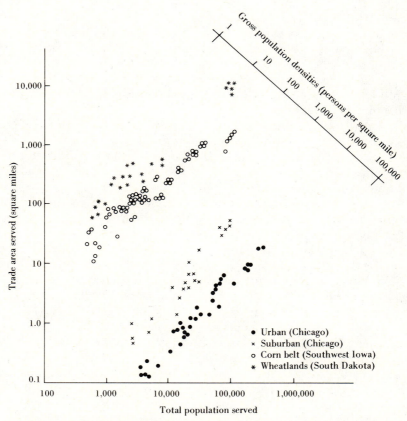

Figure 1.23 Relationship between size of market area and population density. *Source:* B. J. L. Berry and H. G. Barnum (1962), "Aggregate Relations and Elemental Components of Central Place Systems," *Journal of Regional Science* 4: 35–68, fig. 2. Reproduced by permission.

remarkably regular. Market areas in the wheatlands of South Dakota are extensive and serve a relatively small population at low densities. At the other extreme, the pattern of solid black dots representing centers in the highly urbanized Chicago region indicate very small trade areas serving large numbers of people at very high densities. Thus the evidence appears to bear out the importance of taking into account variations in population densities in explaining the spacing of central places.

SPATIAL ORGANIZATION OF AGRICULTURAL PRODUCTION

So far we have assumed that the role of the dispersed population was to support the economic activities in the urban centers through their demands for goods and services, that is, to provide *inputs* to the central place system. But some of the members of the dispersed population—those operating farms—also provide *outputs* of the agricultural products demanded by the residents of the central places. Indeed, the continued existence of such urban centers depends, at least in part, on the supply of agricultural goods. Hence we have a further process to consider, involving the production and exchange of agricultural goods. How will such agricultural production be organized on our homogeneous plain, a plain on which most of the "usual" agricultural variables (soil quality, climate, terrain) are held constant and on which the friction of distance is the sole factor?

One basic characteristic of agricultural production compared with the production of manufactured goods or the provision of services is that it uses relatively large amounts of land. Therefore, the focus of attention when we consider agricultural production is how different units of land are used. The farmers on our plain have a choice of what crops to grow or what kind of livestock to raise, though overall their choices are governed by the kinds of agricultural goods that are demanded by the inhabitants of the central places. But given such demands, how will the production of agricultural goods be organized spatially? In general terms, we can say that different units of land will be devoted to the uses that can achieve the highest return per unit of land. The mechanism by which the scarce land resource is allocated in this way is *economic rent*. It is important to note that *rent* in this context has a specific meaning that differs from that implied by the everyday use of the term. Economic rent is the surplus income that can be obtained from one unit of land above what can be obtained from an inferior unit of land; for the most part, it is measured against land at the margin or limit of cultivation. This marginal land is just capable of producing a return large enough to cover the cost of bringing it into production. The term *rent* in popular usage refers to contract rent, which is an actual payment that tenants make to others for the use of their land or property. The two concepts are related but not synonymous (see Chisholm, 1962).

For our present purposes, we can define *economic rent* as a measure of the level of return that the market at large (all the potential bidders for land) would expect a particular piece of land to produce. It is basically a measure of the advantage, as the bidders see it, of one piece of land over another. This implies, of course, that pieces of land differ in some respect and that such differentiation is reflected in higher or lower returns per unit of land. In our simplified model, the only basis for such differentiation is the friction of distance; that is, the only advantage that one piece of land can have over another piece of land is its location in relation to the market for agricultural products. For this reason, we will use the term *location rent*.

Each of the central places located in an orderly hierarchical manner in our simplified landscape may be regarded as a potential market for agricultural goods. Obviously, the centers vary in importance as agricultural markets according to their size and hierarchical status. The highest-order A centers in Figure 1.9 are likely to be far larger centers of demand for agricultural products than are the C-level centers. However, different-sized market centers pose problems that obscure the essential features of our discussion, so at this stage, we will concentrate on just one highest-order central place and show how agricultural production is likely to be located in the area around it.*

Spatial Organization of One Agricultural Product

The market price of any product is, as we have seen, set by the prevailing relationship between demand and supply. Thus we could construct demand and supply schedules for each of the agricultural products demanded by the central place population in exactly the same way as we did for products such as sausages. The market price is the price received by the farmer when he sells his product at the market. His net return is this price less his costs of production and the cost of transporting his product from his farm to the market. As production costs are assumed to be the same everywhere, the only factor that can influence the farmer's net return is the cost of transportation.

If all farmers could be located at the market, they would all get the same net return. But this is obviously impossible. Farms cannot be piled on top of each other in skyscraper fashion. Farming uses a lot of land, and where this land is located with reference to the urban market is the critical factor determining the net advantage that one piece of land has over another. The greatest advantage belongs to land immediately adjacent to the market center, and the advantage declines as distance from the market increases. Precisely how this occurs in the case of a single crop can be calculated by using the following simple formula:

---

* This kind of approach originated with J. H. von Thünen, one of the founders of economic location theory. Von Thünen was a German estate owner who in the early nineteenth century began to investigate with great scientific precision the best way of organizing agricultural land use. His findings were published as a book, *Der Isolierte Staat [The Isolated State]* in 1926 (Hall, 1966), which has had an enormous impact both on economic thinking in general and on geographic analysis of agriculture in particular.

$$LR = Y(m - c) - Ytd$$

where LR = location rent per unit of land
        Y = yield (quantity produced) per unit of land
        m = market price per unit of product
        c = production cost per unit of product
        t = transport rate per unit of distance
        d = distance of the unit of land from the market

Let us apply this formula to the production of an imaginary milk producer by substituting numeric values for the letters:

Y = 100 gallons per acre
m = $5.00 per gallon
c = $3.00 per gallon
t = $0.05 per mile

At the market itself, the value of location rent for milk would be

$$LR = 100\ (\$5 - \$3) - 100\ (0.05 \times 0) = \$200$$

As no transportation costs are involved, the net return to the farmer is at its maximum. However, for the farmer producing milk on a piece of land 10 miles from the market, the situation is rather different. Yield, market price, and production costs are the same as those of the farmer located at the market. But payment has to be made to ship the milk to market. Location rent, therefore, is

$$LR = 100\ (\$5 - \$3) - 100\ (0.05 \times 10) = \$150$$

If we follow this pattern outward from the market, the relationship between location rent for milk and transportation costs takes the form shown in Figure 1.24. This shows that the margin of production of milk given these yield, price, and cost figures will be at a distance of 40 miles from the market. At that distance, the farmer's net return is completely obliterated by the costs of transporting his milk to market. The farmer cannot therefore remain in business without taking a loss.

Variation in location rent with distance is usually shown by the kind of downward-sloping curve of Figure 1.25. It indicates that the closer a unit of land is to the market, the more desirable it is. This will stimulate competition to exploit its use. The situation can be understood more easily if we distinguish between the dairy farmer who seeks to use a particular piece of land and the owner of that land. Land will be allocated among dairy farmers on the basis of competitive bidding. This location rent curve can therefore also be regarded as a *bid rent curve* (Alonso, 1960) because it gives an indication of how much dairy farmers would be prepared to pay for a unit of land at various distances from the market. In the present example, the cash advantage that a farmer located 10 miles from the market has over a farmer located 20 miles away is $50. It would therefore benefit the more distant farmer to bid up to $50 to acquire the land closer to the market. But it would be useless to bid more because any figure above this amount would be higher than the "ceiling rent" at that location, resulting in a loss.

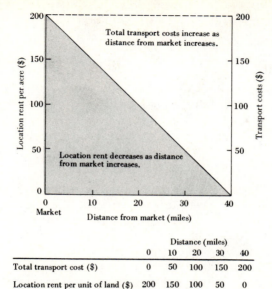

	Distance (miles)				
	0	10	20	30	40
Total transport cost ($)	0	50	100	150	200
Location rent per unit of land ($)	200	150	100	50	0

Figure 1.24 Relationship between location rent for an agricultural product and transportation costs to the market.

Spatial Organization of Several Agricultural Products

Just as in our discussion of goods and services produced in urban centers we progressed from the consideration of one good (sausages) to other goods, so, too, we can examine the spatial organization of agricultural production of other crops and products on our isotropic plain. The basic principle remains the same, though the specific values or price and cost will vary for each product.

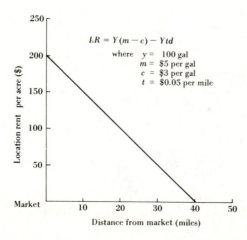

Figure 1.25 Hypothetical location rent curve for milk.

Each agricultural product demanded by the population will be affected by four factors:

1. Its specific market price, depending on the supply and demand relationship for that good
2. Its specific transportation rate, which will vary according to the nature of the product—its bulkiness, perishability, and general transportability
3. Its basic cost of production, which is assumed to be constant in space for any one good
4. Its specific yield per unit of land

Thus every product will have a different location rent curve. Its *height* (the value of location rent at the market) depends on the difference between the product's market price and its production costs; its *slope* (the rate at which the value of location rent declines with distance) depends on the transportation characteristics of the product.

Suppose that we introduce two more products (in addition to milk): for example, potatoes and wheat. What will their location rent curves look like, and more important, how will this affect the relative location of production of all three products? Figure 1.26 shows how the location rent curves for the products overlap. The relative heights of the three curves at any given distance from the market determine which of the three land uses will be adopted at that distance. For example, a farmer located 20 miles from the market could achieve the following location rent values:

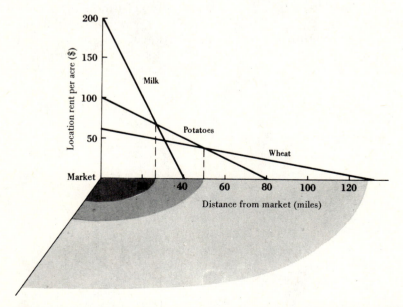

Figure 1.26 Location rent curves and production zones for milk, potatoes, and wheat.

> Milk: $100 per acre
> Potatoes: $70 per acre
> Wheat: $55 per acre

Clearly, the land will be devoted to milk production because this is the land use yielding the highest location rent at that location.* In fact, all farmers located between the market and a distance of some 26 miles from the market would be milk producers because the location rent for milk is higher than that of its nearest rival, potatoes, over the entire distance. But at a distance of 26 miles, milk production would be displaced by potato production; this land use would then dominate over all distances from 26 miles to 40 miles, beyond which wheat would be the crop grown until the margin of cultivation is reached at approximately 126 miles from the market. Beyond that distance, agricultural production would cease under the prevailing conditions of demand and supply.

Figure 1.26 also shows that if we envisage the land use zones on the horizontal axis being rotated around the market point, the result will be an agricultural land use pattern of concentric rings or zones. Each zone accommodates the type of land use yielding the highest location rent. Thus the spatial organization of production of the agricultural goods demanded by the population of our highest-order central place takes the form of a series of concentric zones around the central place. The order or sequence of products is determined by the height and slope of their location rent curves. The transition from one land use to another occurs where their respective location rent curves intersect.

Although we have considered the possibility of producing only one type of crop at a particular location, exactly the same principles apply to combinations of crops. In such cases, the location rent curve applies to the total combination and not simply to the individual elements. Figure 1.27 shows a hypothetical example. A combination of milk, potatoes, and wheat has the location rent curve AX; that for a combination of potatoes, wheat, and barley is shown by BY; and a combination of potatoes, barley, and beef is represented by CZ. The combination yielding the highest location rent at any given location will be the prevailing type of land use. Figure 1.27 also shows that the same product may occur in several zones, though in different combinations (potatoes are present throughout the area in the diagram). Thus the introduction of crop combinations rather than single crops does not alter the basic concentric pattern of

---

* Note that the farmer could still make a profit if either potatoes or wheat were grown instead of milk. To grow either potatoes or wheat would incur an *opportunity cost,* which is the difference between the return actually achieved and the maximum that could be achieved. Thus the farmer's opportunity cost for growing potatoes instead of raising dairy cattle is $30 ($100 − $70), and for growing wheat it is $45. An opportunity cost, therefore, is a measure of the cost of one alternative in terms of the alternative not taken up. But as all our producers aim to achieve optimal solutions, each will adopt the land use generating the highest location rent.

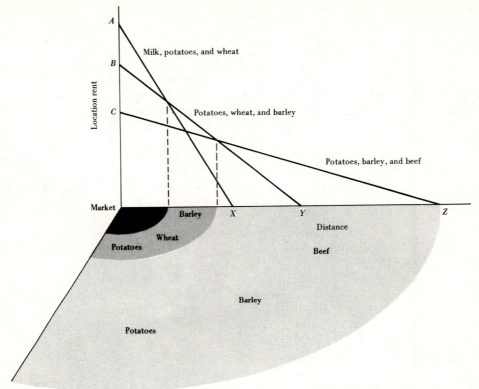

Figure 1.27 Location rent curves for combinations of agricultural products.

agricultural production; it simply means that the number of possible concentric rings is increased to the number of possible combinations demanded by consumers.

Effect of More than One Market Center on the Spatial Organization of Agricultural Production

We deliberately isolated one central place from the hierarchical system in order to see more clearly how agricultural production would be spatially organized around that center. We did this to avoid obscuring the basic allocating mechanism—location rent—and its operations. But what happens when we put our single-market center back into its rightful place as only one center in the hierarchy? It certainly makes life more complicated for the location theorist because every central place is potentially a market for agricultural goods. Suppose, as in Figure 1.28, we introduce the complication of just one additional market. Both centers have the inner land use zones organized concentrically around them, as in the single-market example. However, the outer zones are displaced and take on an elliptical shape because of their

Figure 1.28 The effect of two market centers on agricultural land use zones. *Source:* E. S. Dunn (1954), *The Location of Agricultural Production* (Gainesville: University of Florida Press), fig. 16. Reproduced by permission.

orientation toward two markets rather than one. The dotted line marks the boundary between the two competing supply areas. The existence of more than two market centers produces a more complex picture, as Figure 1.29 illustrates; products of the inner rings are oriented toward the individual towns, as in the simpler case, but those of the outer rings are oriented toward the entire cluster of centers. Thus the existence of several market centers undoubtedly produces a more complex picture, but it does not alter the basis on which the spatial pattern of agricultural production is founded. The land use at any particular location is the one that yields the highest location rent at that location. This allocative mechanism still operates even though the existence of several markets makes its operation more difficult to visualize.

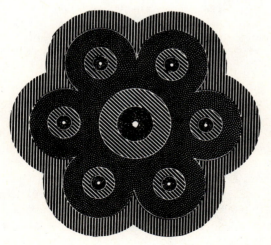

Figure 1.29 The effect of multiple market centers on agricultural land use zones.

THE AGRICULTURAL PRODUCTION SYSTEM IN THEORY:
A SUMMARY VIEW

For the agricultural production system, the objects are the dispersed farm units and the nucleated central places. Inputs to the agricultural system take the form of money income to the farmer derived from the sale of his products, production factors (farm machinery, fertilizers, etc.), and manufactured goods obtained from the central places.

Outputs from the agricultural system are the crop and livestock products. The transformation stage in this case involves the use of capital, labor, and technical skills in manipulating biotic processes to satisfy consumer demands. The functional structure of the system derives from the continuous interaction process as the demand for crop and livestock products is met by supply. Should it stop, the system ceases to exist.

Under the constraints of our model, the spatial organization of the system depends, first, on the specific products demanded and, second, on the extent to which this demand is eroded by distance. For some products, spatial inter-action—the exchange of town-based inputs for crop or livestock outputs—is restricted by the resources lost to transport costs. For these, the spatial ampli-tude of the interaction cycles is small, and in the model, the only feasible locations are close to their markets. For others, the friction of distance bears less heavily, and their range of feasible locations in relation to mar-kets is wider. The precise spatial allocation of land uses is in the long run achieved through a competitive bidding process among prospective land users.

Given all the assumptions of our model and the absence of any disruptive forces, the steady state of the system would be achieved with a regular pattern of zonal land uses. These would be simple and concentric in the single-market case, but for a hierarchical network such as that generated earlier for the central place system, the pattern, even under the constraints of the model, would be extremely complex.

So far, we have treated the central place and the agricultural production systems of the model as though they were separate entities. This is convenient for the purposes of exposition. More realistically, however, they are inextrica-bly bound up with one another. The two combine in a symbiotic relationship. Together they comprise a higher-order, two-sector economic system, the one interacting with the other to satisfy the consumption needs of society in the model landscape. The inputs of one subsystem can now be identified as the outputs of the other. For instance, agricultural produce shipped to the central places (or the money income derived from it) is exchanged for town-based production factors or central goods, and vice versa. The economic system developed up to this point in the simplified model depends there-fore on "two-region, two-sector" trade and interaction to meet consumer demands. Again we can pause to ask whether the theoretical pattern exists in reality.

THE AGRICULTURAL PRODUCTION SYSTEM IN PRACTICE: EMPIRICAL EVIDENCE

Zonation of Agricultural Production Around Individual Urban Centers

According to the von Thünen model, the growing of crops and the raising of livestock will be located in a regular series of concentric zones focused on urban markets. Such zonation is produced by the influence of transport costs on the location rent yielded by units of land at increasing distances from the market together with the sensitivity of different products to the cost of transport.

Von Thünen based his theory on careful observation and measurement of agricultural practices in northern Germany in the early nineteenth century. Thus his suggested spatial organization of agricultural production—Table 1.8 gives a detailed picture of the contents of his land use zones—had a strong empirical content. It is clear from the variety of evidence assembled by Chisholm (1962) in particular that a pattern very like this was a dominant characteristic of agricultural production in the past. Chisholm quotes a contemporary source in describing the location of agriculture around London in 1811 and shows that despite the wide variety of soil types, agriculture around London was arranged in a series of concentric zones. Harvey's (1963) study of the historical development of the Kentish hop industry also revealed that regularity in the land use pattern may override physical differences. In the case of this single crop, there was clear evidence of a decline in the density of hop acreage with increasing distance from the central core area of production.

Such regularities in the agricultural land use pattern are a good deal less evident in today's more complex world, particularly in the highly industrialized economies, but it does appear that some remnants of the preexisting zonal pattern can still be identified. The most widely occurring remnant is the survival of parts of the innermost zone of agricultural production—intensive market gardening and liquid milk production—around urban centers. Despite increased concentration of these activities in areas of greatest comparative advantage, many major cities retain considerable horticulture and dairying on their urban-rural periphery. Jean Gottmann's classic (1961) study of "Megalopolis"—the highly urbanized seaboard of the northeastern United States—amply confirmed this. Agriculture in Megalopolis was highly specialized, with particular emphasis on market gardening, dairying, and poultry husbandry. These goods were produced at a very high degree of intensity, giving one of the highest productivity-per-acre levels for these goods in the whole of the United States. Our theoretical discussion of bid rent curves suggests that farm property values in such a situation should be high, and this was indeed the case. What is more, these high values are directly related to location with respect to the market:

Table 1.8   VON THÜNEN'S AGRICULTURAL LAND USE SYSTEM

Zone	Percentage of state area	Relative distance from central city	Land use type	Major marketed product	Production system
0	<0.1	−0.1	Urban-industrial	Manufactured goods	Urban trade center of state; near iron and coal mines
1	1	0.1–0.6	Intensive agriculture	Milk, vegetables	Intensive dairying and trucking; heavy manuring; no fallow
2	3	0.6–3.5	Forest	Firewood, timber	Sustained-yield forestry
3a	3	3.6–4.6	Extensive agriculture	Rye, potatoes	Six-year rotation: rye (2), potatoes (1), clover (1), barley (1), vetch (1); no fallow; cattle stall-fed in winter
3b	30	4.7–34	Extensive agriculture	Rye	Seven-year rotation system: pasture (3), rye (1), barley (1), oats (1), fallow (1)
3c	25	34–44	Extensive agriculture	Rye, animal products	Three-field system: rye, etc. (1), pasture (1), fallow (1)
4	38	45–100	Ranching	Animal products	Mainly stock-raising; some rye for on-farm consumption
5	—	>100	Waste	None	None

Source: P. Haggett (1965), *Locational Analysis in Human Geography* (New York: St. Martin's Press), table 6.4. Reprinted with the author's permission.

A farm with good soil usually commands a better price than one of the same size with poor soil, but only if their locations are equal. The rockiest pasture ten miles from Boston is more valuable than the finest black loam in central Illinois. (p. 263)

Despite the development of specialized areas of milk production based on factors other than market proximity (see Chapter 3), urban-oriented liquid milk production is still evident. Durand's detailed (1964) study of the major milksheds of the northeastern United States clearly illustrated this. Within the broad zone of specialized dairy production stretching from Maine to Minnesota, which capitalized on favorable natural conditions to serve a national market, Durand identified pockets of liquid milk production that were related solely to local urban markets. As one would expect in a region of strong urbanization, the milksheds of the major cities generally overlapped and also intermingled with the milksheds of smaller cities, but the basic direct orientation of liquid milk production to urban markets was indisputable. Similar conclusions regarding the persistence of urban-oriented agriculture can be drawn in other developed countries. Most cities have pockets of intensive agriculture on their urban periphery. A study of the general relationship between agricultural intensity and distance from the city in the southeastern United States carried out by Winsberg (1981) found that agricultural intensity was indeed lower in the more remote rural areas than near the cities.

Sinclair's Reversal of von Thünen's Analysis

A rather different view was taken by Sinclair (1967). He suggested that although in developed economies the basic allocating force governing land use is in fact economic or location rent, the major force influencing spatial variation in such rent is no longer simply transport cost to the market but rather the massive urban expansion that has occurred on a scale not envisaged in von Thünen's day. Urban land invariably commands a higher value than rural land, and where the two types of use are in direct competition, urban uses generally win. But land that is *expected* to become urbanized also has a higher value—an "anticipated" value—and this has a considerable effect on the type of land use practiced in rural areas. Recall that in the von Thünen model, land adjacent to the urban market tended to be farmed at the highest intensity. However, Sinclair argued that such land is most likely to become urbanized and thus has the highest anticipated value. Under these circumstances, a landowner or farmer is unlikely to invest large amounts of capital and labor in agricultural production when, by waiting a little while, very large financial gain might be made by selling the land to property developers.

The value of land for agricultural purposes, therefore, according to Sinclair, is *lower* very close to an expanding urban center and increases with distance as the likelihood of urban encroachment declines. This relationship is shown in Figure 1.30 for one type of agricultural land use. The value of land for this type of agriculture is lowest near the urban area but gradually increases

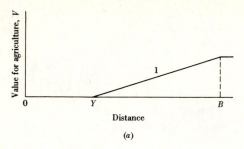

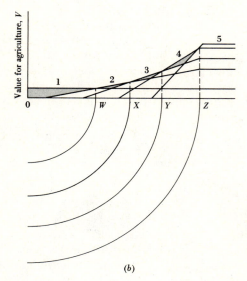

Figure 1.30 Relationship between the value of land for agriculture and distance from the urban center. *Source:* R. Sinclair (1967), "Von Thünen and Urban Sprawl," *Annals of the Association of American Geographers* 57: 72–87, figs. 5 and 7. Reproduced by permission.

outward until at *B,* urban expansion is no longer anticipated and thus has no influence on agricultural land values. In Figure 1.30*b*, a competitive situation among five agricultural types is shown, their competitive position being governed by the steepness of their slopes. These in turn depend on the intensity of agricultural investment.

Using these principles, Sinclair proposed a hypothetical progression of land uses around an expanding urban center in an area such as the specialized feed-grain livestock economy of the Midwest (Table 1.9). Though hypothetical, this scheme does appear to reflect many of the characteristics of agricultural land use around expanding cities in developed economies such as the United States'. In fact, Sinclair backed up his scheme with a variety of empirical evidence. A study by Mattingly (1972) of rural land use around Rockford, near the Illinois-Wisconsin border, produced results that supported Sinclair's hypothesis.

Table 1.9　SINCLAIR'S SUGGESTED LAND USE ZONES AROUND AN EXPANDING URBAN AREA IN A DEVELOPED ECONOMY

Zone	Type of land use
1 (adjacent to urban area)	Land changing to urban use. Subdivision. May be held vacant by speculators. Some "industrialized farming"; poultry, greenhouses, etc.
2	Vacant land. Subdivision not begun. Zone of uncertainty, owners awaiting most profitable time to sell. Land may be leased temporarily for grazing or recreation.
3	Field crop and grazing zone. Low level of intensity. Zone of transitory agriculture.
4	Dairying and field crop zone. Outside "area of anticipation" except at inner margin. But oriented to urban market. Major milkshed.
5	Beyond specific urban influence. Part of nationally oriented agricultural system, e.g., specialized corn belt agriculture.

Source: After R. Sinclair (1967), "Von Thünen and Urban Sprawl," *Annals of the Association of American Geographers* 57: 72–87.

Agricultural Zonation in Developing Economies

In contrast to the situation in developed economies, agricultural land use in less developed countries does seem to be spatially organized in terms of the influence of transportation costs on economic rent. Again, Chisholm (1962) provides a lot of the relevant evidence, though a number of other studies support this view.

Figure 1.31 shows how the percentage of land devoted to each of three types of crops—olives, unirrigated arable produce, and vines—varies with increasing distance from the village of Canicatti in Sicily. The crops grown closest to the village tend to absorb more labor inputs (man-days per hectare). Evidence of similar agricultural zonation has been collected for many of the less developed parts of Europe, for India (Blaikie, 1971), for Brazil (Katzman,

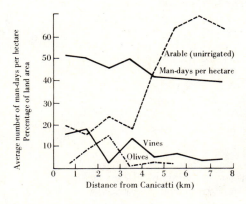

Figure 1.31 Percentage of land area in different crops around Canicatti, Sicily. *Source:* After M. Chisholm (1962), *Rural Settlement and Land Use* (London: Hutchinson), table 6.

1974), and elsewhere, but distance is rarely the sole influence on land use patterns.

Horvath's (1969) study of agricultural patterns around the city of Addis Ababa in Ethiopia is especially interesting because of his suggestion of a direct parallel with von Thünen's forestry zone. Literal interpreters of von Thünen have made much of his apparently anomalous positioning of forestry in the second zone from the city. However, von Thünen was writing in the context of conditions prevailing in the early nineteenth century, when forest products were among the most fundamental and widely used materials by city dwellers. Although this no longer applies in developed economies, it may well still apply elsewhere, as Horvath's work suggests. In Addis Ababa, the importance of timber as the major source of building materials and fuel preserved its position close to the city. This is shown in Figure 1.32 as the zone of eucalyptus forest surrounding the city. Vegetable production, which requires irrigation and a good deal of labor, was located as close to the city as access to water permits. Beyond the forest zone, Horvath identified a zone of mixed semisub-

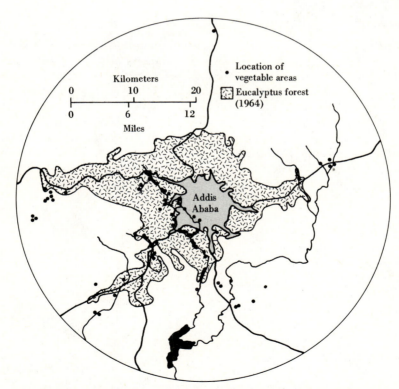

Figure 1.32 Agricultural production around Addis Ababa, Ethiopia. *Source:* R. J. Horvath (1969), "Von Thünen's Isolated State and the Area Around Addis Ababa, Ethiopia," *Annals of the Association of American Geographers* 59: 308–323, fig. 4. Reproduced by permission.

sistence farming with some possible "incipient zonation" within the mixed farming area.

Thus, although the pattern of present-day land use around cities in developed economies differs considerably from our von Thünen–based model, this does not necessarily imply that economic rent is no longer the basic allocating mechanism. It probably is, even though the forces influencing it are no longer dominated by transportation costs from farm to market. By contrast, agricultural land use in less developed countries, where the friction of distance is generally still very high, shows a pattern much closer, though not identical, to that predicted by our location rent model.

Yet such findings do not inevitably mean that the zonation of agricultural production no longer applies to the developed industrial economies. It may not be much in evidence around individual cities in individual countries; but if we alter our lens to the much larger geographic scale of entire continents or even the world as a whole, we may still be able to discern at least some spatial order in the pattern of agricultural production. This is the view of a number of writers, though it is extremely difficult to obtain sufficiently detailed data to confirm such a hypothesis incontrovertibly. The basis of agricultural zonation on a continental or global scale is that such a pattern is oriented toward very large *clusters* of urban markets rather than to individual cities. Figure 1.33 shows one example, that of Western Europe in the early 1950s. A similar broad zonation of agricultural production also seems to exist in the United States, oriented predominantly to the massive urban-industrial complex of the Northeast.

Thus we should not rush too quickly to deny the existence of some distance-based agricultural zonation in advanced industrial economies. It could even be argued, as Schlebecker (1960) does, that on a global scale, the northeastern United States and northwestern Europe form a "world metropolis" (of which New York and London form the axis) around which agricultural production is arranged as a series of gigantic zones.

SPATIAL ORGANIZATION OF LAND USES WITHIN URBAN CENTERS

When we derived the hierarchical structure of central places with its concentrations of functions, we ignored the question of how such functions would be organized within each central place. However, having now explained the way in which location rent "allocates" agricultural land uses in the areas surrounding the central places, we can apply the same basic principle to the internal structure of central places. As Isard (1956) has observed, "In many respects, urban land-use theory is a logical extension of agricultural location theory." The two situations are by no means identical but they have sufficient in common to justify brief consideration here.

We assume that each urban center has within it one focal point—the center—which, on our isotropic plain, is the most accessible location in the

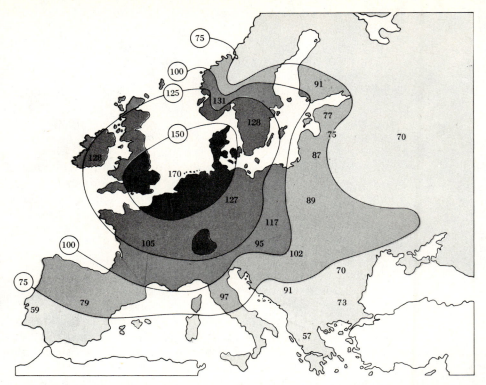

Figure 1.33 Intensity of agricultural production in Europe. The index of 100 was the average European yield of eight main crops. *Source:* After S. Van Valkenburg and C. C. Held (1952) *Europe* (New York: Wiley), map A105.

urban center. Urban land uses will be arranged around that point in exactly the same way as agricultural land uses were arranged around the central place itself, that is, in concentric rings. The basic reason is the same in both cases: land uses compete for the most accessible location and are "sorted out" on the basis of their location rents, which reflect their ability to pay for a particular site. Robert M. Haig (1926) expressed this concept in the following way:

> The center is the point at which transportation costs can be reduced to a minimum. Since there is insufficient space at the center to accommodate all the activities which would derive advantages from location there, the most central sites are assigned, for a rental, to those activities which can best utilize the advantages and the others take the less accessible locations. Site rents and transportation costs are vitally connected through their relationship to the friction of space. Transportation is the means of reducing that friction, at the cost of time and money. (p. 421)

Richard Hurd (1924) expressed the same idea, though rather more concisely: "Since value depends on economic rent and rent on location, and location on convenience, and convenience on nearness, we may eliminate the intermediate steps and say that value depends on nearness." (p. 13). In the simplest terms, therefore, the spatial organization of land uses within the central places on our isotropic plain would look like the general pattern of Figure 1.34. Functions that gain the greatest advantage from locating at the point of maximum accessibility from the innermost zone, with the other uses arranged in sequence according to their location rents. Thus the concentric zonation of land uses, from the center of the city and through urban uses and the various agricultural products to the margin of cultivation, can be attributed to the operation of the following relationships (Alonso, 1960):

1. Land uses determine land values, through competitive bidding among users.
2. Land values distribute land uses according to their ability to pay. This ability depends on the level of location rent accruing to a particular product at a particular location with respect to the market.

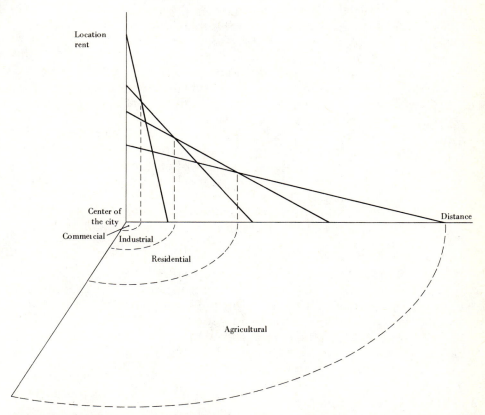

Figure 1.34 Relationship between location rent and the spatial organization of land uses within an urban center.

3. The steeper rent curves capture the central locations: in other words, products that have the most to gain by locating near the market and the most to lose by being farther away end up nearer the center.

SPATIAL REGULARITIES WITHIN URBAN AREAS: EMPIRICAL EVIDENCE

Land Uses Within Urban Areas

In our simplified model, we extended the basic location rent mechanism from its original agricultural context to the organization of land uses within cities. Our expectation based on such theory would be that urban land uses should display a strong concentric regularity around the central point, modified by the extension of zones in a sectoral fashion along major routes. When we try to see how far this pattern actually occurs, however, we face a number of problems. First, we are interested primarily in the arrangement of economic activities, yet, in terms of their share of urban land, these are not the dominant uses: by far the largest user of urban land is the residential sector. Second, cities are exceedingly complex structures that have not only a horizontal dimension but a vertical dimension as well, so that urban land uses tend to intermingle in both dimensions. Third, few if any large cities today have their single most accessible point at their geographic center.

Just as in the case of agricultural location patterns around individual cities in developed economies, we do not find a neat concentric pattern within cities. Such a pattern may well have been present in the past; certainly Park and Burgess's (1925) classic descriptive model of concentric land use patterns in Chicago was a better description of cities at the turn of the century than it is today. Sectoral patterns along major transportation routes are certainly a common feature of modern cities, as are clusters of activities at key locations within the city. Empirical studies of the internal arrangement of economic land uses within cities in developed countries tend to show three important features. First, all cities contain a central business district (CBD), which our rent theory would predict, though especially in North America its importance as a retail center has been declining as peripheral shopping centers have developed. Second, most cities have an industrial zone (not necessarily a continuous one) surrounding the CBD with manufacturing industry extending along particular transportation lines (see Chapter 3) and, more recently, industrial clusters in more peripheral suburban locations. Third, most cities contain a transitional zone close to the center, a zone of incipient speculation similar to Sinclair's urban fringe. In both cases, land tends to be in short-term usage pending possible future development.

Land Values Within Urban Areas

Another way of looking at spatial regularities within cities is to examine the pattern of land values. As we pointed out earlier, land uses tend to be arranged according to their ability to bid for specific locations. The value of land within cities, therefore, should display some spatial order. A number of studies demonstrate some of the regularity we would expect from our theory. The consensus of most of them is that distance from the location of maximum accessibility is an important, though not the sole, factor producing spatial variation in land values.

According to Seyfried's (1963) study of Seattle, for example,

> Market forces result in a structure of site values with the highest value occurring at the location with maximum market accessibility or lowest transportation costs. Thus accessibility tends to centralize site values in a directional sense so that value declines with decreased accessibility at a measurable rate. (p. 283)

Both Knos (1962), in his classic analysis of land values in Topeka, Kansas, and Yeates (1965) in Chicago, identified a similar situation.

Figure 1.35 illustrates the variations in land values in Topeka. Land values precipitously fall within a very short distance of the peak value location. Yeates likewise found that over the entire 1910–1960 period, land values in Chicago were highest in or close to the central business district (the peak land value being at the intersection of State Street and Madison Street). Land values declined markedly with increasing distance from the CBD. But Yeates also pointed out that distance from the center became progressively less important over the 50-year period.

The general configuration of the land value surface is thus far less regular than would be the case if all cities had a single center and equal movement possibilities in all directions. More likely, the actual land value surface resembles that shown in Figure 1.36, where there are a number of peaks and ridges of high values separated by areas of lower values. Nevertheless, at least some aspects of the spatial organization of land values (and therefore of land uses) can be explained by the operation of location rent, though the relationships are much more complex than the simple theory suggests.

MOVEMENT AND INTERACTION IN THE ECONOMIC LANDSCAPE

Bases for Interaction

Interaction—movement of people, goods, or information between places— does not occur in isolation but is stimulated or generated by particular forces. Edward Ullman (1956) suggested that there are three fundamental bases underlying all spatial interaction.* First, for interaction to take place between two places, they must be *complementary* to each other. In other words, there must be a demand-supply relationship between them—one place must want what another place has got, and the latter must be prepared to supply it. We can

---

* Hay (1979) argues that "Ullman's ideas were sufficiently rich to have anticipated inductively many of the more technical advances of the succeeding 20 years" (p. 1).

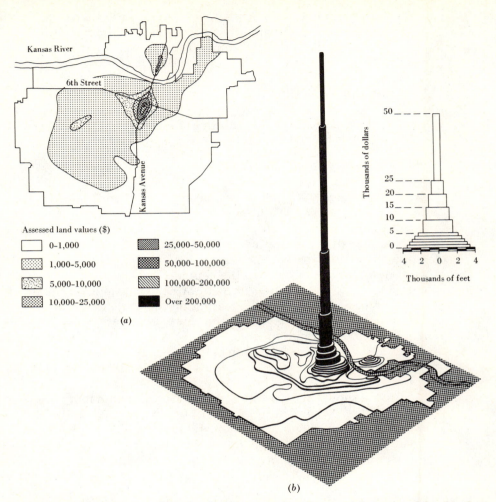

Figure 1.35 The pattern of land values in Topeka, Kansas: (*a*) Assessed land values, 1954–1959; (*b*) Isometric land values. *Source:* D. S. Knos (1962), *The Distribution of Land Values in Topeka, Kansas* (Lawrence: University of Kansas Press), figs. 1 and 2.

illustrate complementarity by referring to Figure 1.14 and examining two of the production centers, for example, G and K. The mix of products at G consists of goods 2, 3, and 6, while at K only products 2 and 5 are produced. A clear complementary relationship exists between these two centers: G can provide K with goods 3 and 6, while K can provide G with good 5.

The concept of complementarity throws additional light on the two types of hierarchies discussed earlier. Figure 1.37 compares the Christaller and Lösch hierarchies in terms of their differing degrees of complementarity. In the rigid Christaller-type hierarchy (Figure 1.37), centers on the same hierarchical level are, by definition, not complementary. They provide exactly the

Figure 1.36 A generalized urban land value surface. *Source:* B. J. L. Berry (1963), *Commercial Structure and Commercial Blight*, University of Chicago, Department of Geography Research Paper No. 85, fig. 3.

same mix of functions and are therefore identical. In a Christaller hierarchy, then, interaction can only occur between one level of the hierarchy and another. Lateral interaction—between centers of the same hierarchical status— is precluded. Thus Christaller's central places are complementary to only a limited extent (defined by the hierarchical marginal good). In a Löschian hierarchy, however, as Figure 1.37*b* shows, interaction can occur in more than one direction within the hierarchy because of its less rigid organization. (As we pointed out earlier, centers of similar order may have different mixtures of economic activities, while low-order places may offer specific goods that higher-order centers—other than the metropolis—do not possess). Allan Pred (1971) suggested that a more suitable solution is one that contains elements of both kinds of hierarchies. In Figure 1.38, for example, in addition to the intercenter links shown in Figure 1.37, there are substantial lateral links between higher-order centers in particular. In each case, the basis underlying such interaction is the complementarity of places, modified, of course, by other factors.

Ullman's second interaction principle relates to the existence of *intervening opportunities*, which can be defined, for our purposes, as alternative sources of supply. For example, one respect in which center K and center G are complementary in Figure 1.14 is that G provides good 3 while K does not. Thus there is a demand at K for center G to provide it with good 3. But this complementary relationship may be negated if, as shown in Figure 1.39, there is an alternative source of supply of good 3 that is nearer to center K. If we imagine the three centers G, J, and K as towns aligned along a highway in the sequence shown, center K will obtain good 3 from the closer town J rather than from G.

Third, according to Ullman, even if there is perfect complementarity and an absence of intervening opportunities, interaction will be reduced, or even be absent altogether, if the cost of such movement—*transferability*—is exces-

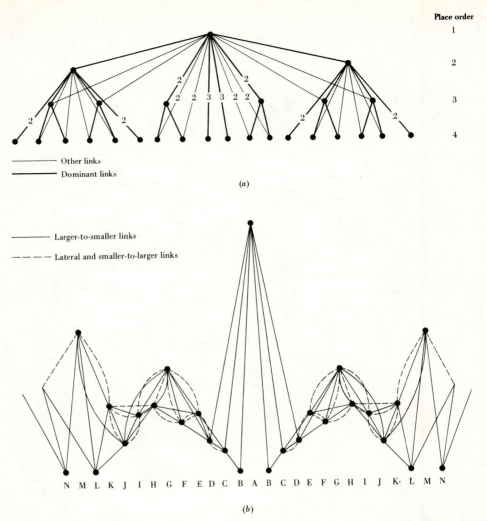

Figure 1.37 Interaction relationships in (*a*) a Christaller-type hierarchy (*k* = 3) and (*b*) a Löschian hierarchy (letters are the same as in Figure 1.14). *Source:* A. Pred (1971), "Large-City Interdependence and the Pre-electronic Diffusion of Innovations in the U.S.," *Geographical Analysis* 3: 165–181, figs. 1 and 3. Reprinted by permission. © 1971 Ohio State University Press. All rights reserved.

sive. This implies more than just distance. Transferability refers to the transport cost characteristics of different goods and is clearly related to the varying movement characteristics of different orders of goods and services and of agricultural products. Some products, as we saw in our discussion of agricultural zonation, are more sensitive to distance than others. We shall return to this point in Chapter 3.

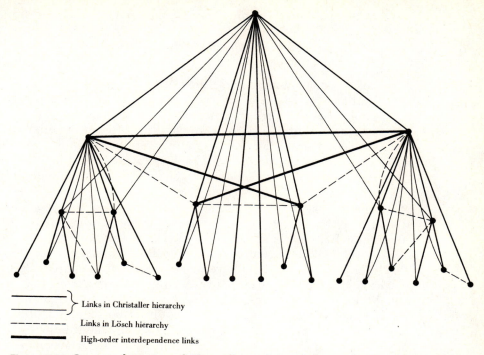

Links in Christaller hierarchy

Links in Lösch hierarchy

High-order interdependence links

Figure 1.38 One combination of Christaller and Lösch hierarchies with links *between* high-order centers. *Source:* A. Pred (1971), "Large-City Interdependence and the Pre-electronic Diffusion of Innovations in the U.S.," *Geographical Analysis* 3: 165–181, fig. 4. Reprinted by permission. © 1971 Ohio State University Press. All rights reserved.

The Gravity Model

If we contend that complementarity represents a force that encourages interaction, whereas both transferability and intervening opportunity represent different aspects of the frictional effects of distance, we can "collapse" Ullman's three principles to two:

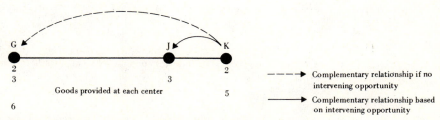

Goods provided at each center

- - - → Complementary relationship if no intervening opportunity

——→ Complementary relationship based on intervening opportunity

Figure 1.39 The modifying effect of intervening opportunity on interaction between complementary places.

1. The *generators* of movement and interaction—the push-pull forces of supply and demand
2. The *restraints* on movement and interaction—the frictional effects of distance

It is useful to do this because it allows us to view movement and interaction in our simplified landscape as a variant on the general physical law of gravity. A few perceptive observers of human behavior, such as Carey and Ravenstein in the nineteenth century, identified a parallel between the migration of people and the basic Newtonian law of gravity. But not until the twentieth century, especially following the work of John Q. Stewart and his "school of social physics," was the *gravity model* systematically applied to social and economic interaction.*

Suppose that we have two urban centers i and j separated by a certain distance d. The gravity concept suggests that the amount of movement or interaction between i and j would, in the first place, be related to the product of the "masses" of these centers (measured, for example, by their population size P). Other things being equal, we would expect, for example, a greater amount of interaction to take place between two very large cities than between two small villages. But other things are rarely equal: in particular, the amount of interaction between two centers is likely to be modified to a considerable degree by the magnitude of the distance that separates them. This, then, is the basic gravity model. Expressed as a simple formula, it becomes

$$I_{ij} = k \frac{P_i P_j}{d_{ij}^b}$$

where I_{ij} = amount of interaction between place i and place j
   $P_i P_j$ = product of the population sizes of the two places i and j
   d_{ij} = distance separating place i and place j
   b = frictional effect of distance; for example, if the value b is 2, this means that the amount of interaction is inversely related to the square of the distance (d_{ij}^2)
   k = an empirical constant

We can interpret the gravity model formula as indicating that the amount of interaction between any two places will be directly proportional to the products of their populations (or some other measure of "mass") and inversely proportional to some power of the distance separating them.†

---

* The gravity model differs from the other models considered so far in this chapter because it is *inductively* based; that is, it is based on empirical observation of regularities in movement. So far, our approach has been *deductive:* we have proceeded by a series of logical steps from a set of initial assumptions.

† Each of the symbols P, d, and b can be expressed in a variety of ways. In fact, the whole question of the interpretation of "mass" and "distance" is complex. Various suggestions have been made regarding the most suitable measure of mass. Population size is most common, probably because of the relative ease of obtaining data, but other suggestions include number of employment opportu-

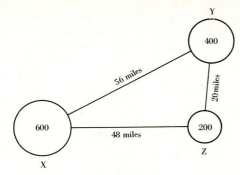

Figure 1.40 Predicted interaction between three hypothetical towns based on the gravity model.

A simple hypothetical example might be the situation shown in Figure 1.40, where there are three urban centers separated from each other by varying distances. If we assume $b = 1$ and $k = 1/3$ (this is simply a "scaling" factor to ensure similarity in the order of magnitude between predicted and observed values), the amount of interaction between these centers would be

$$I_{xy} = \left(\frac{1}{3} \frac{600 \times 400}{56} \right) = 1{,}429$$

$$I_{xz} = \left(\frac{1}{3} \frac{600 \times 200}{48} \right) = 833$$

$$I_{yz} = \left(\frac{1}{3} \frac{400 \times 200}{20} \right) = 1{,}333$$

Thus although X and Y are much larger than Z and would be expected to exert a major attractive force, the actual flow of people is considerably reduced by the 56 miles separating them. For example, if X and Y had been only 20 miles apart, the interaction between them would have been 4,000 instead of 1,429.

MOVEMENT AND INTERACTION: SOME EMPIRICAL EVIDENCE

Movement and interaction of any kind within an economic system is a highly complex phenomenon that is influenced by a whole variety of interrelated variables. Nevertheless, there is general agreement that the intensity of movement and interaction tends to fall off as distance increases.

---

nities, industrial structure, retail sales, per capita income, and vehicle registrations. It is also possible to weight mass in accordance with supposed differences between places in their "propensity to interact." Distance likewise can be measured in a number of ways: straight line, road distance, time, cost, and so on. The exponent applied to distance—b in the gravity formula—is particularly interesting because this represents, in effect, the frictional effect of distance. The higher the value of b, the greater the friction, and therefore the more rapidly interaction falls off with distance. Stewart himself argued that to use other than 1 or 2 was out of step with the model's physical derivation. We shall have more to say about b in Chapter 3.

A detailed review of the gravity model and its development can be found in Haynes and Fotheringham (1984). Important earlier studies include Carrothers (1956) and Olsson (1965).

Zipf (1946, 1949) was one of the first researchers to assemble a large body of empirical evidence on this basis, even though studies using a gravity-type formula date back to the end of the last century. Zipf gathered data on highway, rail, air, and telephone traffic between pairs of cities in the United States and compared them with what would be expected if such flows were directly related to the product of the populations and inversely related to the distance separating them. Figure 1.41 shows his findings for the movement of bus passengers. The actual number of passengers traveling by bus between pairs of U.S. cities is plotted on the vertical axis of the graph, while the predicted value (P_1P_2/D) is plotted on the horizontal axis. Although there is a scatter of points, there is a clear relationship between the two. High predicted values are matched by high actual values, and vice versa. Another example relates to truck trips in the Chicago area (Figure 1.42). Here the distance decay element is more immediately obvious because the intensity of truck trips (measured per 10,000 population) is plotted directly against distance. Again the relationship is not perfect, but the general decline of movement intensity with distance is clearly present.

Studies by Taaffe (1962) and Taaffe and Gauthier (1973) of air passenger traffic in the United States showed how the gravity model ties in with the concept of the urban hierarchy. Taaffe calculated the expected interaction between 100 large United States metropolitan areas using the formula

$$I_{ij} = \frac{P_i P_j}{d_{ij}^2}$$

Each center was categorized as being "dominated" (in air passenger terms) by the center with which it had the highest I_{ij} (that is, expected) value. Exceptions were made if that center was located within 120 miles of the city. Figure 1.43*a* is the map of *expected* air passenger dominance, and Figure 1.43*b* is the map of *actual* air passenger dominance based on observed air passenger flows between the 100 metropolitan areas. Clearly, there was some correspondence

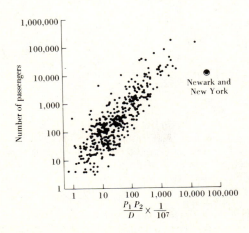

Figure 1.41 The flow of bus passengers between pairs of U.S. cities. *Source:* G. K. Zipf (1949), *Human Behavior and the Principle of Least Effort* (Reading, Mass: Addison-Wesley), fig. 9.15.

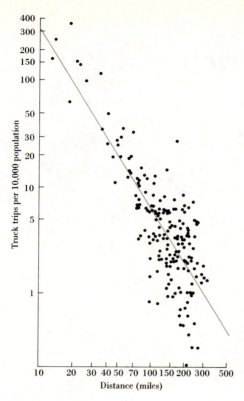

Figure 1.42 Relationship between truck trips and distance in the Chicago area. *Source:* M. Helvig (1964), *Chicago's External Truck Movements*, University of Chicago, Department of Geography Research Paper No. 90, fig. 18.

between the observed and the expected patterns. Equally, however, there were important differences. New York was far more important to the cities of the U.S. manufacturing belt than the gravity model predicted; its air passenger hinterland extended a good deal farther west to incorporate Detroit, Cleveland, and Cincinnati, in particular. In this respect, therefore, Taaffe points out, the gravity model overestimated the frictional effect of distance. On the other hand, some of the interaction linkages between New York City and some southern cities were weaker than predicted.

Although there is a good deal of agreement that the intensity of interaction declines with distance, there is far less unanimity regarding the *rate* at which such falloff occurs. In other words, what is the "frictional value" of distance? As we saw earlier, this frictional effect is incorporated in the gravity model by fitting an exponent (b) to the distance measure. For example, Zipf's study used an exponent of 1, whereas other studies have used exponent values varying from a little less than 1 to around 3. Almost certainly, there is no uniquely correct exponent of distance. It is likely to vary from one type of movement to another; from place to place, depending on such factors as the efficiency of the transportation network and degree of congestion; and from time to time because as transportation improves the frictional effect of distance is likely to

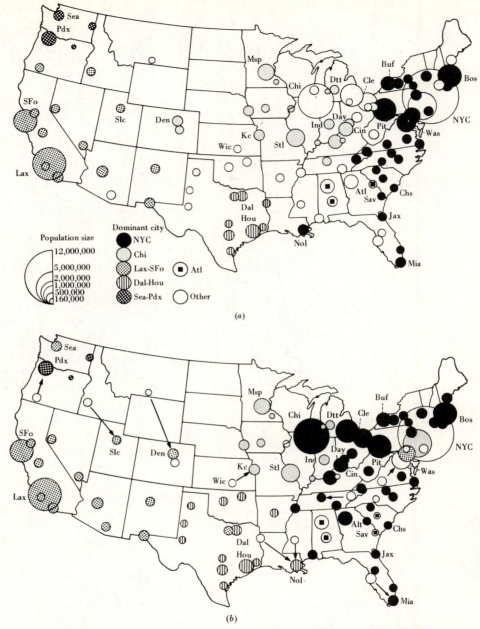

Figure 1.43 Expected and actual air passenger dominance in the United States, 1962: (a) The city circles are shaded according to the city expected to be dominant according to a modified gravity model in which a distance exponent of 2 was used and all cities within 120 miles of each other were excluded from the calculations; (b) Shading shows actual dominance. *Source:* E. J. Taaffe and H. J. Gauthier (1973), *Geography of Transportation* (Englewood Cliffs, N.J.: Prentice-Hall), figs. 3.8 and 3.9. © 1973. Reprinted by permission.

decline (see Chapter 3). Clearly, the frictional effect on a journey between, say, New York City and San Francisco was a great deal higher in the days of the '49ers than it is today.

A SUMMARY VIEW OF THE SIMPLIFIED ECONOMIC LANDSCAPE

The most important conclusion to be made is that even if a perfectly uniform land surface were to exist, economic activities would still tend to be organized spatially in a clearly differentiated manner due to the effect of distance. Such organization is, in theory, both complex and orderly. Its particular form and structure are derived from the patterned nature of economic behavior as goods and services are produced, exchanged, and consumed within the specific assumptions on which our model is based.

Over space, this patterned behavior takes the form of interaction: the movement of consumers to central places, of goods to consumers, and of agricultural products to central markets. The sensitivity of these movements to the attenuating effects of distance, expressed in transportation cost, provides the key to the particular spatial order displayed. The spatial economic system is then a complex of interacting elements, producers and consumers, farms and central places connected by continuous flows of goods, services, people, and information. It functions as an integrated whole, and although its various parts play an individual role in the operation of the system, no part is wholly independent of the others. A change in the functional role of one part will have an important effect on the operation of all the other parts and on the functioning of the total complex. The *space economy* generated under such idealized conditions therefore represents a system by which people seek to satisfy their wants from the means of production at their disposal in the face of the spatial disparity between those wants and the resources necessary for their satisfaction.

Empirical evidence reveals that there is a considerable degree of order and regularity in the spatial organization of economic systems, although such regularity is a good deal less than the theoretical models suggest. Central places in particular and urban centers in general tend to be hierarchically organized, and this is reflected in both functional structure and population size, as well as in the flows of consumers to centers of different hierarchical levels and, to a lesser extent, in the regularity of their spatial patterns. Similarly, there is a degree of spatial regularity in the pattern of agricultural production and of urban land uses, though, again, this does not correspond exactly to the pattern predicted. Finally, interaction between places does appear to show a fairly strong relationship with distance.

Of course, the degree of fit between such theoretical models and reality is far from perfect. Each of the theoretical works discussed in this chapter was based on some very rigid simplifying assumptions. We have deliberately held all variables, apart from geographic distance, constant in order to concentrate on the operation of that single factor. We must now begin progressively to relax these constraints, in other words, to introduce factors other than distance into our discussion.

FURTHER READING

Berry, B. J. L., H. G. Barnum, and R. J. Tennant (1962). "Retail Location and Consumer Behavior," *Papers and Proceedings of the Regional Science Association* 9: 65–106.

Berry, B. J. L., and W. L. Garrison (1958). "Functional Bases of the Central Place Hierarchy," *Economic Geography* 34: 145–154.

Berry, B. J. L., and W. L. Garrison (1958). "A Note on Central Place Theory and the Range of a Good," *Economic Geography* 34: 304–311.

Berry, B. J. L., and W. L. Garrison (1958). "Recent Developments of Central Place Theory," *Papers and Proceedings of the Regional Science Association* 4: 107–120.

Chisholm, M. (1962). *Rural Settlement and Land Use*. Chicago: Aldine.

Christaller, W. (1966). *Central Places in Southern Germany*, trans. C. W. Baskin. Englewood Cliffs, N.J.: Prentice-Hall. (Originally published 1933.)

Hall, P. (1966). *Von Thünen's Isolated State*. London: Pergamon.

Haynes, K. E., and S. Fotheringham (1984). *Gravity and Spatial Interaction Models*. Newbury Park, Calif.: Sage, chaps. 1–2.

Isard, W. (1956). *Location and Space Economy*. Cambridge, Mass.: MIT Press, chap. 3.

Lösch, A. (1954). *The Economics of Location*. New Haven, Conn.: Yale University Press, chaps. 9–12. (Originally published 1939.)

Mulligan, G. F. (1981). "Lösch's Single Good Equilibrium," *Annals of the Association of American Geographers* 71: 84–94.

Parr, J. B. (1978). "Models of the Central Place System: A More General Approach," *Urban Studies* 15: 35–49.

Preston, R. E. (1985). "Christaller's Neglected Contribution to the Study of the Evolution of Central Places," *Progress in Human Geography* 9: 177–193.

Sinclair, R. (1967). "Von Thünen and Urban Sprawl," *Annals of the Association of American Geographers* 57: 72–87.

Ullman, E. L. (1956). "The Role of Transportation and the Bases for Interaction." In W. L. Thomas (ed.), *Man's Role in Changing the Face of the Earth*. Chicago: University of Chicago Press, pp. 862–880.

Chapter
2

A Heterogeneous Land Surface

RELAXING THE SIMPLIFYING ASSUMPTIONS

The empirical evidence presented in Chapter 1 revealed that the operation of a single variable—distance—even in its most simplified form is responsible for some of the spatial variability in the pattern of economic activities. Clearly, however, the gap between the model situation and the real-world situation is still considerable. To reduce this gap, we need to relax some of the rigid simplifying assumptions on which the theoretical model of Chapter 1 was based. For ease of explanation and to help us to understand the spatial impact of each factor, the constraints will be relaxed one at a time. However, because all of these factors are interrelated, there is no unique sequence in which this operation should be performed.

The fundamental concern of Chapter 1 was the relative location of economic activities in space. Given a homogeneous plain with resources of uniform quantity and quality available everywhere, the problem was to determine the distances that separated economic activities. The resulting pattern was one of spatial order and regularity in relative location. In the central place system, producers of the same good were uniformly spaced, while in the agricultural system, production patterns followed an ordered spatial zoning.

In both cases, the only variable costs to each producer were those derived from the movement of evenly distributed input factors and of final products to an evenly distributed market. Implicitly, of course, there were also cost differentials derived from the lower unit costs of large-scale enterprise; without them, there would be no incentive to shift from a subsistence economy in the first place. Apart from this, however, no allowance was made for other variable costs of production.

In the real world, things are very different. Many other factors enter the determination of production costs in space. Transportation costs for both input factors and final products are highly variable over space and vary widely with the type of medium used, the nature of the terrain and the type of good carried. Factors of production like labor, capital, and technical knowledge are not available everywhere, nor are they all infinitely and equally mobile. Perhaps the first thing to strike the geographer about the unreality implicit in the model is the fact that material resources are in no sense of equal quality or ubiquitous in location. We shall begin dismantling the constraints of the model with this question of variable material resources, taking each of the other "unrealistic" constraints in their turn in subsequent chapters.

SPATIAL VARIATIONS IN RESOURCE QUALITY AND AVAILABILITY

For the present, then, we retain all the constraints of the model which we set up in Chapter 1 with the exception of that dealing with spatial variations in resource quality. We now want to look at the way natural resources of all kinds influence the location of economic activity, and this takes us on to look at some of the traditional variables of economic geography.

The spatial distribution of industrial raw materials and sources of energy exerts its most powerful locational influence in the mining and manufacturing industries. The model constraints have given us little opportunity to explore these economic activities so far. Indeed, one of the main conclusions of Chapter 1, in which the simplified model was tested against conditions in the real world, was that there was a relatively poor correspondence in the case of manufacturing. The most obvious, though by no means the only, reason for this was the model's assumption of a uniform availability of materials of homogeneous quality. In reality, most industrial raw materials and energy sources are restricted in their spatial availability. They are localized rather than ubiquitous (available everywhere), as the model assumed them to be. Thus the Lösch-Christaller principles around which the simplified model was constructed are more applicable to the location of tertiary activities and the manufacturing industries that use such ubiquitous materials as do exist, such as air (oxygen manufacture) or water (soft drinks). For them, transportation costs are most important from the point of view of the distribution of the good or service, but most manufacturing industries must use some localized materials, and thus the cost of assembling material inputs from a variety of sources is a significant component of this total cost structure. What effect, then, will the procurement of spatially localized materials have on location patterns? We can begin to answer this question by drawing upon some of the ideas of Alfred Weber (see Box 2.1).

WEBER'S ANALYSIS OF THE MINIMUM TRANSPORT POINT

In the absence of spatial differences in basic production costs, Weber observed that manufacturing plants will locate at the point where total transportation costs are minimized. He suggested that transportation costs are, in effect, determined by two factors:

1. The *weight* of the materials to be assembled together with the weight of the final product to be shipped to market
2. The *distances* over which the materials and the product have to be moved

---

Box 2-1   **Alfred Weber**

Alfred Weber, a German economist, first published his classic work *Über den Standort der Industrien* in 1909. This followed a long tradition of German interest in the economics of location. It was begun in 1826 by von Thünen, whose work formed the basis for our discussion of agricultural location in Chapter 1. The translation of Weber's book into English in 1929 as *The Theory of the Location of Industries* assured him of wider recognition as the acknowledged pioneer of this aspect of location theory. In any case, however, Weber's work was far more comprehensive and rigorous than that of any of his predecessors. The particular format of our treatment in this book denies us the opportunity to treat Weber's theory as a whole, and you are encouraged to consult, for example, Daggett (1968) and Smith (1981) for a more rounded view. Gregory (1982) has attempted to throw new light on Weber's contribution. In brief, however, what Weber set out to do was find the answer to the question, What causes an industry to move from one place to another? In seeking his answer, he made use of the model approach of the classical location theorists, setting up, as we did in Chapter 1, a series of constraints that assumed a flat plain, conditions of perfect competition, and so forth. What he was looking for were the general factors that influenced the location of manufacturing industries.

In essence, Weber saw the location of industry as a response to two interconnected sets of forces: *primary causes* for the regional distribution of industry (regional factors) and *secondary causes* of the redistribution of industry (his agglomerating and deglomerating factors). At the regional level, his general regional factors were the costs of transportation (our particular concern in this chapter and in Chapter 3) and labor costs (which are taken up in Chapter 4). The secondary factors by which Weber saw industry to be redistributed within the regional context are taken up under the heading of scale economies in Chapter 5.

The combination of these two elements permits him to come up with a simple index of cost, the *ton-mile*. The locational problem is then simply to find the point where the total ton-mileage is minimized for the particular production-distribution process.

The key that determines the suitability of a particular location for a manufacturing activity is the total number of ton-miles accumulated at that site. If every possible site is examined in this way, the site that accumulates the lowest total of ton-miles for assembling materials and getting the product to market is the best location (that is, it is the minimum-transport-cost location). In looking at the ton-mileage calculations for a variety of situations, however, it becomes obvious that certain key principles of location can be extracted, making copious calculation unnecessary. These give some general indications as to the best sites for potential classes of industry, depending on certain characteristics of the materials they use.

1. *Ubiquitous materials*, available everywhere, exert no separate locational force on manufacturing. For industries involved only in assembling such materials, the principles of marketing alone condition locational choice, and the best location will be at the market site. (This is the simplified model case).*

2. *Localized materials* exert a specific influence on location. They are of two types, depending on the ratio between the weight of the material and the weight of the product produced from it. For *pure localized materials*, the entire weight enters into the product. For example, yarn for the manufacture of cloth can be regarded as a pure material because no loss of weight occurs in the manufacturing process. *Gross localized materials* do suffer a loss of weight in the process of manufacture. Perhaps the most widely quoted example of this is the manufacture of sugar from sugar beets. The weight of raw sugar extracted by this process is only one-eighth of the weight of the raw material that goes into the process of extraction.

3. The *material index* (MI), calculated as follows:

$$MI = \frac{\text{weight of localized materials used in the industry}}{\text{weight of the product}}$$

shows the extent to which a particular industry will be material-oriented or market-oriented. A material index greater than 1 indicates a tendency toward a material site location. In this case, the sum of the

---

* Weber assumes the market for manufacturing to be concentrated at one or a number of discrete points in space rather than dispersed. Isard (1956) suggests that this is not conceptually different from the market assumed in the central place model. It merely represents a special case derived from a change in the scale at which the market is treated. For the central place model, the household is the smallest point of reference; for Weber's analysis of manufacturing, the scale of the smallest unit is at the higher level of, say, the village or town.

weights of the localized materials used in the industry is greater than the weight of the product. In industries where ubiquitous materials are involved, however, it is possible that the product may have a weight greater than that of the localized materials. Take the case of brewing. The heaviest material used by far is, to all intents and purposes, ubiquitous; that is water. Compared to the weight of the beer itself, then, (mostly water), the weight of the localized materials such as hops, sugar, and chemical additives is small, and the resultant material index will be less than 1. A material index of less than 1 indicates a tendency to favor market location.

For industries using only pure localized materials or where the balance of gross materials and ubiquities gives a material index exactly equal to 1, then, in theory, location can be at the materials site, the market, or any point in between. For reasons connected with transportation rates, a subject taken up in the next chapter, intermediate locations are in fact less likely to result in this case than locations fixed at either the market or the materials site.

Having established some of the basic principles of Weber's approach to the location of manufacturing, let us consolidate our understanding with the help of a simple example. Take the case of a manufacturer who requires only one raw material and sells his entire output at a single market (Figure 2.1). Location depends simply on the nature of the material used in production. If it is gross (weight-losing), the production site will tend toward the material source. For example, suppose the manufacturing process is the refining of sugar beet, a material that, as we saw earlier, contributes only about one-eighth of its weight to the final product. A location at any point other than the raw material source would involve paying unnecessary transportation costs for the shipment of the seven-eighths of each beet that are not required in the manufacturing process. The same argument would apply in the case of any other gross material. Why pay transportation costs to ship unwanted material? Only pure materials, whose entire weight enters the final product, could be processed at a location other than the source. Thus in the single-market, single-material situation, all production processes with material indices greater than 1 will be located at the sources of their materials, while those using pure materials will, in theory, be free to locate at any point along the lines connecting their markets and material sources.

Clearly, the two-point case is too simplistic; very few, if any, manufacturing processes use only a single material input. However, even where the production process is more complex, the same basic principles apply. Figure 2.2 illustrates the situation in which a manufacturer requires one raw material and an input of fuel, both derived from separate localized resources, and ships

Market   *Y*   *Z*   Material source

Figure 2.1 Simple location problem: a firm using one material and selling to a single market.

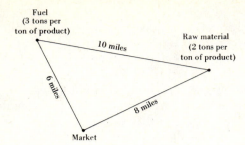

Figure 2.2 Location of a firm using two inputs and selling to a single market.

the product to a single market. By looking separately at each of the three extreme points of the triangle and calculating the ton-mile total at each of them, we can establish which of them is the "best" location.

The total ton-mileage that would be incurred in manufacturing 1 ton of the product at each of the locations is as follows:

LOCATION AT THE MARKET

Fuel cost: 18 ton-miles (3 tons over 6 miles)

Raw material cost: 16 ton-miles (2 tons over 8 miles)

As the product would be manufactured at the market, no transportation cost is incurred in marketing the product.

Total transportation cost at market: 34 ton-miles

LOCATION AT THE FUEL SOURCE

Raw material cost: 20 ton-miles (2 tons over 10 miles)

Marketing product: 6 ton-miles (1 ton over 6 miles)

No transportation cost on fuel because manufacture would be at the fuel source.

Total transportation cost at the fuel source: 26 ton-miles

LOCATION AT THE RAW MATERIAL SOURCE

Fuel cost: 30 ton-miles (3 tons over 10 miles)

Marketing product: 8 ton-miles (1 ton over 8 miles)

No transportation cost on the raw material because manufacture would be at the raw material source.

Total transportation cost at raw material source: 38 ton-miles

In this case, the fuel source is the minimum-transportation-cost point. Only by locating the plant there could the manufacturer avoid the heavy costs involved in moving around those 3 tons of fuel that he needs for every ton of product.

The material index for this industry is 5; that is, every ton of product requires 5 tons of localized materials (3 of fuel and 2 of raw material). We should have known, then, without the ton-mile calculation that the industry would have a strong tendency to be material-oriented. It is also a general rule that where a single gross localized material input exceeds the sum of all the others, the least-cost location will be the source of that input. A quick glance could therefore have told us that the fuel source was the best location and that any deviation from it would incur costs that no profit-maximizing manufacturer would be willing to bear. There was in this case no possibility of an intermediate location.

An intermediate location is possible, however, when the ton-mile cost of any single input (or output) does not exceed all others. In such circumstances, the locational polygon is useful. Weber described the locational problem in terms of a "struggle" between the corners of the polygon (each corner representing an input source or a consumption point). The outcome of the struggle, however, still depends on the material index. One method of determining the minimum-transportation-cost point is to use a mechanical model such as the *Varignon frame,* in which distances and weights in the locational polygon are simulated by appropriately scaled weights and pulleys connected by wires. The point where the connected wires come to rest at a point of balance is the optimal location. Figure 2.3 illustrates this diagrammatically. The respective weights represent the strength of the attractive force of each corner of the polygon.

Table 2.1 shows for various combinations of pure and gross localized and ubiquitous materials the particular locational tendencies that would arise. Again this provides a shortcut to judging the location that a given industry

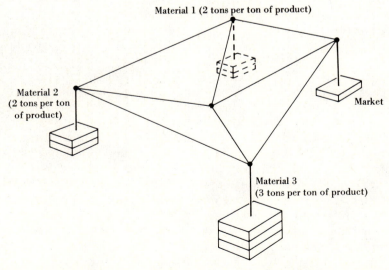

Figure 2.3 A mechanical solution to the multipoint location problem.

Table 2.1 INFLUENCE OF WEBER'S MATERIAL TYPES ON THE LOCATION OF A MANUFACTURING PLANT

Types of materials used	Material index	Location[a]		
		Material source	Intermediate location	Market
Ubiquitous material only				+
1 pure material	1	=	=	=
1 pure material + 1 ubiquitous material	<1			+
> 1 pure material	1			+
> 1 pure material + > 1 ubiquitous material	<1			+
1 gross material	>1	+		
1 gross material + ubiquitous materials	<1			+
> 1 gross material	>1	→		
Gross materials + pure materials	>1	→		
Gross materials + pure materials + ubiquitous materials				→

[a] + indicates definite location; = indicates equally possible locations; → indicates a tendency for production to be attracted toward a particular type of location.
Source: After Weber (1909). Hamilton (1967).

might choose without going through the arithmetic of the ton-mile calculation. The definitive locational types, marked by plus (+) signs, are interestingly biased toward the market location. Only two definite material locations occur: those where there is either a single gross material or where there is one gross material and a ubiquitous material. The reason for this is, of course, that a number of gross materials may neutralize the pulls of each other at the corners of a locational polygon, leaving an intermediate location as the resolution of these forces. This will tend toward a material location but may not be sited directly at one. Weber has often been criticized for overemphasizing the attractive force of raw materials in his model, and the number of pluses in the market column shows how frequently a market location is a final solution for many combinations of raw materials.

There is little doubt that Weber's distinction between ubiquitous and localized materials and between pure localized materials and gross (weight-losing) localized materials has had a profound influence on the economic geographer's approach to industrial location problems. But how efficient in reality is the material index in predicting the orientation of the manufacturing industry? The most comprehensive test of the index was undertaken by Smith (1955). Basing his pioneering analysis on 65 British industries using 1948 census data, Smith investigated the extent to which weight-losing materials were tied locationally to their raw materials.

In the case of primary industries, the initial processing stage of manufacturing, the relationship between high-weight-loss materials and location of production was strong. Thus, for example, sugar beet (MI = 8), manufactured dairy products (MI = 6), and pig iron manufacture (MI = 3 to 4) were all strongly material-oriented. Table 2.2 summarizes Smith's findings for industries located entirely at the material source, partly at the material source, and at other locations. Weber's claim that industries with a material index of greater than 1 tend to locate at material sources was validated. Of all 22 industries sited at material locations, not one had an index of less than 1. Of those "partly at materials," 9 out of 12 had a material index above 1. For industries defined as "not located at materials," Smith's findings are however less clear, but so is the basis for the classification. The group of eight industries with a material index between 2 and 5 classed as "not located at materials" invites caution in applying Weber's simple principles to real-life situations. As Wilfred Smith (1955) pointed out, "The material index provides us with a tool of analysis but it is a blunt tool and is effective only at the very extremities of the classification" (p. 109).

One of the problems with the material index is that by standardizing inputs and outputs as a ratio, it loses the power to distinguish the sheer size of the quantities of materials that some industries have to move. For those already heavily committed to materials in the sense that they have a high material index, this is no problem. There are, however, many middle-range industries whose relatively low material indices fail to indicate strongly enough how far they are committed to a material location because of the scale of the raw material movements they need to undertake. Smith (1955) suggested combining an index of the weight of materials per operative with Weber's material index to make this clearer: "Loss of weight has significant locational effects only when it is combined with large weight per operative, for variations in transport costs are substantial enough to affect location only if weights handled are large" (p. 111). He goes on to make the point that for some industries, like engineering, which produce large amounts of waste in processing, the material index itself is misleading, and this in part explains the tendency shown in Table 2.2 for some industries with high material indices not to be located at material sites. Though some engineering industries have a high propensity to

Table 2.2   OBSERVED RELATIONSHIP BETWEEN WEBER'S MATERIAL INDEX AND LOCATIONS OF MANUFACTURING IN 65 BRITISH INDUSTRIES, 1948.

Location	Material index				
	>5	2–5	1–2	<1	Total
At materials	2	4	16	0	22
Partly at materials	0	4	5	3	12
Not located at materials	0	8	12	11	31
Total	2	16	33	14	65

Source: After W. Smith (1955). "The Location of Industry." *Transactions of the Institute of British Geographers* 21: 7, table 1.

generate scrap in their operations, the weight of material per operative in these industries is small. Using the two indices in combination makes it easier to see why engineering industries in general are not tied to materials, despite the fact that some of them have high material indices.

Further evidence of the relevance of Weber's formulation to the location of certain industries is provided, for example, by Kennelly (1954), Craig (1957), and Lindberg (1953). One conclusion of Kennelly's classic study of the location of the Mexican steel industry was that the industry was located in accordance with Weber's transport orientation principle and that in this case, weight and distance were the principal factors of transport cost. Also, the Weberian distinction between ubiquitous and localized material was found to be valid. On the other hand, Kennelly found the material index to be inadequate because it emphasized the relative weight of materials and products at the expense of their relative locations. Similarly, Craig's investigation of location factors influencing the development of steel centers in the United States revealed that its basic transport orientation could be formulated in Weberian terms. Lindberg made a detailed study of the Swedish paper industry. He found that although the industry was not, as commonly believed, oriented toward materials, it was undoubtedly located so as to minimize the cost of transporting materials and product.

On the intraurban scale, too, there is abundant evidence of the operation of Weberian locational principles in the nineteenth-century cities of North America and Western Europe.* What Scott terms the "large-scale materials-intensive manufacturing activities" were pulled by Weberian forces to specific kinds of location:

> In an economic system in which the intercity movement of goods by rail or water was cheap compared to the intracity movement of goods, such industries had to minimize the costs of assembling input commodities and established themselves near the transport terminals which were the effective location of imported materials. Such industries became concentrated near transport nodes and formed the early manufacturing heart of 19th century cities. (Webber, 1982, p. 205)

The kinds of industries involved in such locational orientation included cattle slaughtering, blast furnaces, foundries, and brick production (Scott, 1982).

Thus Weber's work shows how the existence of localized materials might affect the location of manufacturing industry. In particular, it points out how the need by some industries for gross localized materials will draw them to resource sites. This distorts the logic of a central place location under the predictions of the simplified model for some, but by no means all, types of manufacturing.

---

* Both Scott (1982) and Webber (1982) provide detailed discussions of manufacturing location and change within cities.

ISARD'S SUBSTITUTION FRAMEWORK

Walter Isard (1956) took Weber's basic theoretical logic and greatly enhanced its scope and flexibility by placing it in the context of *substitution* analysis. Although this added little to the theoretical implications of the model, it did make it a more powerful predictive tool, and a review of Isard's technique at this stage will greatly simplify later discussion. The fundamental basis of Isard's approach can be illustrated by looking again at the simple two-point location problem shown in Figure 2.1. Let us assume that location can occur at any point along the straight line connecting the material source and the market, as would be the case if the material were pure. There are thus two variables in this locational situation: distance from the market and distance from the material source. This relationship can be plotted graphically, as in Figure 2.4, to yield a *transformation line,* which expresses all possible substitution re-lations for the two variables. The market and the material source are 7 miles apart. Thus the two axes have a length of 7 units. At the extreme points, a location at the market is 7 miles from the raw material source and, of course, 0 miles from the market. Conversely, a location at the material source is 7 miles from the market and 0 miles from the material source. On the transformation line that connects these two extreme points, all the possible combinations of distance adding up to 7 miles are shown. At point Y, location is 6 miles from the material source and 1 mile from the market. At Z, it is 6 miles from the market and 1 mile from the material source. Each mile over which shipment must be made represents a necessary transport cost input that the producer will have to bear. Location at Y, then, involves "6 miles' worth" of transport input on materials and "1 mile's worth" of transport input on the finished product. Of course, the two-point problem, especially for a pure material, is a trivial one since we already know that location at any point along the straight line is equally costly and that the "7-mile" transformation line for distance inputs covers all the possibilities. It does, however, help us to get the transformation

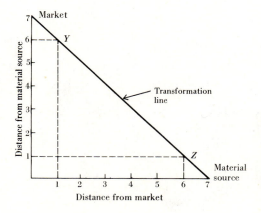

Figure 2.4 A transformation line for the two-point location problem. *Source:* W. Isard (1956), *Location and Space Economy* (Cambridge, Mass.: MIT Press), fig. 16. Copyright © 1956 by the Massachusetts Institute of Technology. Reprinted with permission.

principle established in simple terms before we go on to make it more complicated.

Suppose now that we add a second raw material to the producer's input
requirements. Now not just two but three sets of substitution relations must be
taken into account. Take the case set out in Figure 2.5. Transformation lines for
distance inputs can now take the following form:

1. For any fixed distance from the market, there will be a transformation
   line showing possible substitutions between the distance from
   material A and the distance from material B.
2. For any fixed distance from material A, there will be a transformation
   line that shows substitutions of distance from material B against the
   distance from the market.
3. For any fixed distance from material B, there will be a transformation
   line that shows substitutions of distance from material A against distance from the market.

These cover all the substitution possibilities. The reason for fixing one of the
distance variables each time is simply because the graphical approaches used

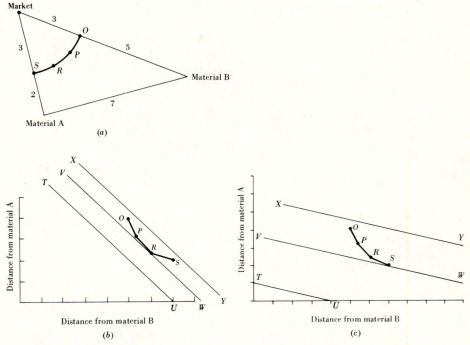

Figure 2.5 Solution of the three-point case using a transformation line. *Source: (a), (b),*
W. Isard (1956), *Location and Space Economy* Cambridge, Mass.: MIT Press, figs. 17
and 18. Copyright © 1956 by the Massachusetts Institute of Technology. Reprinted by
permission.

in generating transformation lines demand that the problem be stated in two-dimensional form. In effect, we "creep up" on the solution to the problem by looking at *pairs* of distance inputs from two of the corners of the triangle while we give the third corner a controlled series of fixed-distance values. In a sense, then, we search the inside of the triangle by giving the fixed corner a series of arbitrary distance values at a radius of, say, 1 mile, 2 miles, and so forth, stopping at each stage to look at the transformation relationship between the two corners, whose distances are free to vary.

The overall minimum-cost point is found by identifying the point at which total mileages are at their lowest for each of the three individual sets of transformations. In effect, the minimum-mileage point for each pair of transformations forms a partial solution, and the partial solutions are used as "stepping-stones" in finding the optimum (or minimum-transport-cost) location. Figure 2.5*a* shows a transformation example for the first of the possible cases of substitution. Here distance from the market is fixed at an arbitrary 3 miles. If we assume, to make things simple, that 3 miles is known to be the "best" distance from the market, we can follow the process through. Figure 2.5*b* shows the transformation line for the variables labeled "distance from material A" and "distance from material B" (with distance from the market fixed to the distance represented by the arc *OS*). In this example, it is assumed for the sake of simplicity that location can take place at only one of four discrete points *O*, *P*, *R*, *S* along the arc (perhaps because these are the only points located on direct transportation routes between materials and market). The optimal location will be the one that incurs the minimum total cost of both material A and material B to some point 3 miles from the market. If we assume that material A and material B are either both pure materials or have equal weight-losing properties, there is no problem. The best location is simply that with the *least accumulated mileage*. On the graph, it will be the point on the transformation line nearest the origin—*R* in the case shown in Figure 2.5*b*. The lines *TU*, *VW*, and *XY* are known as *iso-outlay lines* (see Box 2.2), and they would help us find the point nearest the origin if it were not immediately obvious. In this particular case, the iso-outlay lines simply connect the vertical and horizontal scales of the graph in a one-to-one ratio. One unit on the vertical axis joins one unit on the horizontal axis, two joins two, and so on. This is consistent with the particular situation where material A has equivalent weight per ton of produce to material B. Their price ratio in terms of the costs of moving them is 1 : 1. The lines drawn on the diagram are just a selection from an infinite number of lines rising upward and to the right in a one-to-one ratio. It is easy to visualize the process in terms of a parallel ruler being moved upward and to the right from the origin of the graph until it just touches the point marked by *R*.

In the specific case under consideration here, let us remind ourselves that what we are trying to do in Weberian terms is to minimize the number of ton-miles moved by a particular production unit. So far in the case of substitution analysis, we have said a good deal about miles but very little about tons. Only in the case where there are pure materials or where there is no weight-loss differential between materials can we get away with this. To avoid getting

Box 2-2   **Iso-outlay Lines**

Iso-outlay lines show the possible alternative combinations of goods or resources (or in our specific case, transportation services) that can be obtained for a *given value of total expenditure.* The concept appears under a variety of titles in economics. Price-ratio lines, budget lines and consumption-possibility lines all derive from the same idea, differing only in context, viewpoint, or terminological choice.

To make things clearer, let us take a simple example. Suppose that in spending the surplus money we have at the end of the month, we are presented with a straightforward two-way choice. For the same amount, say, $60, of spending money, our preferences are for either textbooks, which sell at a fixed $20 each, or compact disks, selling at $10 each. Our consumption possibilities within this budget are not difficult to determine. We could buy three books, six disks, or any finite combination of the two items (no half books or half disks). For the $60 level of outlay (iso-outlay), we could draw up the following schedule of choices:

CONSUMPTION POSSIBILITIES FOR A BUDGET OF $60

Books at $20.00 each		Compact disks at $10.00 each		Total outlay ($)
Amount	Cost	Amount	Cost	
0	0	6	60	60
1	20	4	40	60
2	40	2	20	60
3	60	0	0	60

If we were to draw a simple graph of the available choices with the number of books on one axis and the number of disks on the other, we would derive the iso-outlay line for $60. The slope of the line would reflect the price ratio between the two types of good. In this case, 2 : 1 is the ratio of the price of books to the price of disks. Graph the books-disks iso-outlay line and experiment with other sets of binary (two-variable) choices.

involved in dollars and cents, and in accordance with Weber's basic ideas, we have so far assumed that the costs that a producer bears are measured solely in ton-miles. Thus his outlays on production are ton-mileage amounts. Bearing this in mind, let us return to the example of Figure 2.5 under slightly changed circumstances. Suppose that material A contributes four times as much weight to each unit of the product as B. For a given total outlay (measured in ton-miles) by the producer, material A would have a price-ratio four times that of material B.

Let us begin by setting a budget on the amount our producer could afford to spend in ton-mile terms. Suppose that he can afford the money equivalent of only 4 ton-miles. Figure 2.5c shows that he could have 4 full units of material B and no material A (the point U). At the other extreme, he could have just 1 unit of material A and no B (the point T). He could also choose any combination that lies between these limits (the line TU). The iso-outlay lines for this situation are shown in Figure 2.5c. Their slopes now reflect the additional weight that must be given to material A in the transport calculation. Our producer can afford to carry material A over fewer miles than material B if he is to stay inside his budget. The whole range of iso-outlay lines above and below 4 ton-miles now slopes parallel with the new iso-outlay line TU. Going back to Figure 2.5a, it can now be shown that the "best" location has shifted from R to S, that is, closer to the more expensive (in ton-mile terms) material source and slightly farther away from the cheaper material source.

SMITH'S SPACE-COST CURVE

Smith (1966) provides another technique that will be used in subsequent chapters. Like Isard's substitution method, it can be helpful in relatively complex situations, but it can also be introduced here in its most elementary form to demonstrate the impact of raw material and finished goods shipment costs on the location of industry.

The basis of Smith's method is the *isodapane* technique developed by Weber himself, primarily to examine the impact of a second factor, the cost of labor, on the locational pattern that he had derived purely by minimizing transportation costs. We shall be taking up the case of labor costs in Chapter 4.

Weber's isodapane is a line drawn through all those points in space that have equal total transportation costs from the point of view of a given production unit. It is best visualized as a contour map, except that in this case the contours join places with equal total transportation costs instead of heights above sea level. It can be interpreted in exactly the same way as a contour map, with its troughs and basins of low transportation costs, peaks or ridges of high ones, and plateaus or plains where there is little spatial variation.* Two stages are involved in the construction of an isodapane surface:

1. Isotims are plotted around each separate supply or market point. An *isotim* is a line connecting points of equal transportation costs on each material and on the finished product. It shows how the cost of transporting each individual component increases as distance from *its* minimum-cost point increases. Thus isotims are drawn around each point in the locational situation (materials and market). Where the

---

* Lindberg's (1953) Weberian analysis of the locational evolution of the Swedish paper industry makes good use of the isodapane technique. He shows, in a series of isodapane maps, how the spatial structure of the industry developed over a 100-year period.

costs of transportation are the same in all directions and cost increases away from each point are strictly proportional to distance, the isotims are an equal distance apart and take the form of concentric circles running out like ripples from each supply and market point. They form the construction lines used in the second stage of the operation. In Figure 2.6a, only two-point sources are used, a material site and a market, but depending on the strength of one's eyesight and one's patience, a large number can be incorporated.

2. The total transportation costs for assembling the materials and shipping the product to market for a series of convenient points over the "search" area are summed. Likely points are the intersections of a series of isotims. If the isotims are set at 1-mile intervals and we are shipping only 1 ton of material or product, the total ton-mile cost at each intersection becomes very simple. The number of isotims or rings are counted from the material source. This is the ton-mileage incurred in getting enough of the material together to make 1 unit (ton) of the product. The number of rings from the intersection to the market then gives the ton-mileage that must be added to get 1 unit (ton) of the product to market. For example, in Figure 2.6a, the intersection of the $7 isotim around the material source and the $8 isotim around the market gives a total transportation cost at that point of $15. The same total cost is clearly obtained at the intersection of the $9 and the $6 isotims and so on, thus defining the position of the $15 isodapane. The process is continued for all isotim intersections until a total-transport-cost surface is obtained.

If there were a unique point at the bottom of the cost surface "basin," this would obviously represent the minimum-transport-cost point under the assumptions made. The case we have chosen for ease of explanation does not provide us with one, but under other assumptions of the weights to be moved, with nonuniform transport, or with more sources, a point solution could result.

For Smith, the isodapanes are interpreted more generally as "cost isopleths" or "cost contours"—lines of equal total cost, including costs other than transportation. Only in the version as we have it at this stage are they synonymous with isodapanes, since we have constrained all cost factors other than those allowed by the Weberian example. From such cost isopleths Smith derives two important concepts.

The first is the *space cost curve* (Figure 2.6b), which is simply a section drawn through the cost contour map (Figure 2.6a). The lowest point of the curve, where one exists, is the least-cost location. In some cases, of course, like the one in our example, several locations on a "cost plain" are equally likely. The steepness of the slopes on the surface give some notion of the sensitivity of the industry to locational or space-derived costs. Those with a high sensitivity to weight and distance would have steep slopes; those with a low sensitivity would be shallow.

Smith's second concept, derived from the space cost curve, is that of

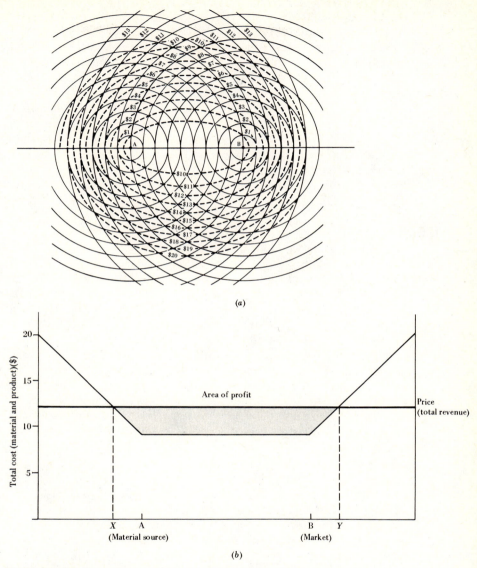

Figure 2.6 (*a*) Transportation cost surface; (*b*) space cost curve.

*spatial margins to profitability.** The manufactured product will be sold at a price that for simplicity here is assumed to be constant in space. At some point

---

* The concept of spatial margins to profitability was first introduced by Rawstron in 1958. In Smith's (1987) view, "Rawstron's exposition of spatial limits to the area in which profitable production is possible was a brilliant feat of intuitive insight. . . . It would be hard to imagine a more original geographical contribution to a field dominated, for the most part, by economists" (p. 30).

on the total-cost surface will be a contour that coincides with this value. Everywhere above this level will have costs that exceed this delivered price. This contour represents the spatial margin to profitability under the conditions of the model (X, Y in Figure 2.6b). Inside the margin, profits will be made; outside it, losses. Weber's least-cost point now becomes, more realistically, only one point within a varying zone where profits could be made at a given price. Producers could miss the least-cost point slightly but still survive. Other factors could also pull them to new sites within the margin. Something of the imprecision of real-world locational choice is introduced into the model by this notion. Although Smith does not add anything fundamentally new to basic Weberian theory, his method, like Isard's substitution framework, adds analytic power to the theory, as we shall subsequently demonstrate.

LOCALIZED INDUSTRIAL RAW MATERIALS AND THE SPACE ECONOMY: SUMMARY AND HISTORICAL PERSPECTIVE

Let us now recapitulate and turn our attention to the probable impact of industry's need for localized resources on the form of the space economy as we have developed it thus far. Allowing that the primary resource base of the economic system will be composed of various industrial raw materials unevenly distributed over its geographic environment, we can now see how, for some elements of manufacturing, there will be a strong pull to locations away from existing central places. The effectiveness of these pulls in determining the final location will depend on their total weight in the cost structure of the industry in question. Industries with a large proportion of their total costs incurred in acquiring and processing gross, localized raw materials will be powerfully drawn to resource sites. Those processing semifinished goods or with a relatively low proportion of gross materials in their total cost structure will behave more like the central service functions of the Lösch-Christaller model. They will attempt to maximize their access to a market while minimizing distribution costs. Both material-oriented and market-oriented industries comply with the basic requirement of all economic activities to locate where the maximum possible number of customers can be served at the lowest possible cost. The spatial heterogeneity of the resource base now means that in pursuing this end (of maximizing profits), the solution to the locational problem varies from industry to industry, depending on the particular input and market requirements.

There are, therefore, important locational differences between producers of manufactured goods and suppliers of services. Manufacturers assemble a variety of predominantly localized inputs, process them, and distribute them to primarily punctiform markets (i.e., wholesalers and retailers in central places). They must therefore "look both ways" to procurement costs and distribution costs. A whole spectrum of industry types spans the range from those in which resources dominate location to those dominated by considerations of access to markets. Services, by contrast, where they do require nonlabor inputs, employ essentially pure materials (e.g., the goods handled by wholesalers and retailers). For them, the localization of the material source has little impact. They perform their specific service function and market the result. This goes to sources of demand that range widely from highly localized seats of government through the entire hierarchy of central places to the dispersed domestic consumer. Here the primary locational concern is access to the market. The spectrum of service industry types covers the range from those tied relatively loosely to a specific market to those tied totally. In general, however, the choice of location begins and ends in some central place within the overall hierarchy.

The locational outlook of manufacturing and services is thus different (although the difference is primarily one of emphasis). This is important when it comes to reviewing the nature of the space economy with localized manufacturing added to it. As we have seen, many manufacturing industries with a high degree of market orientation would offer little disruption to the central place–dominated space economy. They would seek to minimize the distance to at least one punctiform market by locating there. Industries processing inputs with high material indices would be different. They would find locations at raw material sites and to perform their operations would attract labor and so population. This would distort the regular central place lattice (Figure 2.7). A manufacturing town of this kind would itself provide a market for central goods and services, drawing to it tertiary services, primarily of low order, that could gain a profit by supplying its needs.

For example, the very high volume of coal needed for most heavy industrial processes in the nineteenth and early twentieth centuries ensured that many such industries located on the coalfields. Similarly, the iron and steel industries were pulled to sources of iron ore. The result was the generation of new urban centers based on resource use. Such urban areas invariably acquired lower-order services to provide for their local populations, but their

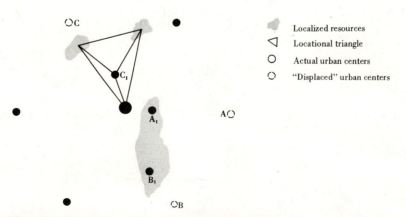

Figure 2.7 Distortion of the spatial pattern of urban centers due to resource localization.

service status was often a good deal lower than their population size would suggest.

Figure 2.7 provides a simple example of the distortion of the uniform pattern of urban centers by this kind of resource localization. Towns A_1 and B_1 are examples of towns based on direct use of a high weight-losing material. The location of the resource precludes the development of urban centers A and B at their "expected" locations in the corners of the hexagon. C_1 is an example of a town based on an industry whose minimum-transport-cost point is intermediate between materials and market. The consequence of resource localization, therefore, is to produce specialized industrial centers in the settlement pattern. These tend to be "out of step" with the spatial uniformity of central places in the Lösch-Christaller model.

In the real world, however, the nature of demand and the technology of production are constantly changing. The nature of the resource base and its distortional impact on settlement and population distribution also change. Resources are exploited for a period and then are either worked out physically or become too costly. Alternatively, a shift of demand may lead to the abandonment of a previously exploited raw material. A new technique may significantly change its material index, weakening the localizing pull of the resource on the other factors needed for production. The resource complex available to any economic system changes constantly in tune with the evolution of the system itself. As a result, the space economy at any point in time carries within it structures derived from a previous stage in its functional evolution. Sometimes these structures carry sufficient momentum from their early growth to exert a long-standing impact on the future development of the space economy. Take, for example, the coalfield areas of Pennsylvania or the Black Country of Great Britain. Long after the basic resources on which they were founded passed the peak of production, their growth and development continued. Other raw material sites fail to achieve the necessary momentum and are short-lived. They distort the regularity of the space economy only in an ephemeral way, as in the case of the abandoned mining towns of Arizona and New Mexico.

Technological change—particularly the replacement of one material or energy source by another—is a key influence in such spatial change. As one example, Smith's space cost curve can be used to demonstrate the locational impact of substituting a ubiquitous energy source (electricity) for a highly localized source (coal). In Figure 2.8, two stages are identified. In stage 1, the industry in question uses a single raw material with coal as an energy input (Figure 2.8a). The source of the material is at the location shown by P, and the space cost curve derived from a section through its space cost surface is the curve ACP. The source of coal is at Q, and the space cost curve rises steeply away from this point. Coal has the stronger localizing pull.

When total costs for both inputs are added together (Figure 2.8b), the overall cost curve ATC_1 "favors" coal but generates an intermediate low point at the place O. For the total situation, the spatial margins to profitability (where costs just equal price) are at E and D. Thus profitable operation is possible over a relatively limited portion of space.

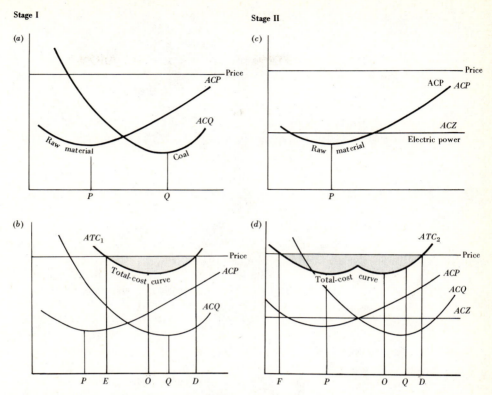

Figure 2.8 Locational impact of supplementing a highly localized fuel with a ubiquitous one. *Source:* After D. M. Smith (1966), "A Theoretical Framework for Geographical Studies of Industrial Location," *Economic Geography* 42: 95–113, fig. 12. Used with permission.

Moving to the new era of fuel technology (stage 2), electric power becomes available. It is as near to any (reasonable scale) producer as the nearest plug socket and has a flat space cost curve (ACZ) (Figure 2.8c). The space cost surface for the material is retained as before (ACP). The new space cost curve for fuel, then, offers two posibilities (Figure 2.8d). Coal can still be used (ACQ), or electricity can be adopted (ACZ). For any producer in the system, the overall space cost curve now looks different (ATC_2). The new spatial margin to profitability expands to cover the area between F and D. Coal retains its pull in the area around Q, but electricity has opened up a new area over which profits are possible. Of course, this has its most cost-effective impact around the source of the material at P. It is not now necessary to move fuel, and since production savings are made by not having to move the material, a new dip in the total-cost surface appears around P. On balance, the new combination of electric power at the resource site is marginally cheaper, and P becomes the overall least-cost location, replacing O, which had this status in stage 1.

Times change, and so does fuel technology. What we have shown is how this may alter the resource complex available to industry and how this in its turn produces a basis for spatial change (though it does not make it happen). We referred earlier to the "distortional" impact of localized materials on the central place–type space economy. Some locations probably from the old farming–central place network are favored over others, and a clustering of industry-based settlement takes place. This pattern in its turn becomes the "normal" one of an industrial society, and other forces, which we shall examine later, then come into play to add further distortions. The previous phase is not entirely obliterated, however, and at any point in time, the economic structure of the space economy reflects the accumulated results of previous events.

SPATIAL VARIATIONS IN THE QUALITY OF AGRICULTURAL RESOURCES

We have suggested that the localized pattern of particular material resources "distorts" the central place–derived pattern of settlement. Now we turn our attention to another "distortion." This time it is the way in which the spatial variation of agricultural resources alters a farming space economy developed (as ours was in Chapter 1) on the basis of location rent alone.*

By definition, agriculture is concerned with the use and, where possible, improvement of "the natural genetic and growth processes of plant and animal life, to the end that these processes will yield the vegetable and animal products needed and wanted by man" (Zimmerman, 1951, p. 148). Quite clearly, therefore, the quality of land (comprising all its physical attributes, including climate) is of fundamental importance. In particular, the complex interaction of three basic elements—climate, soil, and topography—plays a large part in influencing the spatial pattern of agricultural production.

Crops and livestock have particular fundamental physical requirements, especially certain levels of temperature, moisture, and nutrient supply. Not only is the provision of these uneven in space, but also the actual requirements differ from one crop or livestock type to another. Each has its optimal requirements. There is, too, a considerable range around the optimum within which growth is possible, for most crops have a degree of tolerance of suboptimal conditions. Nevertheless, this tolerance has its maximum and minimum limits, and this helps, first, to determine the absolute limits of crop and livestock production and, second, to influence the proportions in which the various crops are grown.

We can set this situation graphically within the framework suggested by McCarty and Lindberg (1966). Two sets of effective limits on agriculture are defined. The first of these are *physical* limits. They are set at the points where

---

* Seeing the real-world variety in the pattern of world resources as a distortion is, of course, a function of nothing more substantial than the particular way we have chosen to structure our analysis, that is, beginning with a study of the impact of the distance variable.

for a given crop or husbandry activity, production becomes physically impossible. The second set of limits apply to *profitability*. These are set by economic considerations, but physical conditions also exert a powerful impact in their determination. We know something of the nature of economic limitations already since we have already met them in the simplified model. We will return to them in due course, but first let us look at physical limits and their impact on the spatial pattern of agriculture.

Figure 2.9 illustrates the physical "optima and limits scheme" set out by McCarty and Lindberg. Two key variables are employed: temperature and moisture. Over a particular area (in the center of the graph), there is an optimal combination of both inputs for a particular crop. Outward from this, however, the restrictions imposed by the physical limits of the two variables make conditions less and less favorable. Ultimately, under extreme conditions, physical limits are reached that make production impossible. Of course, it is always possible, even within the context of physical limits, to provide artificial growing conditions. Technically, it would be possible to grow strawberries at the South Pole given sufficient market incentive. Costs, however, rise with a degree of steepness that depends on the particular product as the optimum is left behind. At the margins of the square figure in Figure 2.9, they rise to infinity, meaning that a real technological barrier exists. In effect, the optima provide the farmer free of charge with the maximum benefits that nature can provide in terms of temperature and moisture. To move away from the optimum imposes costs and recourse to scarcer resources.

For all practical purposes, the economic production of agricultural commodities is spatially restricted. Large areas of the earth's surface are useless agriculturally. Under prevailing technological conditions, they do not provide the critical physical requirements for crops and livestock. However, these outer limits are by no means fixed. Changes in agricultural technology—the development of special and general fertilizers, the evolution of more suitable crop varieties such as hybrid corn, and the introduction of types of livestock suited to particular environmental conditions, such as drought-resistant breeds

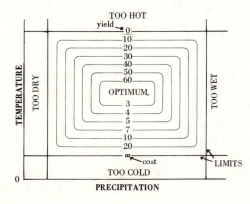

Figure 2.9 An optima and limits scheme for the impact of physical factors on agricultural production. *Source:* H. H. McCarty and J. Lindberg (1966), *A Preface to Economic Geography* (Englewood Cliffs, N.J.: Prentice-Hall), fig. 3.3. Copyright 1966. Reprinted with permission.

of cattle—may have the effect of extending the margin of agricultural production.

Outer limits to agricultural production are controlled very largely by climatic conditions. It is generally accepted that plant growth ceases below a daily mean temperature of 4°C to 5°C. This is the base or threshold temperature at which growth begins to take place. This threshold value varies somewhat from one crop type to another. For satisfactory growth to occur, there must also be a sufficently lengthy period during which temperatures exceed the threshold, and the longer this period, the greater the growth, other things being equal. This fundamental temperature requirement is either completely or partly unfulfilled over very large areas in the high latitudes and elevations. Apart from the polar regions themselves, more than half the area of Canada and Siberia are too cool for agricultural production on any significant scale.

Similarly, extensive areas are deficient in moisture, precluding agriculture in all but the most rudimentary form. Moisture deficiency is a product not only of a shortage of precipitation but also of the relationship between precipitation and evaporation. The lack of sufficient moisture to ensure growth has been estimated to preclude or severely restrict crop production in no less than one-third of the United States and Asia, one-half of Africa, and two-thirds of Australia. It is important to realize that both extremes of temperature limit the amount of available moisture. On the one hand, consistently high temperatures encourage rapid evaporation; on the other, subfreezing conditions lock water in an inaccessible form. At the other end of the precipitation scale, excessive moisture also imposes limits on agricultural production. Very high and prolonged rainfall leads to the waterlogging of soil and the virtual asphyxiation of plants. Under certain topographic conditions, the vital plant nutrients are leached or washed out of the soil, rendering it impoverished and infertile.

The extreme effect of the climatic forces, then, is to preclude or very greatly restrict agricultural production in large areas of the world. Within the feasible limits, of course, physical conditions help to modify land use patterns. This is partly because, as we have pointed out, certain combinations of physical conditions are more suitable for some crops than others. In this respect, soil quality is a particularly important differentiating factor. The varying demands of crops for soil-based nutrients means that some soils favor certain types of crops and produce higher yields. Thus in the southern United States, the higher cotton yields are generally associated with nutrient-rich black soils, while the soils of the Bluegrass Basin of Kentucky have produced yields of tobacco 75 percent greater than those obtained from adjacent but less suitable soils. Local topographic conditions can also modify land use patterns. Apart from the altitudinal influence on temperature, precipitation, and so on, the slope of the land surface may be significant, especially in the case of modern, large-scale, highly mechanized farming operations.

Let us turn at this point to consideration of the profitability aspects of the "optima and limits" approach to agriculture. As far as the individual producer is concerned, the major aim is still to maximize the profits and will therefore involve choosing the combination of crops and livestock most likely to achieve this end. Particularly favorable or unfavorable physical conditions will influence the broad range of choice, affect productivity or yield, and thus modify unit production costs, for environmental factors operate primarily through spatial variations in costs and revenues. In other words, favorable natural conditions, such as more fertile soil, will increase the economic rent accruing to that unit of land.

We have already explained the concept of economic rent in the case of the simplified model of Chapter 1. We described it then as basically a measure of the advantage, as the bidders see it, of one piece of land over another. Under our simplifying assumptions, at that stage, land could achieve this advantage only in one way: it could be better located in relation to the market for its products than other parcels of land.

For this reason, we termed the particular form of rent in the model *locational rent.* Now we have introduced another way one piece of land can have an advantage over another: physical-environmental properties. These, as we pointed out, are offered as free benefits to certain crops or pastoral activities. They are "gifts of nature." In combination with location, these special attributes of land give it a value to bidders who want to make use of it in production. They would compete with each other to "hire the services" of a piece of "good" agricultural land (of course, its "goodness" would depend on what they had in mind). What they would be prepared to bid for, say, a fertile acre of flat land on Long Island would depend on how many bidders were in the market for it and what return they would expect to get from hiring it. The bidder who finally succeeded in hiring it (in a perfect world run according to the laws of classical economics) would be the one who would expect to get the highest return from doing so. The *economic rent* of the piece of land would be measured in terms of what the highest bidder in the market would expect to get in return for its use.

Extending the argument from a single piece of land to land in general with all its locational and physical attributes, we can see yet again how economic rent is a device for allocating parcels of land to various uses. Good land near the market will fetch high returns. Poor land in the backwoods will fetch low ones. Land that is "useless" will fetch no returns at all and will have no bidders and therefore no economic rent. Economic rent measures the *relative advantage* of land as a factor for producing the goods that a market wants. It is also an *allocative device*, ensuring in a perfect world that the best land goes to uses with the highest returns and so on down the line. At this stage in our discussion, then, we have given a more realistic basis for the valuation of land. In the simplified model it was location alone. Now it is location plus fertility, climatic attributes, flatness, and so forth.

Having extended and revised our understanding of the economic rent concept, let us turn to look further at the profitability limits as they apply to the use of land that is within the boundary of physical limits. Figure 2.10 illustrates the way in which rents contribute to the determination of economic limits. (Interestingly, the exact reverse is also true; economic limits set the base value for rents.) The productivity characteristics assigned to land by the two variables, temperature and moisture, in Figure 2.9 are now translated into unit costs of production. The "free rider" effects that nature provides in the zone of optima are now converted into cost savings (compare Figure 2.10 with Figure 2.9). Movement away from the physical optimum promotes a rise in unit costs of production. This is shown by a graph that resembles the space cost curve in Figure 2.6*b*. If a price is set for a particular product, this will determine the level of return possible for that product. Suppose that this price is set at 7. No part of the producer region with unit costs above 7 will offer a net return (profits less costs). The isoline marking the margin to profitability (but this time not a directly spatial one) will be that at which nature's bounty gives not less than 7 units of cost saving to a prospective hirer of land. Land outside this production cost margin will have no economic rent, as the diagram shows. This is because it offers no return to those who hire it. From the isoline for 7 inward, however, rent is acquired by land for the particular use, and it rises to a peak where 4 units of rent are returned in the center of the optimum zone.

Logically, the diagram in Figure 2.10 offers a direct analogy to the bid rent lines that allocated land use with reference to locational advantage in the von Thünen model. We must be careful to note, however, that the "regions" of Figure 2.10 are not regions in the spatial sense. They are regions of temperature and moisture conditions as determined by the axes of Figure 2.9. Similarly, the margins to profitability are not spatial ones but the temperature and moisture margins. Bearing this in mind, however, it can be seen that in the competitive struggle by bidders to hire land with good temperature and moisture advantages, uses that give the highest returns to moisture and temperature optima will capture the "inner" segments of the diagram. Uses with lower

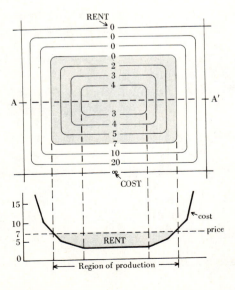

Figure 2.10 Productivity translated into unit costs and rent for land under agricultural production. *Source:* H. H. McCarty and J. Lindberg (1966), *A Preface to Economic Geography* (Englewood Cliffs, N.J.: Prentice-Hall), fig. 3.4. Copyright 1966. Reprinted with permission.

returns to temperature and moisture will be outbid and will occupy locations of lower quality.

What we have to do now is put the attributes of physical resource quality and location together. We want to see how the addition of differential resource quality influences the regular order that location rent alone gave to the pattern of land use in the simplified model. At first sight, it would seem that little spatial order could possibly remain because the quality of the natural environment varies spatially in a largely nonsystematic manner. This is in direct contrast to the effect of distance, which, by its relatively predictable influence on transportation costs, tends to create a more regular spatial pattern around the major spatial reference points, urban markets.

As Dunn (1954) has pointed out, however, all is not lost:

> The lack of regularity in the distribution of these [environmental] attributes would seriously distort the regularity of the land-use patterns. However the phenomenon of the systematic progression of crops would not be altered. All crops could earn more on the more productive land, but, for any given degree of productivity, the crop that would earn the most would be the crop which had the highest marginal rent-line established by the economic influence of distance. The systmatic nature of crop progression would be preserved because the more productive land would be more productive for all uses. The principal location influence would be a distortion of the regularity of the pattern by extending the crop boundaries in the areas of high productivity and restricting them in areas of low productivity. (p. 67)

A hypothetical example of such distortional effects is shown in Figure 2.11. Shaded area A to the east of the market town presents an area that is unsuitable for agricultural use. Shaded area B is an "inlier" of crop 3, which is otherwise produced at a greater distance from the market. However, area B is particularly suited to the production of crop 3 and gives much higher yields at lower cost. This increases the economic rent of crop 3 above that of crops 1 and 2 in this restricted area and allows crop 3 to be substituted for crops 1 and 2.

Variations in land quality, therefore, are important primarily for their influence on the spatial pattern of economic rent in the course of agricultural production. Such variations may also be significant for other forms of economic activity. Manufacturing industries requiring very extensive sites for efficient plant layout will be influenced, to some extent, by the availability of sufficiently large areas of flat land. The extent to which this factor is significant depends, of course, on the importance of site costs in the industry's cost structure (in most cases, this is not very great). Similarly, climatic influences may in certain cases be a consideration for the manufacturing process. The often-quoted example of the advantages of the California climate for aircraft manufacturing may be cited. But as in the case of the relationship between the physical environment and agriculture, the impact is generated primarily through economic costs. On a larger scale, variations in the physical environment may have an impact on population distributions and therefore on settlement patterns. In the days when settlements had to be supplied with food from their local area, there was some relationship between land quality (hence crop

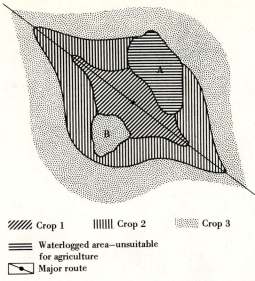

▨▨ Crop 1   ⫿⫿⫿ Crop 2   ⠿⠿ Crop 3

▤▤ Waterlogged area—unsuitable
for agriculture

◺◹ Major route

Figure 2.11 Distortion of agricultural land use patterns by variation in the physical quality of the land.

yield) and population density. Over time, of course, this relationship has ceased to exist, but the inertial element has meant the perpetuation of such patterns, for, as with industrial raw materials, the forms generated by the functioning of earlier agricultural systems under different conditions of demand and technology exert a strong impact on agricultural patterns in the space economy at any subsequent point in time.

Thus the impact of the physical environment on the location of agricultural production is seen as one controlling cost factor among many. Its effect on a space economy such as that of the simplified model of Chapter 1 will be to distort and complicate conditions on the isotropic plain, making it more realistically heterogeneous.

FURTHER READING

Dunn, E. S. (1954). *The Location of Agricultural Production.* Gainesville: University of Florida Press, chap. 5.

Gregory, D. (1982). "Alfred Weber's Location Theory," *Progress in Human Geography* 6: 115–128.

Isard, W. (1956). *Location and Space Economy.* Cambridge, Mass.: MIT Press, chaps. 4 and 5.

Kennelly, R. A. (1954). "The Location of the Mexican Steel Industry." In R. H. T. Smith, E. J. Taaffe, and L. J. King (eds.), *Readings in Economic Geography.* Chicago: Rand McNally, pp. 126–157.

Lindberg, O. (1953). "An Economic-geographic Study of the Swedish Paper Industry," *Geografiska Annaler* 35: 28–40.

McCarty, H. H., and J. B. Lindberg (1966). *A Preface to Economic Geography*. Englewood Cliffs, N.J.: Prentice-Hall.

Scott, A. J. (1982). "Locational Patterns and Dynamics of Industrial Activity in the Modern Metropolis," *Urban Studies* 19: 111–142.

Smith, D. M. (1966). "A Theoretical Framework for Geographical Studies of Industrial Location," *Economic Geography* 42: 95–113.

Smith, W. (1955). "The Location of Industry," *Transactions of the Institute of British Geographers* 21: 1–18.

Webber, M. J. (1982). "Location of Manufacturing Activity in Cities," *Urban Geography* 3: 203–223.

Weber, A. (1909). *Theory of the Location of Industries*. Chicago: University of Chicago Press, chap. 3.

Chapter 3

Transportation and the Spatial Organization of Economic Activities: Routes, Networks, Transportation Costs

Throughout our discussion so far, we have been concerned with *movement* and its fundamental role in influencing how economic activities are arranged on the earth's surface. We have gradually relaxed some of our initial assumptions, in particular the notion that movement was both equally possible and equally likely in any direction on our plain. But we have still retained some restrictive assumptions concerning both the form of transportation routes and the structure of transportation costs. In this chapter, we examine the effect on the spatial organization of economic activities of removing both of these constraints.

THE LOCATION OF TRANSPORTATION ROUTES AND NETWORKS

We saw in Chapter 1 that even on a homogeneous plain, movement would tend to become channeled into specific lines or routes. But transportation routes do not exist in isolation; they are organized into networks with varying degrees of interconnection. Such networks are the physical expression—the trace left on the earth's surface—of the flows of materials, people, and information that bind the economic system together. The question we pose in this section, therefore, concerns the nature of the forces that influence the *spatial* form of the transportation routes and networks that play such a fundamental part in our daily lives.*

It is obvious that variations in the form of the land surface will have an impact on the location of transportation routes and networks. Mountain ranges and extensive plains, river valleys and expanses of swampland provide differential resistances to movement. Such variations in physical geography led many geographers in the past to examine the location of particular routes almost entirely in terms of the influence of such physical conditions. It seems obvious, perhaps, when looking at a map showing a mountainous region dissected by river valleys, that the highways or railroads "choose" the "natural routeways." It is tempting to suggest that the river valleys represent the raison d'être of the railroads and highways that pass through them. But as Appleton (1963) so aptly pointed out, "A so-called 'natural routeway' is a means and not an end. It affords the opportunity for communication; it does not create the demand" (p. 21).

In fact, Appleton goes on to observe that in the development of transportation routes in the Great Australian Divide, for example, the most obvious physical route was, in several cases, *not* used. Meinig's (1962) comparative study of railroad development in the Columbia basin of the western United States and in southern Australia showed that many alternative routes other than the obvious ones were surveyed and considered by the developers. In the case of underdeveloped countries, too, it seems that variations in the physical landscape can be relatively unimportant in determining the development of transportation facilities (Taafe, Morrill, and Gould, 1963).

To find the fundamental reason for the location of a transportation route, we have to return again to demand and, especially, to the principle of complementarity. As we saw in Chapter 1, interaction occurs between places if they have some kind of demand-supply relationship. If two urban centers are complementary but not connected by a transportation route, it is likely that such a route will be constructed to meet the need. It is in the context of construction costs, therefore, that variations in the land surface are influential. The costs of traversing difficult terrain have to be set against the benefits that such construction would bring. The cost of a transportation route consists of two basic elements:

1. *Fixed or capital costs.* These are the costs involved in actually building the route. Land may have to be purchased (in heavily urbanized areas, this may well be the largest single component in the cost of building a route); uneven ground may have to be leveled; waterlogged ground may have to be drained and filled. Cuttings may have to be

* Hay (1973) provides a comprehensive discussion of the spatial form of transportation networks.

blasted through solid rock. Fixed costs are closely related to the length
of the route.

   2. *Variable or operating costs.* The cost of a transportation route does not
   end when it is built and opened to traffic. There are recurrent costs that
   vary according to both the length of the route and the volume of traffic
   flowing along it. Routes have to be maintained and kept in good repair.
   They may have to be plowed or salted in winter to keep them clear,
   and maintenance staffs have to be employed to carry out these and
   other tasks.

These two cost elements may well vary from place to place in terms of their
relative importance. In some cases, fixed costs may be extremely high and
operating costs relatively low, or vice versa. Thus the relationship between
fixed and variable costs goes some way toward explaining the precise form of
transportation routes and networks. To some extent, we can view the route
problem as one involving two extreme sets of goals. One goal is to build the
route or network as cheaply as possible—what Bunge (1966) called the "least-
cost-to-build" motive. The other extreme is to build the route or network with
the aim of keeping the costs to the user as low as possible—the "least-cost-to-
user" goal. The former objective would result in a network as small as possible,
while the latter would result in a highly connected route network in which
every place is directly connected to every other place.

Let us suppose that we wish to build a transportation network in an area
that contains five towns (Figure 3.1*a*). What form should the network take? The
answer clearly depends on our basic objectives; in particular, do we wish to
keep construction costs as low as possible, or do we wish to build a network
offering greatest convenience to people who are going to use the network and
perhaps thereby increase total revenue? The five towns could be connected
into a network in many different ways. Figures 3.1*b* and 3.1*c* show the two
extreme alternatives that correspond to the least-cost-to-user and least-cost-to-
build principles. In Figure 3.1*b*, we see the network that gives maximum

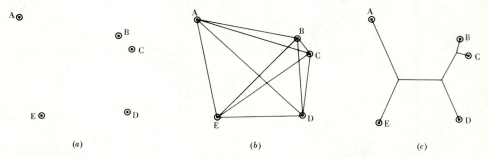

(a)   *(b)*   *(c)*

Figure 3.1 Alternative ways of connecting five settlements by a transportation network:
(*a*) Relative locations of the settlements; (*b*) "Least cost to use" solution; (*c*) "Least cost
to build" solution. *Source:* W. Bunge (1966), "Theoretical Geography," *Lund Studies in
Geography*, Series C, 1, fig. 7.10 and 7.14.

benefit to the user in that each of the five towns is directly connected to every other town. There is no problem, therefore, of having to change trains to pick up an airline connection to the next town. Figure 3.1*c* shows the solution that meets the criterion of least cost to build because it minimizes the total route length. But this is clearly far less convenient to the user.

Bunge suggested that the railroad pattern of North America in the 1950s could be at least partly understood in terms of these two basic patterns (Figure 3.2). The least-cost-to-user network was characteristic of the Northeast and the Midwest, where the large metropolitan centers are spatially clustered and transportation demands are greater. Elsewhere in North America, urban centers are relatively scattered, involving greater distances and smaller traffic volume. In such conditions, the dominant consideration would tend to be construction costs, and a least-cost-to-build network would be expected.

In locating a transportation route, therefore, a balance has to be struck between making the route short enough to keep construction costs down and making it long enough to connect places generating a large amount of traffic and revenue. We can thus identify two forces that help to make transportation routes deviate from a straight-line connection between origin and destination. *Positive deviations* result in making a route longer in order to increase revenue. Figure 3.3 shows a hypothetical example in which the problem is to locate a transportation route in an area where there are eight towns. The two major towns, X and Y, both generate revenues of 10 units, while the others vary in their revenue-generating capacity. The problem is that increasing the length of the route to take in other centers raises construction costs. Which is the best route? In Figure 3.3*a*, the net benefits of the straight-line route connecting only X and Y are 10 units. Lengthening the route to connect two additional places raises costs but greatly increases revenues (Figure 3.3*b*). However, connecting every town does not give the greatest net benefit; this is achieved by connecting six of the centers and bypassing two of the smallest ones (Figure 3.3*c*).

Negative deviations of transportation routes are produced mainly by the effects of a heterogeneous land surface in differentially raising or lowering construction costs. This effect is not simply restricted to the more obvious physical phenomena such as mountain ranges, river valleys, glacial spillways, and marshland areas. Political boundaries, differences in the cost of acquiring land, and other socioeconomic phenomena all contribute toward the "distortion" of transportation routes and networks. Even airline routes are not simply straight-line paths between airports: they, too, become distorted by variations in atmospheric pressure and wind strength and direction.

As transportation routes pass through space, therefore, they encounter varying degrees of "resistance." Lösch saw this as being analogous to the way in which light rays become refracted or bent as they pass from one medium to another, just as a stick partly immersed in water appears bent rather than straight. Consider Figure 3.4, which depicts a transportation situation in which goods have to be moved over two different media, land and sea. Each of these has a different resistance to movement in the sense that transport costs per mile

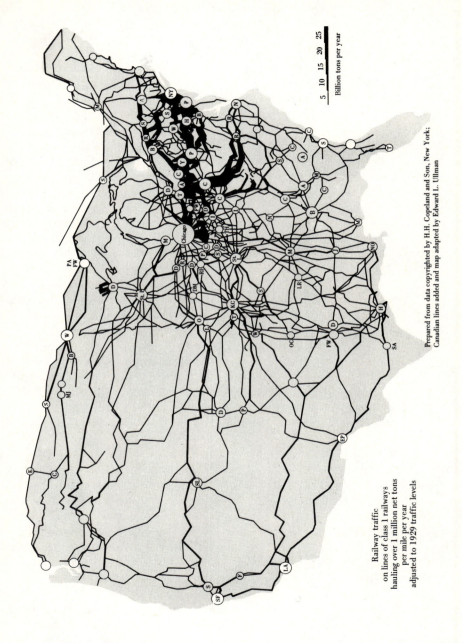

Railway traffic
on lines of class 1 railways
hauling over 1 million net tons
per mile per year
adjusted to 1929 traffic levels

Prepared from data copyrighted by H.H. Copeland and Son, New York;
Canadian lines added and map adapted by Edward L. Ullman

5   10   15   20   25
Billion tons per year

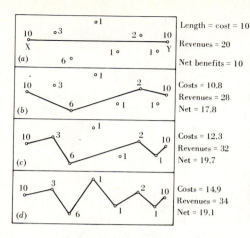

Figure 3.3 Alternative ways of locating a transportation route to maximize net benefits. *Source:* R. Abler, J. Adams, and P. Gould (1971). *Spatial Organization: The Geographer's View of the World* (Englewood Cliffs, N.J.: Prentice-Hall), fig. 8.42. Reprinted by permission.

are higher over land than over sea ($1.50 per mile compared with $0.75 per mile). What will be the optimal route for moving the goods from *A*, which is a port, to *B*, which is an inland location in a different country? The diagram shows three alternative routes using different combinations of land and sea distances:

1. Route *ANB* is the straight-line route crossing equal distances over land and sea. At the prevailing transportation rates the total transportation cost will be

$$(20 \times 0.75) + (20 \times 1.50) = \$45.00$$

2. Route *AMB* maximizes the distance traveled over water (the lower-cost medium) and minimizes the distance traveled over land. Total transportation costs will be

$$(30 \times 0.75) + (10 \times 1.50) = \$37.50$$

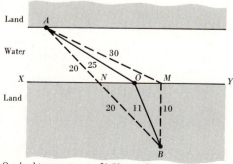

Overland transport cost = $1.50 per mile
Oversea transport cost = $0.75 per mile

Figure 3.4 Application of the principle of refraction to transportation routes: movement over land and water.

3. Route *AOB* is the compromise route, with total costs of

$$(25 \times 0.75) + (11 \times 1.50) = \$35.25$$

The least-cost route, therefore, is the one that crosses the coast at *O*.

Lösch (1954) also suggested that such "refraction by lenses" occurs in both natural and human conditions. In Figure 3.5, the problem is to transport a consignment of goods as cheaply as possible from *A* to *B*, two cities separated by a high mountain barrier. The excessive cost of crossing the barrier rules out the possibility of the straight-line connection. Lösch saw the shaded area in Figure 3.5 as being exactly like a "biconcave lens." "The greater the refractive index of the lens—that is, the more it resists the passage of a beam of light—the more will the beam be deflected" (p. 186). He referred to the changing pattern of trade between the eastern and western coasts of the United States in the nineteenth century as an example. Initially, overland transport costs were so exceptionally high that much of the trade was via the long sea route around Cape Horn. Subsequently, the degree of refraction was lessened by the construction of the Panama Canal; today, east-west movement is refracted hardly at all by overland barriers.

We can quite easily extend this basic idea by considering a case where an origin and destination point are separated by several areas of varying resistance to movement. In Figure 3.6, there are four such areas that affect the level of construction costs. The darkest shaded areas represent the highest construction costs per mile. These areas could be seen as different types of physical terrain—a mountainous region with intervening valleys across the desired line of movement—or as a land value surface in which the cost of acquiring land for the route is much higher in some places than in others. Whatever their exact nature, the presence of such areas of differential resistance is to deflect the optimal route from the straight-line path between P_1 and P_2. It is interesting to note that in Figure 3.6, the route with the lowest total cost is in fact the longest (route 3).

Transportation routes are not only distorted by variations in the land surface itself; boundaries, especially political boundaries, also have a very considerable influence. In some cases, the effect on the network structure is quite spectacular. One of the best-known cases is that either side of the United

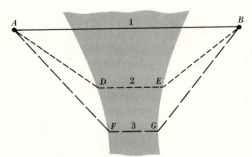

Figure 3.5 Refractive effect of a mountain barrier. *Source:* After Lösch (1954), fig. 51.

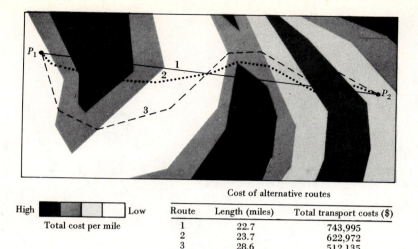

Cost of alternative routes

				Route	Length (miles)	Total transport costs ($)
High			Low	1	22.7	743,995
				2	23.7	622,972
Total cost per mile				3	28.6	512,135

Figure 3.6 Effect of areas of different construction costs on the location of a transportation route. *Source:* C. Werner (1968), "The Law of Refraction in Transportation Geography," *Canadian Geographer* 12: 28–40, fig. 6 and table 1. Reproduced by permission.

States–Canadian border. In the approximately 700-mile stretch of country shown in Figure 3.7, only eight railroad lines cross the border, though more than 20 approach it and terminate before reaching the border. Two major east-west routes run parallel to each other on either side of the border. As Wolfe (1962) pointed out, "If the boundary between Canada and the United States were removed from the rail network map, one would have little difficulty in putting it back again in about the same place" (p. 184).

Figure 3.8 shows an even more striking example of the effect of political boundaries on transportation networks. The line *XY* on the map marks the boundary between the provinces of Quebec and Ontario. To the east of the

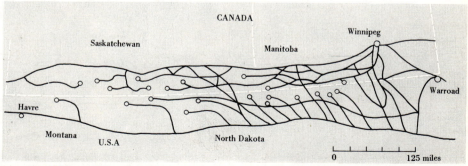

Figure 3.7 Effect of the U.S.-Canadian border on railroad networks. *Source:* A. Lösch (1954), *Die räumliche Ordnung der Wirtschaft* [*The Economics of Location*] (Stuttgart: Gustav Fischer), fig. 85.

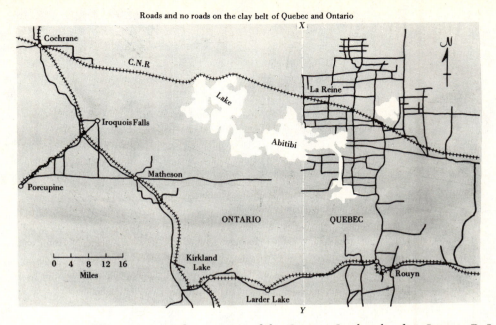

Figure 3.8 Road patterns astride a section of the Ontario-Quebec border. *Source:* R. I. Wolfe (1962), "Transportation and Politics," *Annals of the Association of American Geographers* 176–190, fig. 4. Reproduced by permission.

line, there is a dense network of roads; to the west, there are virtually none until the railroad is reached.

One likely result of such disruption and distortion of transportation networks at political borders is a reduction in the intensity of movement and interaction between one side of the border and the other. In the extreme case, the barrier may be almost total, allowing no transportation routes to cross from one side to the other. More usually, the barrier is permeable; that is, it acts as a kind of filter through which some, though not all, movement passes. The two cases shown in Figures 3.7 and 3.8 are of this type. In the particular case of the Quebec border, an interesting study of interaction was carried out by Mackay (1958). Mackay used the gravity model to analyze the flow of telephone traffic between, on the one hand, three Quebec cities—Montreal, Quebec City, and Sherbrooke—and, on the other, 50 Canadian and 20 U.S. cities. By predicting the intensity of telephone traffic from the gravity model and testing the values against actual telephone traffic, Mackay was able to measure the effect of the border on such interaction. He found that telephone traffic across the border was consistently lower than expected on the basis of the actual distance between cities. In the case of interaction between Quebec cities and Ontario cities, the Ontario cities received between one-fifth and one-tenth of their predicted telephone calls, while U.S. cities received only one-fiftieth the number of the predicted calls. As Mackay pointed out, the border in effect increased the distance between places, by between five and ten times for interaction within Canada and by 50 times for interaction between Quebec and the United States.

For a number of reasons, therefore, both physical and human, transportation routes and networks have a complex spatial form. Insofar as all economic activities are locationally related to transportation facilities, the intricacies of the network result in location patterns that are considerably removed from the regularity presented in Chapter 1. The relative positions of urban centers and the distortions in the size and shape of market and supply areas and urban and agricultural land use zones produced by simple linearity of transportation facilities (see Chapter 1) become infinitely more complex, though the underlying locational principles remain.

THE STRUCTURE OF TRANSPORTATION COSTS

A particularly important assumption on which the simplified economic landscape of Chapter 1 was based was the structure of transportation costs.* These were assumed to be strictly proportional to distance; in other words, each additional unit of distance added an equal increment of cost to total transportation costs (Figure 1.1). Movement of a good over a distance of 100 miles therefore would cost twice as much as movement of the same good over 50 miles.

In reality, transportation costs are rarely exactly proportional to distance, for several reasons. First, all transportation media, whether railroad, truck, airline, waterway, or pipeline, incur a certain level of fixed costs that do not vary with the length of journey. We have mentioned some of these costs in connection with the construction of transportation routes, but there are also "handling" costs that are part of fixed costs—costs of picking up and loading freight and of billing customers. Such handling costs are also unrelated to the length of journey and have to be incurred whether the goods are moved 1 mile or 1,000 miles. These unvarying-with-distance costs are collectively termed *terminal costs* (T in Figure 3.10). Although terminal costs are not dependent on the length of journey, they greatly influence transportation costs because as length of haul increases, terminal costs can be spread to a greater extent than on a short journey. The result is that the cost per mile tends to decline with increasing distance. Figure 3.9 shows this for a situation in which terminal costs are $1.00 and line haul (movement) costs are 5 cents per mile. The total cost per mile (terminal plus line haul costs divided by the distance traveled) falls very rapidly and then flattens out. Thus the cost per mile if the journey is only 1 mile is $1.05, whereas it is 25 cents per mile for a journey of 5 miles and

---

* Transportation cost structures are extremely complex; in this section, we do no more than review some of the more spatially significant features. For more detailed and expert treatment, see Sampson and Farris (1980).

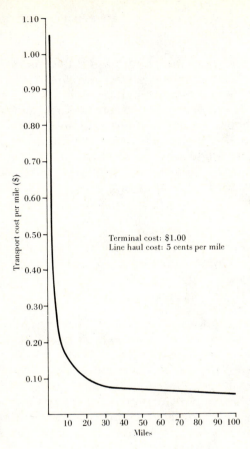

Terminal cost: $1.00
Line haul cost: 5 cents per mile

Figure 3.9 Relationship between transport cost per mile and length of journey.

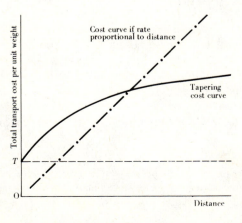

Figure 3.10 Comparison of transportation costs that are proportional and less than proportional to distance.

only 6 cents per mile for 100 miles. In terms of *total* transportation costs, the effect is as shown in Figure 3.10: the cost curve "tapers off" with increasing distance.

For the most part, tapering transportation costs are the result of spreading a fixed terminal cost over a greater distance, but there are other influences too. One is that line haul costs themselves are not always exactly proportional to distance; another is that rates for longer distances are kept lower than they might otherwise be in order to encourage longer-distance movement. Figure 3.10 also shows how tapering cost curves compare with those where transportation costs are strictly proportional to distance. In effect, short-distance movements are more expensive and longer-distance movements are less expensive when tapering transportation costs are in operation.

The general form of transportation costs, therefore, is for them to taper off as distance increases. However, the degree of tapering varies greatly from one type of transportation to another, depending on their respective terminal and line haul costs. Figure 3.11 is an idealized representation of transportation cost curves for three types of transportation: highway, railroad, and waterway. Highway or truck transportation costs are seen to be only slightly less than proportional to distance. This is because fixed costs are low, possibly only about 10 percent of total costs. One reason for this, of course, is that truck operators do not maintain their own highways, unlike railroad operators, who incur track maintenance costs. Such contributions as are made toward highway costs (e.g., through gasoline taxes) vary largely according to the use made of them. Conversely, line haul costs are high for truck operators, so trucking costs increase more or less directly with distance.

Railroad and waterway costs, by contrast, are characterized by far higher terminal costs but lower line haul costs. Both railroad and waterway networks are coarser than highway networks, providing fewer terminal facilities, which generally involves costly transshipment of goods from truck to freight car (or barge) and finally to truck again. For example, in 1962, the route mileage of

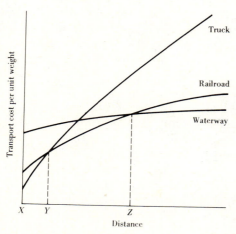

Figure 3.11 Idealized transportation cost curves for three transportation media: truck, railroad, and waterway.

highways in the United States was ten times greater than railroads and 100 times greater than navigable waterways (Barloon, 1965). The difference is even greater today because of rationalization of the railroad network with the closure of large stretches of "uneconomic" routes. Although the development of standardized containers has greatly facilitated transshipment from one type of transportation medium to another, terminal costs on rail and waterway remain high. However, line haul costs are relatively low, particularly in the case of waterways, producing cost curves that are convex upward.

As a result of these varying cost characteristics, each transportation medium offers advantages over different lengths of haul. For short distances such as *XY* in Figure 3.11, truck transportation is cheapest. For medium-length hauls (*YZ*), the railroad is least costly, whereas over very long distances (beyond *Z*), water transport, if available, is cheapest. The extent to which these media can compete for traffic is therefore somewhat limited. If, for example, railroads wish to compete in the short-haul market, they must offer rates over such distances that are comparable with trucking rates, and vice versa. However, such competition is probably restricted to certain types of freight since there is a considerable degree of specialization in the type of freight moved by road and by rail. Table 3.1 summarizes some of the basic characteristics of the major transportation media.

Although most transportation rates take the general form described here, a number of other influences have a significant geographic impact on transportation costs. We shall look briefly at just three of these:

1. Grouping of freight rates into zones
2. Variations in freight rates according to the nature of the freight being moved
3. Variations in freight rates according to traffic characteristics

Grouping of Freight Rates into Zones

Two aspects deserve our attention. The first relates to the practice, prevalent until the 1950s in the United States, of allowing different rate *levels* to exist in different parts of the country. For the first half of the twentieth century, in fact, railroad freight rates were grouped into five major regions. Figure 3.12 shows how the *average* rate varied from region to region compared with the eastern or official region. The disparities were quite remarkable; for example, railroad freight rates in the southwestern region were, on average, 61 percent higher than those in the eastern region. In part, these differences reflected the principles discussed earlier—areas of high demand generate a dense transportation network, and such high demand may also be reflected in lower freight rates. The operation of this regional differential between 1887 and 1952 gave the Northeast a pronounced rate advantage, with an average cost of shipping goods significantly lower than the rest of the nation. Though no longer operative, these regional differences in freight rates continued to exert a spatial impact on economic activities for some time because of the stimulus they gave to economic activity.

More significantly today, however, is the almost universal practice of charging freight rates in *zones*. In theory, every separate journey would be charged a specific rate that varied precisely according to its length. But this is time-consuming and costly to administer. Consequently, most transportation companies operate a zonal rate structure. In the case of railroads, for example, a common practice is to group stations into areas and to charge a single rate within that area. Most commonly, rates are set in relation to a "control point" in each area; frequently this is the largest center.

Closely associated with the zoning of origin and destination points is the common practice of quoting freight rates in *steps* of varying width. In Figure 3.13, not only does the length of the individual steps increase over greater distances, but also the steps for interstate freight movement are longer than those for intrastate movement. The diagram shows a case in which interstate rates are not quoted on distances of less than 40 miles (the average size of interstate zones). Between 40 and 100 miles, the zones increase in steps of 5 miles; between 190 and 240 miles, the steps are 10 miles wide; and beyond 240 miles, each step represents 20 miles. In the case of intrastate traffic, the steps are 5 miles wide for movements of less than 100 miles and 10 miles wide beyond that distance. The stepped-rate pattern, with steps becoming wider with distance, retains the tapering principle and favors longer-haul movements.

Figure 3.14 illustrates freight rate zoning on a larger, transcontinental scale. The rate pattern shown in Figure 3.14*a* shows the *blanket* or zonal rate for shipping lumber from western Oregon and Washington to various parts of the United States; the profile in Figure 3.14*b* is for shipment of the same commodity to the Gulf Coast via the southern route. This graph is especially interesting because it shows a combination of blanket or zonal rates and mileage rates. Over the first few miles, the rate per 100 pounds increases very rapidly but then takes on a blanket level of 40 cents per 100 pounds for several hundred miles. This is again interrupted by a lengthy stretch of short-stepped progressions before the level of about $1.40 per 100 pounds is reached at roughly 2,100 miles. From that point, the rate remains exactly the same. Again, therefore, we have a freight rate structure that tapers off with distance but in a more complex way.

Variations in Freight Rates According to Commodity Characteristics

The characteristic shape of transportation rates over distance is, as we have seen, a stepped, convex-upward curve. However, the *level* of transportation costs varies markedly according to the general characteristics of the commodities involved. Just as the tapering of freight rates has its origin in the existence of fixed costs, so these same costs encourage transportation operators to vary their rates in order to cover these costs. The result is a considerable degree of discrimination between commodities in the level of freight rates and the existence of rates that are above or below those justified by the actual costs of transportation. A degree of *cross-subsidization* may be practiced whereby

Table 3.1 CHARACTERISTICS OF MAJOR TRANSPORTATION MEDIA

Mode	Costs	Unit cost per mile (Rail = 1.0)	Distance	Characteristic goods	Distinction	Drawbacks
Railroad	Capital-intensive: large initial investment (incl. right of way). Profitability rests on intensity of use: 350,000 to 500,000 tons/mile/year is operational margin. Terminal costs high.	1.0	Increasing effectiveness with length of haul. Large shipments. cheaper by long or short haul.	Minerals; unprocessed agricultural products; building mats., chemicals. Passengers minor.[a]	Large volumes of bulk goods in comparatively short time at low costs.	Cost and time of assembling units.
Waterways	Investment low, especially where natural waterways utilized. Terminal and handling costs several times line haul costs.	0.29	Increasing effectiveness with length of haul.	Marine: semifinished and finished products. Inland: bulk raw goods—coke, coal, oil, grain, sand, gravel, cement. Passengers negligible.	Low freight rates; slow speed; spec. of goods carriage.	Slow speed.

Mode	Cost characteristics		Distance characteristics	Commodities	Advantages	Disadvantages
Motor transport	Fixed costs negligible. Operates on small margins—operating costs high; vehicle turnover high.	4.5	Short haul, less costly than rail. Wide area coverage.	Perishable goods; lumber. Passengers important.	Light loads, short distances, short time. Flexible and convenient. Improved service. Minimizes distribution costs.	Inadequate capacity for moving heavy volumes, bulk materials. High costs of long hauls. High vehicle operating costs.
Air transport	Fixed costs low. Investment in stock very high. Terminal, takeoff costs high.	16.3	Long hauls, economy with distance.	Passengers dominant. Perishable, lightweight, high-value goods.	Speed.	Very high costs.
Pipelines	Fixed costs high. Large economies through diameter of pipe. Costs increase almost directly with distance. Viscosity adds costs.	0.21	Long haul in bulk.	Crude oil and petroleum products in large volume.	Bulk movement.	Restricted commodity use. Regular flow and demand needed. Large market.

[a] In Europe, passenger revenues usually exceed freight revenues.

Source: Michael Eliot Hurst, A Geography of Economic Behavior: An Introduction. © 1972 by Wadsworth Publishing Company, Inc., Belmont, California 94002. Reprinted by permission of the publisher, Duxbury Press.

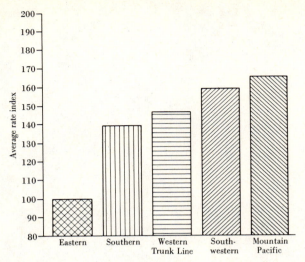

Figure 3.12 Relative freight rate levels by major freight rate region, United States, to the 1950s.

"losses" on transporting some commodities can be offset by "gains" on others. But why should such rate discrimination between commodities exist? Two reasons seem to be especially important. One is the fact there are differences in the cost of providing a transportation service for various kinds of freight. The other relates to the ability of various types of traffic to pay high transportation charges.

Differences in the Cost of Providing a Transportation Service A number of factors contribute to differences in the cost of providing a transportation service:

1. *Loading characteristics.* Of particular importance is the weight density of the commodity (weight per cubic foot). Consequently, light

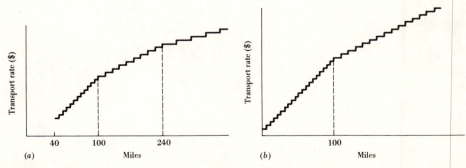

Figure 3.13 Stepped freight-rate curves: (*a*) Interstate; (*b*) Intrastate.

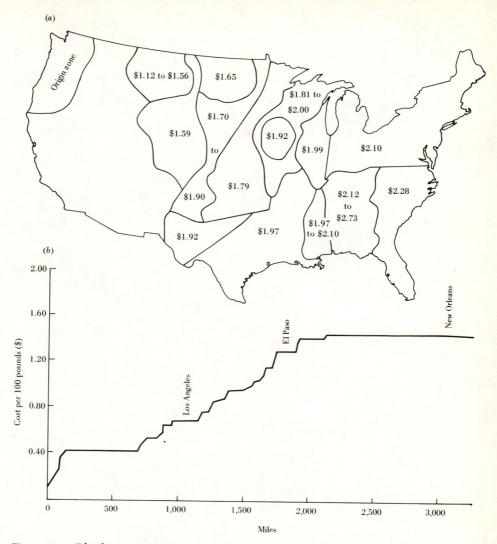

Figure 3.14 Blanket rate freight profiles on the shipment of lumber from Portland, Oregon. *Source:* R. J. Sampson and M. T. Farris (1975), *Domestic Transportation: Practice, Theory and Policy,* 3d ed. (Boston: Houghton Mifflin/Bureau of Business Research, University of Oregon), figs. 19 and 20.

though bulky articles are likely to incur higher freight charges than heavy, compact articles. This explains the common practice of charging more favorable rates on "knocked-down" than on "set-up" items. The decentralization of the automobile industry from Detroit owes a good deal to the fact that there was a significant freight saving in shipping packaged components over the shipping of finished automobiles. Similarly, articles that load compactly are preferable, from a rate viewpoint, to articles of odd shapes and large dimensions.

2. *Size of shipment.* Up to a point, there is a direct relationship between size of shipment and level of transportation costs. Large, single consignments of goods permit economies in administrative, terminal, and in some cases line haul costs per unit weight. A significant distinction is between shipments in carload (CL) and less-than-carload (LCL) lots. Average costs tend to be very much lower the larger the quantity shipped.

3. *Susceptibility to loss or damage and risk liability.* There is considerable variation between commodities in their susceptibility to loss or damage. As a result, freight rates must often be calculated to cover such contingencies. The more fragile and perishable a good—such as delicate instruments and components, on the one hand, and vegetables and fresh fruits, on the other—the higher the rate is likely to be. Often the type of packaging influences the rate level, glass containers being more susceptible to breakage than cans. Closely related to this is the need for certain commodities to be transported in special equipment, such as refrigerated or insulated cars, or by special services—for example, rapid transportation is necessary for perishable goods. Finally, the amount of the transporter's liability must also be taken into account.

Elasticity of Demand for Transportation It is generally accepted that goods of high unit value are better able to "bear" relatively higher transportation charges than goods of lower unit value. This is related to the good's elasticity of demand for transportation, which is less for goods of high unit value. Such variations allow the transportation agency considerable flexibility in rate making. High-value goods are generally charged at least the full cost of the service, including overhead charges, while goods of low unit value may be charged rates that do not meet the actual costs involved. (See Box 3.1.)

Variations in Freight Rates According to Traffic Characteristics

Competition Between Transportation Media Where only one effective form of transportation exists, the operator can set rate levels to cover costs, and in the absence of government intervention, there is considerable scope for establishing high rate levels. But as we have seen, the major transportation media each have the advantage over certain distances. If, for example, railroads wish to compete with trucks for traffic on short hauls, they must keep their rates down to a comparable level. A similar effect occurs because of competition between different routes of the same transportation medium, particularly railroads. The general effect of competition, then, is to lessen the rate differences between direct competitors. There are numerous examples of this in the literature of economic geography; for example, the presence of the

Box 3-1   # Elasticity of Demand

* *Elasticity of demand* measures how much demand for a good or service changes in response to change in the price of that good or service (both usually being measured in percentage terms). At the most basic level, we can recognize three major types of elasticity:

1. *Unitary elasticity of demand.* Where the percentage change (increase or decrease) in demand is exactly the same as the percentage change in price, we speak of elasticity of demand being *unitary.* (For example, this would apply where the price of a good decreases by 5 percent and demand for the good increases by 5 percent.)

2. *Elastic demand.* Where the percentage change in price produces an even greater percentage change in demand, such demand is said to be *elastic.*

3. *Inelastic demand.* Where the percentage change in price produces an imperceptible change in demand, we speak of *inelastic* demand; that is, it changes very little in response to a change in price.

In reality, of course, elasticity of demand generally falls somewhere along a continuum.

New York State Barge Canal had a pronounced effect on railroad rates between Buffalo and New York City. Thus freight rates have been observed to be lower from Chicago to New York than from Chicago to Philadelphia. The opening of the St. Lawrence Seaway in 1959 resulted in a decline in rail freight rates on commodities most likely to be affected by low water transport rates. But the improvement of a transport route will not in itself reduce freight rates to the users of the route if there is no competition between *suppliers* of the transport service. The effect will be simply to lower the costs to the suppliers of the service and increase their profits.

Traffic Density   Where demand for transportation is concentrated, with heavy volume of traffic over certain routes, there is justification for setting lower rates: the greater the traffic density, the lower the unit cost of transportation because the fixed costs can be spread more extensively. Conversely, areas or routes of light traffic may charge higher rates. The variations between freight rate territories discussed earlier reflected such differences in traffic density, as did the network patterns.

Direction of Haul over a Particular Route   A common practice is for freight rates to be lower along a route in the direction of light traffic flow. Such a difference is based on the fact that the railroad cars, trucks, or other vehicles must in any case be returned to the point from which the major traffic originates. Where a return journey has to be made, it costs little more to carry a freight load than to return empty. Thus more favorable "back-haul" rates can be offered, and these have been significant in a number of cases, a notable example being the development of iron and steel plants in the upper Great Lakes region. Here coal was shipped by water transport at low back-haul rates and used with local ores.

TRANSPORTATION COSTS AND THE LOCATION OF ECONOMIC ACTIVITY

The introduction of more realistic transportation costs and, especially, of the tapering characteristics of many freight rate structures inevitably modifies our view of the spatial organization of economic activities. The existence of tapering freight rates compared with rates that are proportional to distance encourages longer-distance movements, as Figure 3.10 showed. This greatly alters distance relationships in our economic landscape; long-distance movements become relatively cheaper, and short-distance movements are less advantageous. We can examine the locational implications of this revised view of transportation costs under three broad headings: first, the effect on supply and market areas; second, the effect on the actual location of production units; and third, the ways in which producers may themselves operate different spatial pricing policies by manipulating their transportation rates.

Effect on Supply and Market Areas

Long-haul economies make possible sales and purchases at greater distances than would be feasible if transportation costs were proportional to distance. The result is that market and supply areas for all kinds of products—agricultural, manufacturing, and central place goods—become spatially extended. In the case of agricultural production, for example, tapering transportation costs modify the shape of the location rent curve. Instead of being linear, the rent curve becomes *curvilinear,* implying that location rent declines more rapidly close to the market than at greater distances. The result is to spread production zones, especially the outer ones, thus extending the boundary of agricultural production as shown in Figure 3.15. Clearly, nonlinear transportation costs in general do nothing to destroy the overall pattern of zonation; they simply modify the size of the zones.

However, if different products have different levels of transportation costs, as well they might, depending on such characteristics as perishability, fragility, or bulkiness, the zonal pattern may be quite drastically altered, and some production zones may disappear altogether. In Figure 3.16, for example, the relatively high transportation costs incurred by product 2 give a steeper location rent curve, which is never higher than that for product 1. Product 2 will therefore not be produced. As a result, the transition point between one crop and another also changes, so that product 1's zone of production is extended outward while product 3 can be produced rather closer to the market.

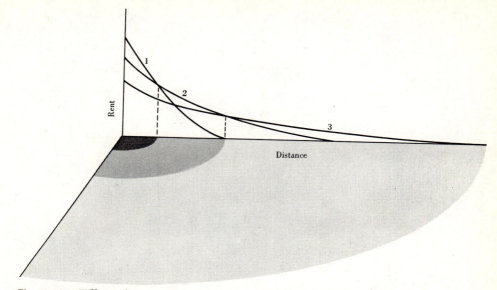

Figure 3.15 Effect of tapering freight rates on agricultural rent curves.

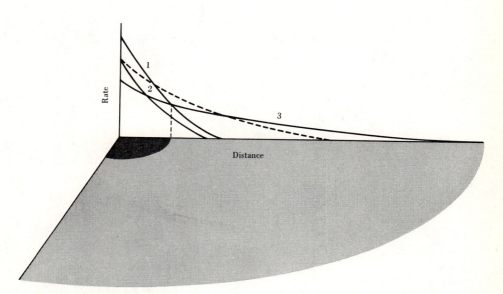

Figure 3.16 Effect of different levels of transportation rates on agricultural land use zones.

Long-haul economies will also tend to extend the market areas of producers. In the absence of competition, this would simply result in an extension of the outer boundary, a shift outward of the range of the good. In the more usual situation in which producers are competing for sales, the effect of tapering transportation costs will depend, as in the agricultural supply area example of Figure 3.16, on the *relative* transportation costs of each competitor. Suppose that as in Figure 3.17, there are two producers of the same good. If transportation costs were linear (that is, proportional to distance), the market area boundary would be exactly halfway between the two producers, at M. But suppose not only that transportation costs are not linear but also that the producer at A pays a lower transport rate than the one at B. This could occur, for example, if the producer at A can ship its products in larger consignments. The result could be that A may push the market area boundary to M' and even isolate B's market area $(M' - M'')$ by being able to charge a lower delivered price beyond M'' even though consumers in that area are closer to B than to A. Such extensions to market and supply areas are not, of course, spatially uniform but vary according to the cost advantage of different routes. The result is not only larger market and supply areas but also boundaries to such areas that are highly irregular. In addition, the grouping of freight rates into distance zones contributes toward the overlap of market and supply area boundaries, thus reducing the clear spatial monopoly that would otherwise exist.

Effect on the Location of Production

In Chapter 2, we examined the locational problem facing a producer of manufactured goods who had to assemble materials that were available only at certain locations and to distribute the product to a market. We saw that production could be located either at the end points of the locational figure or at an intermediate location, depending on the size of the material index in Weber's approach to the problem or on the specific substitution relationships in Isard's analysis. With tapering transportation costs, however, an intermediate location becomes distinctly less attractive because the cost of two separate short hauls is likely to exceed the cost of one long haul from either a material source or a market location (two sets of terminal costs would be incurred rather than one). We can illustrate this in a number of ways to correspond with our discussion in Chapter 2. Figure 3.18 shows the simple two-point locational problem in

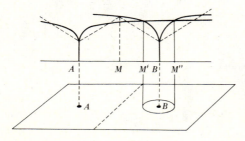

Figure 3.17 Effect of tapering freight rates on the market areas of two competitors.

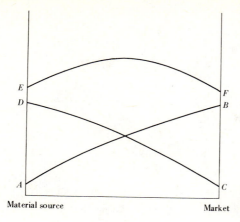

Material source   Market

Figure 3.18 Tapering freight rates encourage location at terminal points.

which both material and product have a tapered freight rate structure. The curve *AB* represents the cost of transporting the material, and the curve *CD* represents the cost of transporting the finished product. Adding the transportation cost on each at every point between market and materials source (curve *EF*) shows that an intermediate location is out of the question because, in every case, it would be more costly in transportation terms than location at either market or material source.

Tapering transportation costs can be accommodated quite easily into the isodapane technique. The effect is basically to increase the spacing between each set of isotims progressively, as in Figure 3.19. This is the same situation as shown in Figure 2.6, except that now transportation costs are higher on the material than on the product (shown by the closeness of the isotims around the material source). The least-cost production point under these conditions is at A (the material source), while the cost surface itself is also modified with repercussions on the extent of the spatial margins to profitability $(X - Y)$.

The existence of stepped and tapering freight rates certainly makes Isard's substitution analysis rather more complex at first sight as Figure 3.20 suggests. This graph shows the substitution relationships between two materials, M_1 and M_2. Two tons of M_2 and 1 ton of M_1 are needed to produce 1 ton of the final product. The graph contains three iso-outlay lines (lines of equal total expenditure on transportation) of $24.00, $26.40, and $30.00. In the comparable Figure 2.5, such iso-outlay lines were straight because they reflected linear transportation costs. In Figure 3.20, they are both curved and stepped, each step representing a distance zone. The derivation of the optimal location is the same as in all such analyses—it is the point where the transformation line just touches the lowest iso-outlay line. At first sight this would seem to be location *D* on the $30.00 line in Figure 3.20. In fact, it is location *B*. This is because a tapering cost structure produces "tails" on the iso-outlay lines that converge with the axes of the graph. For example, the stretch *LM* represents the tail of the $30.00 iso-outlay line. *BG*, however, represents the tail of the $26.40

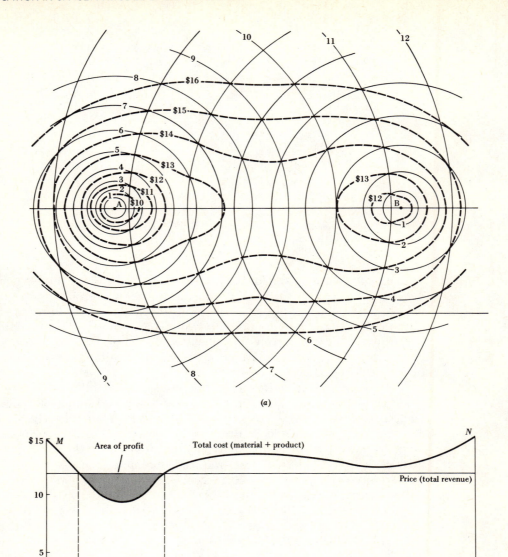

Figure 3.19 Effect of tapering freight rates on isodapanes.

iso-outlay line; hence *B*—the location of material M₁—is on a lower iso-outlay line than *D* and is therefore the minimum-transportation-cost point.

Tapering transportation costs, then, tend to encourage location at terminal or end points—material sources or market locations—rather than somewhere between. However, there are many examples of manufacturing activities that

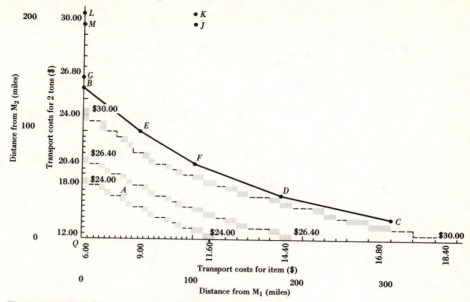

Figure 3.20 Isard's analysis and realistic freight rate structures. *Source:* W. Isard (1956), *Location and Space Economy* (Cambridge, Mass.: MIT Press), fig. 22. Copyright © 1956 by the Massachusetts Institute of Technology. Reprinted with permission.

have been established at neither material sources nor their main market centers. As subsequent chapters will show, there are several non-transport-related reasons for this, but as far as transportation costs are concerned, such intermediate locations may arise in two major ways.

First, and most important, is the case where different transport systems converge. Where goods must be transshipped from one type of transportation to another, additional terminal and handling costs are incurred that destroy any long-haul advantage. These extra costs can be avoided by locating production at the *break-of-bulk point*. If we examine the two-point case again but introduce the additional complexity of different transport media, the effect of their convergence can be seen (Figure 3.21). Between *A* (material source) and *B* (transshipment point), movement is by water transport; between *B* and *C* (the market) rail transport is used. The differing cost characteristics of the two media are shown by the differences in the cost gradients. The cost of transferring the good from water to rail at *B* produces the marked upward jump in both procurement and distribution costs. Consequently, it is cheaper to establish production at *B*. This simple example demonstrates why ports and rail terminals in particular tend to have considerable importance as manufacturing centers, especially in the processing of bulk materials. A similar phenomenon may occur at political boundaries, where crossing from one country to another may result in the journey's being regarded as starting afresh at the border even though no actual break occurs. The result, however, is an increase in transpor-

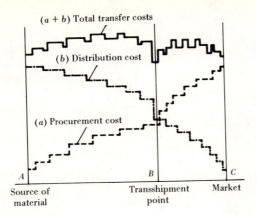

Figure 3.21 Effect of the convergence of two transportation media on total transportation costs. *Source:* E. M. Hoover (1948), *The Location of Economic Activity*. (New York: McGraw-Hill), fig. 3.8. Copyright © 1948. Used with permission.

tation costs that can be avoided only by establishing production at the artificial transshipment point.

A second, and far less important, cause of intermediate location in the transport context has been the granting of "in-transit" privileges by transportation agencies, particularly in the past when railroad companies were eager to attract traffic to their routes. The objective of the in-transit privilege is to remove the disadvantage of intermediate locations, which, as we have seen, would generally involve the combined greater costs of two short hauls. Under this arrangement, the material can be shipped from its source to an intermediate point, processed, and shipped onward to its market destination at the through rate. In the United States, flour milling has been perhaps the most notable example of this practice, but it has also been present in the iron and steel industry. For example, in 1909, the Wheeling and Lake Erie Railroad permitted the stopping of steel in transit at Canton and Toledo for fabrication. This facilitated the introduction of new fabricating plants there to compete with those in the Pittsburgh area (Fulton and Hoch, 1959).

The nature of transportation facilities and the use made of them by different types of economic activity helps to explain some of the locational arrangement of manufacturing functions within, as well as between, urban areas. We noted in Chapter 2 the tendency for certain kinds of manufacturing activity, particularly the large-scale material-intensive industries, to be tied to major transportation terminals during the nineteenth and early twentieth centuries. In particular, differences in the medium of transportation—whether road, rail, or waterway—seem to be especially important because the media themselves differ in the extent to which they exert a locational constraint. For example, industries using only truck transportation to serve local or regional markets will not be particularly constrained in their locational choice by transportation considerations simply because of the fineness of the road network and the flexibility of truck transportation. Those serving a more geographically extensive market may well be attracted to strategic locations such as interchanges on the interstate highway.

Spatial Pricing Policies

In our discussions so far, we have assumed that the price that a consumer pays for a good is related directly to the consumer's location in relation to the good's point of origin. This form of pricing is known as FOB (free on board) or ex-works pricing, whereby a price is quoted at the point of production. The cost of transportation is then added to this ex-works price. Obviously, the existence of transportation costs imposes limits on the spatial extent of a firm's market, while at the same time it gives a degree of spatial monopoly to a producer over the area in which the delivered price is lower than that of competing producers.

FOB pricing is nondiscriminating in the sense that each consumer pays a price in exact accordance with that consumer's location relative to the point of supply (factory price plus transportation cost). In Figure 3.22, for example, two firms, X and Y, both charge FOB prices for an identical product. Where both factory price and transportation costs are the same, they share the market between them equally (the market area boundary in Figure 3.22*a* being shown by the line *MC*). If firm X reduces its delivered price uniformly, from *ABC* to *DEF*—for example, by lowering profit margin in order to increase sales volume—then the firm can invade producer Y's market area as far as *N*. Such an advantage is likely to be short-lived, of course, because Y will probably retali-

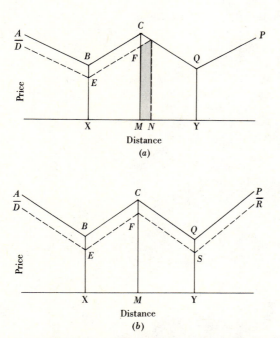

Figure 3.22 Competition for markets: the result of change in the delivered price. *Source:* M. L. Greenhut (1956), *Plant Location in Theory and Practice* (Chapel Hill: University of North Carolina Press), fig. 23. Reprinted with permission.

ate by reducing its delivered price by at least the same amount, and the market area boundary will revert to M (Figure 3.22b).

However, a producer need not adhere to FOB pricing, especially if the aim is to increase its sales at greater distances. *Discriminatory pricing* can be practiced by manipulation of transportation costs in a number of ways. In each case, the price paid by a consumer is not directly related to distance from the point of sale as it would be if FOB pricing were followed; in other words, there is a degree of spatial discrimination.

The aim of discriminatory pricing is to increase total revenue by achieving sales in markets at distances beyond those that could be served using the FOB pricing system. The basis of such discrimination is the spatial monopoly imposed by distance and the extent to which elasticity of demand for the good varies as a result of the location of competitors. In Figure 3.23, the boundaries of X's market area under FOB pricing are M, N. Beyond these points, consumers are supplied at a lower delivered price by Z and Y. The price to consumers within X's market area is a point along CDE, depending on their distance from X. But X *could* charge prices as high as CK and EK without losing sales to Z or Y. Clearly, X's scope for raising prices is greatest close to its location and least at the outer margins of the market area, where potential competition from other suppliers is greatest. In other words, the elasticity of demand for good ZX is greater farther from the seller's location.

In order to extend sales spatially, therefore, a seller must be able to charge a delivered price to distant consumers that is lower than that of competitors. This can be achieved by "absorbing" some or all of the freight cost to the distant consumer. Instead of operating an FOB pricing policy, a firm may operate an *equalized* delivered price system in which a uniform price is charged over a wide area. As Figure 3.24 shows, this, in effect, discriminates against consumers near the point of production or sale and benefits more distant buyers. In other words, local customers pay more and distant customers less than they would if delivered price varied directly with distance. A uniform delivered price system, which may extend over the entire national market in certain cases, is justified if it encourages sales to distant areas that would not

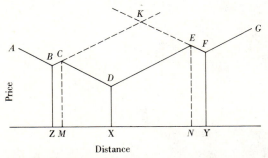

Figure 3.23 The potential for price increase is related to the degree of spatial monopoly. *Source:* After M. Chisholm (1966), fig. 5.

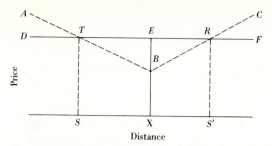

Figure 3.24 Relationship between FOB pricing and uniform delivered pricing.

exist otherwise, provided that the cost of producing additional units of output to such distant consumers is less than the revenue received in selling these units, allowing for the extra cost of freight absorption.

A good deal of controversy exists regarding the extent to which uniform delivered prices operate in practice. On the whole, the system has been most prevalent in consumer goods, especially those advertised and sold statewide or nationwide, rather than in producer goods. In the latter case, much depends on both the relative importance of transportation costs in the industry's cost structure and the distances involved. The small spatial extent of the British market, for example, means that uniform delivered prices are characteristic of a very wide range of economic activities, not merely of those with low transport costs. However, uniform delivered prices are characteristic not only of consumer goods but also of a whole range of commodities.

Probably the most widely known discriminatory pricing policy is *basing-point pricing* (single or multiple), although today it is relatively unimportant and probably restricted to one or two economic activities. But in the past, basing-point pricing was a characteristic of a large number of industries. In the United States, no fewer than 18 major industries used this system at some time until it was finally declared illegal by the federal authorities in 1948. Its use was especially widespread in the steel industry.

Basing-point pricing is, in effect, a rigid pattern of delivered prices systematically followed by all firms in an industry. The system operates in the following way. A base price is fixed at a certain location, and the price quoted to all consumers is this base price plus the freight charge from the base point to the consumer, whether or not the good is bought from the base point or from some other location. In an industry operating on basing-point pricing, all sellers will quote an identical delivered price to any one consumer.

Figure 3.25 shows two production centers. The center at P can quote an FOB price of PS, while RST shows its delivered price at all locations. X is also a producing center, which has much lower production costs (XB) than P and could quote a delivered price ABC. Under a nondiscriminatory pricing system, X would supply all customers to the right of M and P would serve the market to the left of M. Suppose, however, that P is the basing point for the industry. In

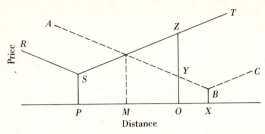

Figure 3.25 The distortional impact of single-basing-point pricing. *Source:* After M. Chisholm (1966), fig. 5.

this case, all consumers must pay the appropriate delivered price along *RST*. For example, *O* represents a consumer who purchases from *X*, the nearest producer. Under the basing-point system, *O* must pay *OZ* instead of *OY* even though the good is transported from *X*. In other words, the consumer must pay for a nonexistent, or "phantom," freight. Quite clearly, the existence of a basing-point pricing system is likely to stimulate the location of users of the good close to the basing point. Only in this way can they reduce the delivered price of the good to them. In the case of the U.S. steel industry, where a single basing point focused on Pittsburgh operated until 1924 and a multiple basing-point system prevailed until 1948, it has been claimed that it greatly helped to keep steel fabricating activity in the long-established areas of steel production. Hence the development of steel production and fabrication in the southern states was inhibited for some considerable time despite the low-cost advantage of the Birmingham, Alabama, site. In fact, in Figure 3.25, point *P* could be regarded as representing Pittsburgh and point *X*, Birmingham.

Thus variations in pricing policies, together with the zonal characteristics of freight rates, tend to reduce the occurrence of distinct market area boundaries. The ultimate goal of the various pricing practices, both nondiscriminatory and discriminatory, is to maximize revenues by increasing sales. Where a nondiscriminatory FOB policy is adopted, with consumers bearing the cost of transportation, there is a stimulus for the seller to be as close as possible to the largest number of buyers. If the market is spatially extensive, firms are likely to be dispersed, other things being equal. Under such circumstances, the location of competitors is of vital importance. On the other hand, where equalized delivered prices are quoted, such locational interdependence is less marked. Location patterns are more likely to differ according to spatial variations in procurement costs (see Chapter 2), costs of production (Chapter 4), and agglomeration economies (Chapter 5).

TRANSPORTATION IMPROVEMENTS AND THEIR SPATIAL IMPACT

An efficient transportation system is, in many ways, the lifeblood of an economic system because it is the means whereby the friction of space is overcome. The need to move goods and people from place to place as rapidly and as cheaply as possible has resulted in major changes in transportation technology. These have, in effect, progressively altered the spatial dimension of the economic system. As a result, we live in a "shrinking" world (Figure 3.26), although the amount of shrinkage is not everywhere the same. We might imagine that the earth's surface is like a sheet of highly flexible material that over time becomes changed and distorted—relatively stretched in some areas and contracted in others.

Let us look more closely at this process using Janelle's (1969) concept of *time-space convergence* as a framework for our discussion. Janelle set out to demonstrate the cumulative, though spatially uneven, impact of transportation improvements on the spatial organization of activities. His starting point is the same as the one we chose at the beginning of this chapter: the demand for

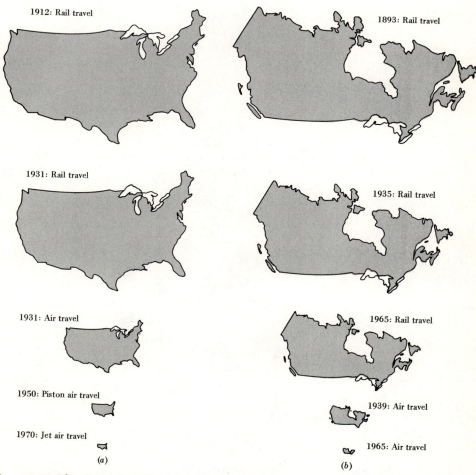

Figure 3.26 The "shrinkage" of (*a*) the United States and (*b*) Canada as a result of innovations in transportation. *Source:* After Dicken and Lloyd (1981), figs. 2.5 and 2.6.

improved accessibility generated by an economic system as a whole or by one or another of its component parts (individual businesses, urban centers, etc.). In Figure 3.27, therefore, demand for accessibility (1) is the driving force. If potential demand is sufficiently large, it is likely that a search for the means of satisfying this demand will occur (2). Searches that prove successful result in a transport innovation (3), which may be of one of two kinds: an entirely new form of transportation, as when the railroad superseded the stagecoach, or an improvement in existing types, for example, paved highways instead of dust tracks or boats powered by steam instead of sails. In either case, the result may be to increase the speed of movement, thus reducing the time-distance separating places or the cost of moving from place to place. Or the innovation may permit the movement of a much larger volume of traffic.

This is not the place to indulge ourselves in a detailed discussion of transportation innovations, but we need to take a brief look at such changes to provide a perspective from which to examine their spatial impact. As many economic historians have observed, the nineteenth-century industrial revolution cannot be considered in isolation from developments in transportation. Indeed, it has been suggested that the industrial revolution, at least in its early stages, might more realistically be called the transportation revolution. Because of the high cost and low level of efficiency of overland transport prior to the development of the steam engine, most industrial and commercial freight was carried by water: hence the early dominance of centers on seacoasts and navigable waterways. Development of the steam engine in the eighteenth century and its application in the early nineteenth century, first to water and later to land transportation in the form of the locomotive, was nothing short of revolutionary. This heralded the succession of transportation improvements that has continued to the present day.

The transportation revolution of the nineteenth century was based on the

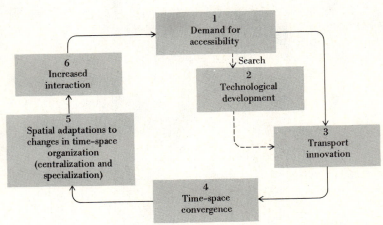

Figure 3.27 Spatial reorganization process in response to transportation improvements. *Source:* After D. G. Janelle (1969), fig. 3.

rapid spread of the railroads. The transportation revolution of the twentieth century is associated with quite different but equally revolutionary transportation innovations. First in importance is the automobile, the current symbol of Western urban-industrial society, and, later, the airplane and a more specialized form of transportation, the pipeline. The automobile in particular has become the most pervasive transportation medium, with a phenomenal growth rate since the turn of the century.

Introduction and development of each successive transportation medium has led to changes in the competitive position of existing types of transportation. This is mainly because each new development generally represented an improvement either in speed or efficiency of service, or both. In addition, as demonstrated earlier, each transport type offered particular advantages, either over certain distance ranges or for particular types of freight. In general, the share of freight carried by truck has increased very markedly since the Second World War, while railroads have experienced a considerable decline except in the movement of certain kinds of freight. Airlines remain more significant as people movers than as freight movers, although some kinds of low-weight, high-value goods do move by air.

Thus the two major trends in transportation innovations have been, first, the development of improved means of transportation and, second, technical developments within the transportation media themselves. The effects of such developments have been further enhanced in recent years by improvements in coordination and linkage between the different transportation media. The development of piggyback and fishyback techniques—standardized containers that are interchangeable between truck, railroad car, and ship—has made an enormous difference to both the times and costs of freight handling.

These successive developments have resulted not only in greatly improved transportation facilities but also in a drastic general decline in the average costs of transportation. For example, the opening of the Erie Canal in 1825 reduced the cost of transportation between Buffalo and Albany from $100 to $10 and ultimately to $3 per ton. Between the 1870s and the 1950s, the real cost of ocean transport fell by almost 60 percent.

The universal decline in transportation costs, combined with the changes toward more efficient use of raw materials in production processes and the substitution of one material for another (for example, electricity for coal), means that transportation has become a relatively less important factor in the overall cost structure of many industries. For some heavy industries, of course, transportation costs remain a very substantial cost element. However, the tendency in recent decades has been for most products to undergo a greater degree of processing. In other words, the *value added* by production to that of the original materials has tended to increase. The greater this degree of value added, the less important proportionally are transportation costs, which, as we have shown, have in any case been falling relatively.

Not only has the cost of overcoming the friction of space declined spectacularly over the past century and a half, but also the time involved has decreased greatly. For example, the construction of the railroad shortened the journey time between New York and Chicago from more than three weeks to less than three days. Figure 3.28 shows the progressive reduction in travel time between Detroit and Lansing, Michigan, because of transport innovations. As a result of such changes, places are effectively closer together than before. In Janelle's terminology, they are characterized by time-space convergence (step 4 in Figure 3.27).

The general result of such developments is an overall reduction in the friction of distance. In terms of the gravity model (see Chapter 1), the distance exponent b has been progressively reduced as transportation improvements have occurred. Insofar as b describes the rate at which interaction or movement declines with distance (it defines the slope of the distance-decay curve), the changing situation resembles that of Figure 3.29. As we move from t_1, a period in which movement was costly and the friction of distance very high, through successive time periods t_2 to t_4, the diminished frictional effect is reflected in gentler and gentler slopes as defined by b, and interaction takes place over greater and greater distances.

More specifically, the result of the kinds of transportation developments we have described briefly in this chapter is to alter greatly the spatial relationships between economic activities. Their *relative* locations have been altered (step 5 in Figure 3.27). This spatial reorganization has three major characteristics:

1. The spatial pattern of production has been transformed from dispersed to concentrated.
2. Differentiation between locations on the basis of their intrinsic qualities has increased.
3. The degree of geographic specialization and the spatial extent of production have both increased.

The production and distribution units (farms, factories, central places), which under primitive or high-cost transport conditions had to be scattered to serve

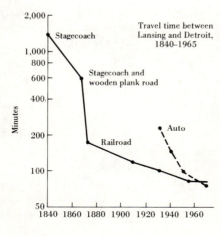

Figure 3.28 Reduction in journey time as a result of progressive transportation improvements. *Source:* D. G. Janelle (1969), "Spatial Reorganization: A Model and a Concept," *Annals of the Association of American Geographers* 59: 348–364, fig. 4. Reprinted by permission.

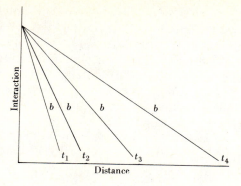

Figure 3.29 Changes in the frictional effect of distance.

distant markets, have tended to become spatially concentrated in areas of greatest advantage. Market areas have therefore become even more extensive as transportation has improved. Similarly, supply areas have expanded. In the case of agriculture, production becomes profitable over a wider area, so the zones of production tend to move outward from the central market. Figure 3.30 presents a striking illustration of this by comparing estimated values of wheat at increasing distances from the market under two different transportation technologies, the railroad and the "ordinary" road in 1852. The value of wheat declined very rapidly if road transportation was used, whereas its value was retained over far greater distances as a result of the railroad. At a 200-mile distance, wheat was worth $19.50 per ton under road transport but $46.50 under rail. Clearly, the margin of cultivation for wheat (and, of course, for other crops) was very greatly extended.

Such expansion of agricultural production zones based on transportation

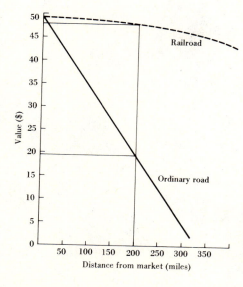

Figure 3.30 The impact of improved transportation (railroad) on wheat values per ton at various distances from the market, United States, 1852. *Source:* Calculated from data in Locklin (1960), p. 3.

improvements partly explains the continental-scale zonation noted in Chapter 1. Peet (1969) has demonstrated how the large-scale spatial expansion of commercial agriculture in the nineteenth century can be interpreted in a von Thünen framework. He shows how the growth in demand for food and raw materials generated by the industrializing and urbanizing "world metropolis" focused on northwestern Europe and the northeastern United States was met by tapping increasingly distant agricultural areas in a way made possible by the transportation improvements of the nineteenth century. Figure 3.31 shows how the average distance from London to regions from which imports were derived increased over the period 1831 to 1913. In the 1831–1835 period, no

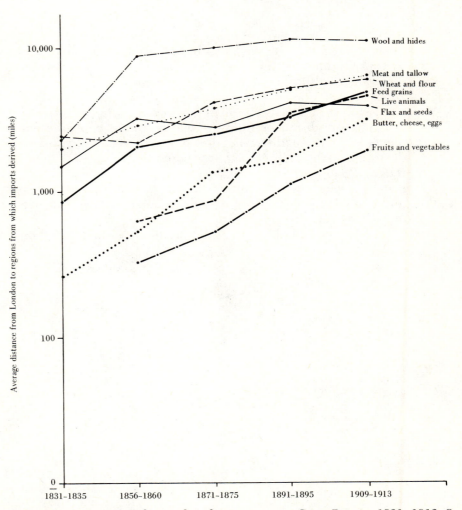

Figure 3.31 Expansion of agricultural import zones, Great Britain, 1831–1913. *Source:* Compiled from R. Peet (1969), table 1.

significant imports of fruits and vegetables or live animals occurred (products originating in Ireland were not classed as imports). By the 1856–1860 period, however, fruits and vegetables were being imported from an average distance of 324 miles from London and live animals from 630 miles on average. As the century progressed, these eight products were derived from increasing distances from London. In 1831–1835, the maximum distance involved was 2,430 miles (for wheat and flour); by 1891–1895, this maximum distance had increased to 11,010 miles (for wool and hides). Thus although the outward expansion of agricultural zones was not regular in every case, the general trend shown in Figure 3.31 was for an overall progressive increase in the distances involved.

Decline in the relative importance of transportation costs therefore implies that other factors increase in significance. In particular, specialization of production on the basis of comparative advantage other than market accessibility emerges, with an accompanying increase in geographic specialization. In agriculture, for example, remote locations with particularly advantageous physical conditions can be exploited, whereas their location relative to the market precludes their optimal use under less favorable transportation conditions. An obvious example is the change that has occurred in the location of intensively cultivated horticultural crops. They were formerly tied to the area immediately adjacent to the market; today they tend to be concentrated in areas offering the greatest natural advantages—often in other countries—with only small pockets remaining at the urban periphery (see Chapter 1). The general effect of improved transportation on the location of agricultural production has been to increase the relative importance of natural environmental factors.

Such spatial changes at what might be termed the interregional scale have been paralleled by changes at the intraregional level. In particular, the spread of automobile transport has induced profound changes within metropolitan areas, primarily by allowing the infilling of interstitial areas between major transportation arteries. Developments in intrametropolitan accessibility have permitted a rearrangement of urban-economic functions resulting in greater growth at the periphery than in the central area.

As Figure 3.27 shows, the result of such spatial reorganization is a higher level of interaction (6). Whereas under more primitive transport conditions most needs could be satisfied locally, the greater spatial concentration and specialization of production mean that this is no longer the case, and interaction must inevitably increase as places engage in trade with each other. This in itself generates new demands for further improvements in transportation as the routes and networks become congested and obsolescent after a period of time, and so a renewed cycle is initiated (step 1 in Figure 3.27).

But even though in general terms the friction of space has greatly diminished over time and continues to do so, the process has not been spatially even: some areas have benefited more than others. Transportation improvements tend to enhance the strategic position of some areas and to diminish that of others. In effect, they change the entire spatial form of the economic system by altering the functional distances between places. Invariably, transportation improvements tend to be greatest in and between places that already possess considerable economic status, generate a high level of demand for transportation, and are likely to benefit most from such improvements. These places tend to "converge" most significantly in time and space.

Developments in transportation thus have a pronounced effect on the spatial dimension of economic systems. As well as increasing the general level of interaction within and between systems, such developments profoundly alter space relationships. The implications of this process for regional economic development are profound because in many cases, areas that are peripheral to the main concentrations of economic growth become increasingly remote and their degree of real integration within the system diminishes. In some cases, of course, transportation improvements have enhanced the accessibility of formerly peripheral regions and increased their degree of integration. Finally, at a more specific level, we should mention the impact of transportation improvements on the overall cost structure of businesses. As transportation costs decline, they become relatively less significant in a firm's cost structure, and other costs—for example, costs of production—become relatively more important. We turn to the spatial form of such costs and their locational impact in the next chapter.

FURTHER READING

Alexander, J. W., S. E. Brown, and R. E. Dahlberg (1958). "Freight Rates: Selected Aspects of Uniform and Nodal Regions," *Economic Geography* 34: 1–18.

Greenhut, M. L. (1956). *Plant Location in Theory and Practice*. Chapel Hill: University of North Carolina Press, chap. 6.

Haggett, P. (1965). *Locational Analysis in Human Geography*. New York: St. Martin's Press, chap. 3.

Hay, A. (1973). *Transport for the Space Economy: A Geographical Study*. Seattle: University of Washington Press.

Isard, W. (1956). *Location and Space Economy*. Cambridge, Mass.: MIT Press, chap. 5.

Janelle, D. G. (1969). "Spatial Reorganization: A Model and a Concept," *Annals of the Association of American Geographers* 59: 348–364.

Sampson, R. J., and M. T. Farris (1980). *Domestic Transportation: Practice, Theory and Policy*, 4th ed. Boston: Houghton Mifflin, chaps. 2–5, 9–12.

Taaffe, E. J., and H. Gauthier (1973). *Geography of Transportation*. Englewood Cliffs, N.J.: Prentice-Hall, chaps. 4 and 5.

Chapter
4

Spatial Variations in Production Costs

COMBINING FACTORS OF PRODUCTION

The production of any good or service requires that the producer assemble a number of basic *inputs*. *Land* will be required on which to place the factory, farm, or office. *Materials* will have to be assembled and *labor* recruited over and above that supplied by the producer alone. *Capital* will certainly be needed, both as money to set up the business and buy the first inputs of materials from suppliers and to purchase machinery and equipment. *Technology* in one form or another will also be required. In simple terms, technology is simply "know-how."* These various inputs required for a business to function are traditionally known as *factors of production* (Figure 4.1). Like the finished goods we buy for ourselves, their price or cost is subject to the laws of supply and demand (see Chapter 1). There are *markets* for production factors: labor markets, capital markets, and so on.† Assuming, for now, that the decision of

---

* Although land, labor, and capital are relatively easily distinguishable from one another and are, in theory at least, discrete measurable factors, technology is difficult to define and even more difficult to separate out and measure. Much technology is "built in" to the tools available to a society. Tools, however, would be more readily identifiable as capital. Even more know-how is built into the particular "social technology," the institutions that different societies provide for getting things done.

† We shall have more to say about how such markets work in Part Two.

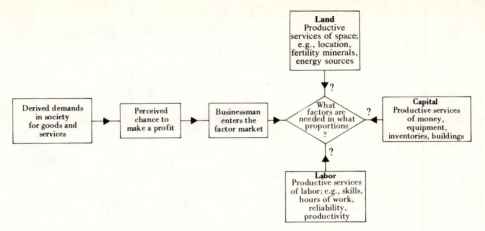

Figure 4.1 Choices in the hire of production factors.

what to produce has already been made, the question of the precise mix or combination of factors to use is of prime importance. This in turn depends on the technique of production to be used. It is almost always possible to vary the precise combination of, say, labor and machinery according to their relative cost. However, the extent of such possible *factor substitution* is not infinite.

Suppose, for instance, that our producer is an aspiring shirt manufacturer. Even for the same output of shirt units per week, choices will have to be made. Should the producer, for instance, invest in a set of computer-controlled cutting machines to cut the cloth, or is it better to take on skilled workers? Should buttonholing be done by machine or worker? Should the shirts be hand-stitched or sewn on a machine? Should allowance be made for floor space and shelving to carry a reserve stock of shirt cloth, or should this be bought as and when required? Obviously, many of these choices will be more critical in relation to different scales of the operation,* but even at the same scale of output, choices of *factor mix* would have to be made. It is also clear that in some activities, *technical substitution* of, say, capital for labor or land for capital is not possible. For any production unit, therefore, there is some notional range of *production possibilities* that implies an ability to substitute one factor of production for another. An economist would express this as a production function:

$$O = f(K, L, Q, T)$$

which is read as, "The output produced will be some function of the amounts and combinations of the capital (K), labor (L), land (Q), and technology (T) used."

---

* Scale of production is discussed in Chapter 5.

If we were to attempt to graph all the possible combinations for substitution from this set, we would, of course, need four dimensions—one for each of the factors. So let us simply subsume both land and technology for the moment into the capital factor. Under these constrained conditions, the logical expression for the production function would be

$$O = f(K, L)$$

and we could draw a two-dimensional graph of the production function (Figure 4.2). The axes represent increasing amounts of the capital and labor factors from the origin, 0. The curves OU_1 and OU_2 show, for two given levels of output, the possible combinations of amounts of capital and labor. We have encountered similar curves before in Chapter 2. In that case, we were using the curves or transformation lines to show possible substitutions of distance from pairs of locations in attempting to solve the manufacturer's distance minimization problem. Now we are looking at the substitution possibilities between capital and labor inputs. To produce 1,000 units of output on the basis of the production function shown in Figure 4.2, our producer could choose, for example, to hire 50 units of capital combined with 10 units of labor. Alternatively, the producer could go to the other extreme and hire only 20 units of capital but 35 units of labor. These would represent the extremes of capital intensity or labour intensity. However, a wide selection of intermediate choices is available. Every combination of capital and labor inputs at any point on the curve between its two extremes offers an opportunity for substitution of one against the other. Clearly, some combinations represent more rational

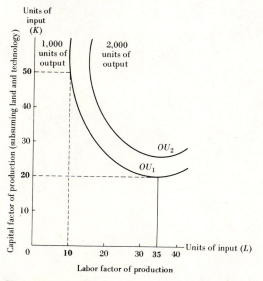

Figure 4.2 Production function for two factors of production: capital and labor.

choices than others. Where the production possibility lines "recurve" toward their extremities, more of *both* factors would be required for the same output. No rational producer would choose such a combination.

We have looked at the possibilities available to the producer in deciding on the particular factor mix. So far, however, we have said nothing about cost. Yet our aspiring producer is confronted not simply with the technical possibilities of substituting the quantity of one factor of production against another but also with the relative costs of the various factors. As a profit maximizer, the concern will be to make the decision that minimizes costs. (We are retaining for the moment the assumption of uniform demand.) The producer will therefore go for the factors that other things being equal, involve the lowest technically feasible outlays for a given volume of output.

Figure 4.3 provides a graphic illustration of what might be involved. Taking the production function for 1,000 units of output from Figure 4.2, we can

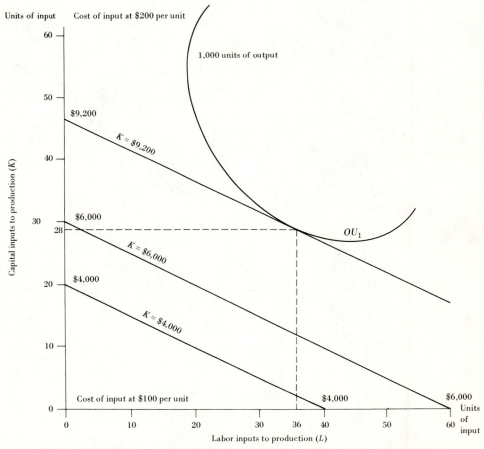

Figure 4.3 Production function and factor outlays for a given scale of output.

show the impact that differential factor costs might have. Costs are added to the graph by means of iso-outlay lines drawn diagonally from axis to axis. Suppose that the relative prices of capital and labor per unit of output are in the ratio 2 : 1. In other words, capital for the producer's particular application is twice as expensive as labor for each unit produced. There will then be an infinite number of iso-outlay lines. The figure shows only a selection of them.

Take first the iso-outlay line of $6,000. If the producer decides that the total budget would be $6,000, there will be choices in the way it can be spent. Applying the logic outlined in Chapter 2, it could, for instance, be spent on capital, and at $200 per unit, 30 of our notional units of capital could be obtained. Alternatively, it could all be spent on labor to get 60 units of labor at $100 per unit. The producer could also combine inputs of capital and labor so long as their total cost did not exceed $6,000. The iso-outlay line for $6,000 shows all the possible combinations at this cost.

However, if the producer had set out to produce a minimum of 1,000 units as shown by the production function OU_1, it is obvious that a $6,000 expenditure would not be enough. The outlay line does not nearly begin to reach the level at which the chosen scale of production is technically possible. For the 1,000 units, the actual level of outlay on production factors could only be represented by iso-outlay lines that intersected the curve of the production function. Since it is desired to minimize costs of production, it would be sensible to choose the one that gave the lowest total outlay on the two production factors in combination. In finding the point at which the production possibility curve intersects the lowest iso-outlay line, the producer would discover that the least-cost combination of factors would be that demanding 28 units of capital at $200 per unit and 36 units of labor at $100 per unit—a total of $9,200. With the production function shown in the curve OU_1, this is the lowest feasible cost at which 1,000 units of output could be produced. Notice that the slope of the iso-outlay lines, which reflects the relative costs of the two factors, tends to encourage the use of more labor—the cheaper factor—and less capital. Were the relative cost positions of the two reversed, the slope would be steeper and would favor capital over labor. In the real world, such simple solutions would be made far more difficult by the fact that the relative cost of factors is not static but ever changing. For example, if labor was a cheap factor for all producers and they exercised their choices to hire more of it, the relative scarcity of labor would increase. Its value would rise, and the price of the factor would rise with it. Soon the price advantage of taking on labor as a substitute for other factors would be significantly reduced, and other factors would become relatively more attractive. At this point, it might be more sensible (to the producer) to substitute more capital for labor.

So we have seen, in outline, the sort of considerations that an aspiring producer would have to take into account in setting up an enterprise. The producer would have to consider what factors were needed to produce the good, what combinations were technically feasible at the chosen scale, and what would be the least-cost combination. Each choice, as we have seen in Chapters 1 and 2, has its spatial ramifications. In the case of the land factor, in its pure sense, we have already shown how the geographic heterogeneity of the landscape influences the location of economic activity in space. The need for appropriate soil or climatic conditions at an economically feasible distance from the market absorbed our attention in the case of agriculture, while for manufacturing we have concentrated thus far on the distance costs associated with assembling the resources needed for a given productive activity. But these considerations tell only part of the story. However good the space-cost models have proved themselves to be in providing a working basis for the understanding of location, they are necessarily limited in their application. Where the influence of space has great importance, say, for economic activities needing large amounts of specific localized raw materials or land of a special type, the models (as we saw in Chapters 1 and 2) give a good fit. Equally, activities serving spatially localized markets in a situation where shipment costs contribute in major proportion to total costs are also appropriate subjects for analysis with models founded on space costs.

For some economic activities, however, space or land costs have never had a special importance in comparison with, for example, labor or capital costs. For the small handicraft manufacturer, space inputs are traditionally small; a loft or workshop is often sufficient. Transportation costs on inputs or the final product also tend to be small. For this kind of producer, the availability of labor with the right skills at the right price or the presence of a friendly financier to help in the setup of the business may be the crucial element in the decision process. The producer will be drawn to locations where such attributes exist. In this sense, labor and capital, though not by nature intrinsically spatial like land or distance, are no less critical to location. Further, as we pointed out in Chapter 3, the distance element in space costs has been drastically reduced with the transportation revolutions of the past 150 years. This being so, in relative terms, nonspace factors of production have increased their importance in the cost structure of businesses of all kinds. Therefore, we now turn our attention to those ingredients of the production function: labor, capital, and the particular forms of technology. The simplified model assumed them to be available everywhere. Let us now remove this constraint.

LABOR

As we have pointed out, some input of labor is fundamental to the operation of all production systems. Its relative importance varies widely, however, from one type of economic activity to another. Nevertheless, a growing body of opinion argues that the particular characteristics of labor make it of special significance as a spatially variable factor of production for all economic activities.*

---

* This view will be developed more fully in Chapter 10.

Up to this point, we have assumed, for the sake of simplicity, that labor has had no significant impact on locational choice. The simplified model assumed that labor was evenly distributed over space and that workers were equally endowed with an infinite variety of labor skills that they could perform with uniform productivity. Of course, nothing is farther from the truth. Labor, defined as the productive services offered by human beings, is highly variable in its spatial distribution. Even if we were to equate labor availability simply with the distribution of population, a glance at any population density map would immediately indicate the extreme spatial variety in the supply of this factor. If we were to take into account differences in skill, productivity, reliability, versatility, and so on, labor supply would take on the most complex patchwork of pattern and variety over space.

The Importance of Labor as a Location Factor

For U.S. manufacturing as a whole, even in these days of mechanization and automation, labor costs are still a significant proportion of business costs in general. Wages and salaries of production workers still account on average for one-fourth of the value added in manufacture for all industries. If nonproduction workers—managers, technical and office staff, and so on—are added, the figure is 46 percent. So even though our society is becoming more dominated by computers and machines and wage and salary payments to production workers have been falling steadily (in 1974 they accounted for 30 percent of value added), the drift of the labor force from blue-collar to white-collar jobs is maintaining the *overall* cost significance of labor as a production factor.

Figure 4.4 shows the main classes of U.S. manufacturing activity ranked by decreasing contribution of the work force payroll to value added. As the figure shows, the range is very considerable. In terms of all workers, wage costs are more important than the U.S. average in eight industries; in terms of production workers, only 11 industries had wage costs above the national average. Wage costs are least important in industries such as petroleum, chemicals, and tobacco products and most important in primary metal industries, timber and wood products, transportation equipment, and textiles.

It is clear, therefore, that there are some industries whose key factor input is labor. There are also some that are all but insensitive to cost variations in this particular factor. When Alfred Weber set out his model of industrial location in 1909, he was aware of this fact. Labor costs ranked at the highest level as a "locational factor." They were given the status of a general regional factor of location along with "the relative price range of deposits of materials and the costs of transportation." He saw all industries as being affected in their location to some degree by labor costs, but he also acknowledged that some have a more specific degree of labor orientation due to the greater importance of the labor input in relation to that of other factors.

Within the terms of his analysis, Weber sought to provide some measure of the degree of labor orientation applicable to various industries. His *index of*

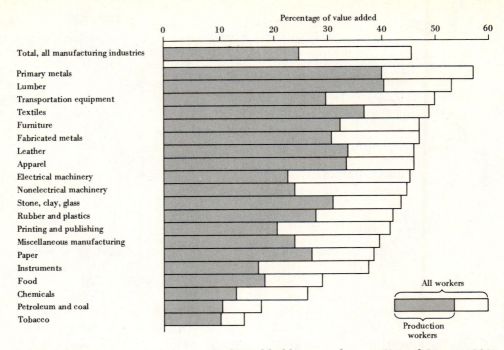

Figure 4.4 Wages as a percentage of value added by manufacture, United States, 1982. *Source: Statistical Abstract of the United States, 1985,* table 1342.

labor cost measured the average cost of labor required to produce a given unit weight of product. Industries with a high index would be more generally sensitive to spatial variations in labor costs than those with a low index. However, the real significance of the labor factor in location depends on its relative weight as compared with the other elements in the production function of the industry that have an identifiable locational impact. In Weber's analysis, the prime component among these was the total weight of materials to be moved for the production of a unit weight of product—that is, the *locational weight.* Thus in Weberian terms, the sensitivity of an industry to spatial variations in the cost of labor as an input factor is simply given by the *labor coefficient,* the ratio of the labor index to the locational weight.

In reality, of course, the determination of the relative importance of labor as a locational factor is far more complex than that.

Spatial Variations in the Cost of Labor

Labor need not be the largest component in an industry's cost structure to be locationally significant. What matters is the extent of the difference in labor costs—or sheer availability—between locations. In the past, specific geographic clusters of workers with particular skills exerted an extremely important locational influence. However, developments in the production process itself—notably the introduction of greater mechanization and automation—have reduced such locational influence.

That there are spatial differences in labor costs in the United States is shown in Figure 4.5. The median level of average hourly earnings for all production workers in 1983 was $8.84. But the interstate range of earnings was from $12.34 in Alaska to $6.68 in North Carolina. In general, earnings tended to be lower in most southern states and highest in the east north central and Pacific groups of states. However, earnings vary not only from state to state but also between urban and rural areas and by size of urban area. In general, earnings tend to be higher in urban areas than in rural areas—this partly explains the differential between northern and southern states. Earnings also tend to be higher in bigger cities than in smaller cities. Of course, much depends on the particular industrial mix in a particular area. Earnings in some kinds of economic activity are higher than in others. Some areas may have a larger proportion of low-paying industries, and this will obviously depress the average earnings level. But even within the same industry, there may be geographic variations in earnings.

However, labor costs are not merely a function of wage levels. The crucial factor is not so much what a producer has to pay for labor but rather the return

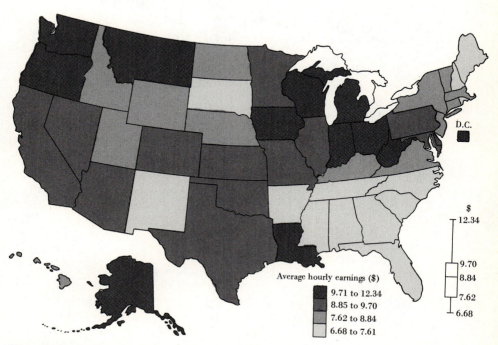

Figure 4.5 Average hourly earnings of production workers in manufacturing, United States, 1983. *Source: Statistical Abstract of the United States, 1985*, table 697.

achieved for a given payment. In other words, we must consider productivity or output per worker. For example, suppose that wages in city A were 5 percent lower than in city B but the productivity of the labor force in B was 10 percent higher than in A. It would clearly benefit the producer to locate in the higher-wage location because of the greater volume of output that would be received.

Low-wage areas, therefore, are not such an obvious attraction to business enterprises as a first, uncritical glance would suggest. Frequently, it costs a business a substantial amount to initiate training programs to meet even the most modest productivity targets. High levels of labor turnover while the local populace tries out the new jobs often means that many more workers are trained than ever produce output. Similarly, high levels of initial absenteeism mean disruption of training programs and manning schedules, with a consequent loss of output. For this reason, many governments, in attempting to attract industry to areas where wages are likely to be low, are still bound to offer further inducements such as tax holidays (exemption from taxation for a limited period) and training and investment subsidies. These help tide businesses over an initial period of heavy outlay, part of which is concerned with the need to raise productivity levels in the labor force to a point where any wage savings constitute a real gain in cost efficiency.

The productivity of labor, then, depends on many things. If we ignore, for the moment, considerations such as the equipment a worker is given or the management skills that direct the worker's activities (both of which are crucial to effective productivity), we might see labor attitudes as a critical variable. When talking of the quality of their labor force, most producers speak in terms of rates of labor turnover, rates of absenteeism, the frequency of industrial disputes, willingness to adapt to new production methods, and the like. Indeed, for a very large percentage of industrial enterprises moving to new locations, the labor attitudes and labor relations aspect tends to come to the fore in discussions of the role of labor as a factor in locational choice. Not that wage rates are not important as a consideration; they are. But to the everyday operation of a business, differentials in "strike proneness" or absenteeism tend to weigh more heavily these days on a decision maker's mind than a few cents an hour differential in cash labor outlays.

We will discuss these issues more fully in Chapter 10. However, it is worth noting at this stage that substantial spatial variations exist in the extent of unionization of the U.S. labor force. Figure 4.6 shows that the degree of unionization of the nonagricultural labor force varied from a mere 7.8 percent in South Carolina to almost 40 percent in New York. The median state value was 22.3 percent. In general, unionization was highest in the traditional manufacturing states of the Northeast and lowest in southern states. By 1980, fully 20 states had enacted "right to work" legislation that makes compulsory membership in a labor union illegal. Not surprisingly, the geographic distribution of right-to-work states roughly parallels the distribution of states with low levels of unionization. All but two of the states in the lowest quartile were right-to-work states, including all the southern states.

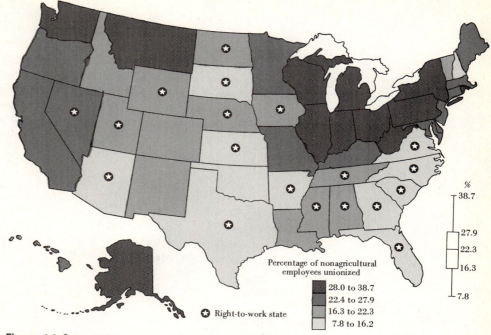

Figure 4.6 Interstate variations in unionization and right-to-work legislation, United States, 1980. *Source: Statistical Abstract of the United States, 1985,* table 709.

The Mobility of Labor

Spatial variations in the cost and availability of labor would only be important locational influences in the short term if labor were perfectly mobile. Under such circumstances, workers would simply migrate from one location to another in search of job opportunities and higher earnings.

But labor is not infinitely mobile, and the labor market does not operate freely under conditions of perfect competition, at least not outside the realms of pure theory. In the real world, while it is certainly true that a good deal of migration during the twentieth century has been from low-wage, mainly rural areas to higher-wage urban areas, the classical notions that "labor reacts to wage differentials by moving to the jobs offering the most favorable terms and conditions" and that "workers compare alternative jobs and choose the ones that are superior" (Gitlow, 1954, p. 62) require substantial qualification. Even where such movements have taken place, they have conspicuously failed to remove spatial wage differentials.

Long-distance labor movements have to be examined in the context of migration as a whole, a complex behavioral process the parameters of which are far from simply economic. In general, labor tends to be relatively immobile, particularly where long-distance movement is involved. In the short run at least, labor is an immobile factor of production. In fact, some types of labor are more mobile than others. In general, male workers are more mobile than female workers; skilled workers are more mobile than unskilled workers.

For a variety of reasons, then, labor supplies and labor costs vary significantly from place to place and exert differential locational attractions. We shall probe more deeply into these underlying reasons in Chapter 10; at this stage, we move on to consider spatial variations in the availability and cost of another major factor of production: capital.

CAPITAL

Every business—even the smallest and the simplest—requires some capital to start up. Once started, further capital is required to keep the business going. Money capital is derived from four possible sources, in broadly increasing order of scale and organization of productive operations:

1. Personal savings
2. Loans
3. Reinvested profits
4. Sale of stock (equity) in the business

Personal savings may well be important in the initial startup of a totally new venture, but in most cases, some or all the launch capital will come from raising a loan from one of the financial institutions. All the regular banks provide business loans, but there are also venture capital firms that specialize in financing potentially profitable businesses. Loans are also a major source of financing for businesses, although as time goes on, more successful firms may be able to finance their investment plans by reinvesting some of their profits or by selling equity in the business itself.

The market for capital thus consists of a complex, interlocking network of financial institutions—banks, finance corporations, insurance companies, government agencies, and so on—that serve to accumulate savings and disburse investment capital to would-be hirers of the factor. Most domestic savings in the modern world come in an institutionalized form. The government "saves" and reinvests household income through taxation. Pension funds and insurance premiums channel savings to insurance companies and their associated banks, which form the largest private-sector sources of financing in the modern era. These represent the effective sources of supply to the developer seeking funds. They have a particular spatial distribution. Similarly, the aggregate demand for capital has a spatial pattern. We can therefore look at the locational impact of capital in terms of the spatial disparity between supply and demand for the factor and its mobility in matching the two at different locations. First, however, let us briefly examine the importance of capital as an input to different classes of economic activity.

The Importance of Capital as an Input Factor

Although all economic activities need capital to function, they vary considerably in terms of how much capital is required. Ignoring obvious variations in capital needs by firms of different sizes (large firms need more capital than small firms), some industries are far more capital-intensive than others. Wholesale and retail trade, finance, insurance, and personal and business services are heavy on people and light on machines. Though the advent of the computer has revolutionized many aspects of the work in these industries, it has not significantly altered their degree of labor bias. By contrast, the mining and manufacturing sectors depend very heavily in general on expensive inputs of capital equipment, and there has been a consistent tendency over the years to substitute capital equipment for the hire of labor services. Within each of the sectors themselves, there is, of course, a wide variation in capital intensity. Figure 4.7 shows for the major classes of U.S. manufacturing the amount of capital invested per production worker in 1971. It is hardly surprising that the petroleum industry was far and away the most capital-intensive of the major classes, with more than three times as much capital invested per shop floor worker as tobacco, which ranked second. Chemicals, the transportation industries, and primary metals were also highly capital-intensive by this measure. Apparel, furniture, leather, and textiles, in contrast, ranked lowest in capital invested per worker.

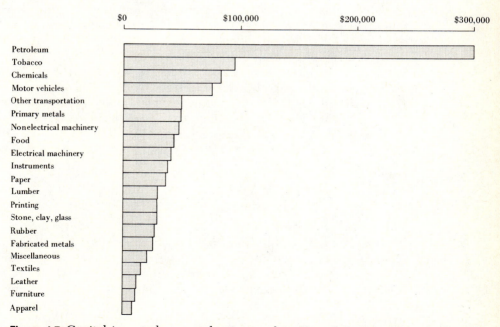

Figure 4.7 Capital invested per production worker, United States, 1971. *Source: Statistical Abstract of the United States, 1975*, table 1258.

Spatial Variations in the Availability of Capital

The accumulation of capital within an economic system depends on the level of savings, the willingness of savers to invest, and the level of net capital inflows from outside. This represents broadly the supply side of the situation. On the demand side we must take into account the level of actual or expected profits in the system and their relationship to profit levels in other systems to which capital might flow.

In some regions—for instance, old, established, successful industrial regions—capital may have been accumulating over a long period to give a large basic stock of physical capital in the sense of plant, equipment, and infrastructural facilities as well as an abundant supply of monetary investment capital from savings out of profits and wages. These regions may also be large sources of demand for capital, the high levels of profits within them absorbing internally generated investment funds and drawing in additions from outside.

Other regions, perhaps old and established but less successful, may still exhibit a large accumulation of physical capital and may, from the fruits of a more successful past, still accumulate large supplies of investment capital. They may, however, be unable to absorb all of this investment capital within themselves and stand out as regions of capital surplus. At the other end of the scale, developing areas, perhaps new frontiers for settlement, may be sources of a limited supply of capital but may at the same time exert a powerful demand for capital inputs through the potential returns to be had from mining and agricultural ventures.

Thus over space as well as over time and from one industry to another, there exist variations in levels of supply and demand for capital as a factor of production. The major allocative mechanism by which scarce supplies are distributed among competing uses (regional, sectoral) is a market system regulated by price, which, in the case of capital, is the *interest rate*. As in the case of labor, given perfect mobility and a free competitive market for capital, geographic differences in interest rates and hence the cost of capital would virtually disappear. Capital would flow from surplus areas to deficit areas until a balance, with equal rates of interest, was achieved. In fact, capital, like labor, is far from infinitely mobile, and free competition is, like the isotropic plain, an idealized concept.*

The Mobility of Capital

The mobility of capital differs according to the specific form of the capital involved. *Physical capital* in the form of plant or equipment is largely immobile as a factor once it is put in place. This is one of the sources of the phenomenon often referred to as geographic or industrial inertia, which is the tendency for an industry to remain in operation at a location when the reasons

---

* For a recent comprehensive discussion of regional capital theory, see Gertler (1984).

that brought it there in the first place have either lost their significance or completely disappeared. Once put to work at a particular place, the value of physical capital lies in its being used as much as possible; every moment of lost production costs money. The cost of taking it out of production to move it to another location is one that any producer would seek to avoid and is therefore a significant barrier to mobility. Thus physical capital, once set in place, becomes a powerful locational force guiding the development of the space economy through its inertial impact (see Chapter 10).

Monetary capital, by contrast, is considerably more mobile. Its movement is restricted more by institutional barriers than by distance or the frictional drag of space in general. The boundaries between nations, economic communities, trading blocs, or currency areas represent steep steps in the path of its flow, tending to keep finance capital within homogeneous financial systems, although the current internalization of capital markets is reducing these barriers. Within the same currency area, trading bloc, or nation, monetary capital is assumed to be highly, or even perfectly, mobile—free to move at no perceptible cost from place to place. In the United States, the revolution in communications technology and the development of a national federal reserve bank system are usually regarded as important causal factors in the emergence of a nationally integrated capital market (Gertler, 1984). But as Estall (1972) has pointed out, there may be a number of impediments to the free flow of investment funds that are not necessarily directly reflected in aggregate figures for the cost of capital at different places. Especially for the small business or for businesses wanting to invest in unusual or nonstandard activities, the physical presence at a given place of some institution willing to finance them may provide a point-source of investment funds that would simply not be available at other places.

Small business is often dependent upon local financing from banks or loan institutions where there are close, often personal contacts between business owner and financier. A move away from the effective range of such personal contacts may make financing for developmental purposes difficult or even impossible to obtain. There may therefore be a strong "distance decay" effect in the mobility of capital for the small business, with financiers willing to put up funds only so long as they can "keep a close eye on them." The presence of "specialist" lending institutions may well provide a source of capital to these industries, which is to all intents and purposes fixed in its effective location. Indeed, particular ethnic or religious traditions in the lending process may effectively restrict certain investment projects to particular quarters associated with such groups. Conversely, investment capital is highly mobile geographically within large multilocational firms, a point we will develop further in Chapter 8.

Estall (1972) points to a further neglected consideration influencing the effective spatial and sectoral mobility of capital. Most new investment funds go into the support and expansion of existing capital investments, which are, as we have seen, themselves constrained by the immobility of physical capital once in place. It has been estimated that up to 80 percent of all new manufacturing investment in the advanced nations goes to support the expansion of existing plants. Clearly, if this is the case, whatever the potential returns to be obtained at new locations, only a limited proportion of total capital stock is available to finance development. In this sense, capital as a production factor is far less mobile than we have hitherto tended to assume.

The activities of national and state governments seeking to promote investment in their less developed regions provide effective testimony to the immobility of capital as a production factor. Almost all of the advanced industrial nations find it necessary to provide investment capital in the form of grants and subsidies paid for out of taxes in regions where they hope to promote new development. This form of capital (derived from income the state "saves" on behalf of households as taxation) is becoming an increasingly important source of investment funds for new development. Indeed, the assistance schemes are designed for just this purpose—to pull new investment to depressed regions against an inertial trend that would reinforce regions previously capitalized. Capital of this order is clearly mobile in an imperfect sense. It is available in some regions and not in others, and in regions where it can be obtained, tariffs may restrict its availability to specific locations.

In summary, social, political, and economic discontinuities tend to discourage the free flow of monetary capital between different economic systems. The most significant of all these discontinuities is that between the advanced and underdeveloped nations. Even within homogeneous systems, however, the free flow of monetary capital cannot be said to be "perfectly" mobile either sectorally or spatially. In some senses it is—for the large, well-known corporation, funds may be as close as the nearest telephone. But for the small firm or the nonstandard capital application, there may be no effective mobility, spatially or otherwise, in the factor. Finally, governments these days exert a powerful impact on the flow of capital for development, regulating its overall levels by policy instrument and making it more readily available in some areas than in others.

In general, then, like labor, capital is localized (or perhaps more accurately, regionalized) in its supply and is not in any broad sense mobile. This means that the price of capital—the interest rate—does vary spatially, and it is to this variation that we now turn our attention.

Spatial Variations in the Cost of Capital

A number of writers have pointed to the existence of spatial variations in the cost and availability of capital and its impact on the evolving economic geography of the United States. Davis (1966) showed how, in the mid-nineteenth century, the financial network of the United States was dominated to such a large extent by the established centers of New England and the Middle Atlantic states that capital movement was unable to keep pace with demand generated by the westward expansion of economic activities. Davis suggests, for example, that an acute shortage of capital in the South prevented the major redevelopment of the region's textile industry until after the Civil War. Even then, local capital was insufficient, and northern capital had to be acquired at high interest rates. This situation was a result of the prevailing condition of the financial market, which consisted not of an integrated national network but of a series of small separate markets. The result was a wide divergence in interest rates: in 1870, for example, there was a 10 percent interest rate differential between New York City and the smaller centers of the West Coast.

At a more localized level, the work of Burgy (cited in Estall, 1972) points to the impact of differential availability of capital and interest rates on the evolution of the New England textile industry in the nineteenth century. The effective source of investment funds was the shipping and commercial enterprises of Boston and the region around Narragansett Bay, with a marked rise in interest rates culminating in a total reluctance to venture funds increasing with distance from the source. This contributed substantially to the early localization of the textile plants in Boston and the Providence-Pawtucket area in the early years of that century. Gradually, with improvements in communication promoting a more realistic appreciation of the risks involved in investing money farther away, Boston capital began to fund new develpments in New Hampshire and Maine. Later, funds were available to the whole of New England, and with time, Boston capital was to become one of the key sources of financing for developments throughout the United States.

It seems reasonable to assume that with time and the spread of the communications network, with effective national integration, and with universally applicable legislation and security services protecting the interests of capital, investment funds would become freely available in the United States. Lösch (1954), however, showed that even in the 1920s and 1930s, there were still significant spatial variations in general interest rates. Figure 4.8a is based on data for 20 financial centers at various distances from New York, the key source

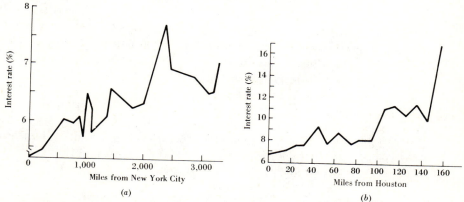

Figure 4.8 Spatial variations in interest rates within the United States: (a) Increase in the rate of increase with distance from New York City, 1919–1925; (b) Increase in the rate of interest with distance from Houston, Texas, 1936. *Source:* A. Lösch (1954) *Die räumliche Ordnung der Wirtschaft (The Economics of Location]* (Stuttgart: Gustav Fischer), figs. 94 and 95.

of U.S. capital. It shows average interest rates on six major types of bank loan for the period 1919–1925. The graph shows that although interest rates in each of the 20 centers were above that for New York and that there was a general increase with distance, the increase was not uniform. Rates were highest in the southern states (average 6.84 percent) and in the Continental Divide region at El Paso (7.68 percent) and Helena (7.73 percent); beyond this region, rates fell again toward the West Coast. The implication is that all the major banking centers were subordinate to New York. But in turn, each major center dominated a surrounding region within which interest rates increased with distance from the dominant center. Figure 4.8*b* demonstrates this phenomenon for the area round Houston, Texas.

Lösch attempted to explain the tendency for interest rates to be higher in the West than in the East in two ways. First, he suggested that demand for capital at the same rate of interest was greater in the West because of the larger range of undeveloped possibilities there together with the need for more credit to bridge over agricultural losses. Second, the total supply of capital, at the same rate of interest, was held to be smaller in the West. This was the result of a number of factors. One was the fact that the supply of local capital was smaller because of a predominance of agriculture, whose profits were lower than those of the major eastern industries. Another factor was the need for western banks to keep larger cash reserves on hand, and a third factor was the higher costs of banking in the West (partly because of the great distance from New York).

Not only was the supply of local capital smaller in the West, but also the supply of *eastern* capital was smaller. Again a number of factors inhibited capital movement despite the high interest rates. Lösch mentions the inaccessibility of the New York market in terms of both distance and knowledge and the need to employ intermediaries. He suggested, too, that the risk increases with distance because creditors cannot have full knowledge of conditions at a distance. Moroney and Walker (1966) attempted an explanation of development differentials in the North and South by reference in part to the comparative capital abundance of the former as compared to the latter. The high cost of capital in the South and the relatively low cost of labor are hypothesized as the basic factors underlying the region's strong attraction on the labor-intensive industries rather than more capital-intensive ones during the 1950s.

But how far do such geographic differences exist today? In Smith's view, "[interest] rates appear not to differ very much in advanced industrial nations, and the cost of financial capital is thus not very influential in locational choice" (1981, p. 48). However, as both Estall (1972) and Gertler (1984) suggest, it is not necessarily cost per se that is important. Interest rates—the cost of capital —may not vary a great deal geographically within countries like the United States, but the willingness of finance capital institutions to lend money for development is certain to vary, as anyone who has ever sought an overdraft or a bank loan will know. For some industries, then, the evaluation of capital as an input factor with an impact on location will depend solely on its availability or the reverse. The first-time entrepreneur wishing to open a business in the ghetto might not find capital available at any price, while General Motors could presumably raise capital almost anywhere at any time. Estall summarizes the position very succinctly:

> We conclude that capital funds are an important geographical variable in the internal location problem. Capital is not equally available at all locations, given identical risks and opportunities; and the constraints that operate to impede its equal availability must, by extension, act also as constraints upon location. (pp. 197–198)

TECHNICAL KNOWLEDGE

No study of the impact of variations in production costs would be complete without a consideration of the role of technology. In its broadest possible definition as "society's pool of knowledge regarding the industrial arts" (Mansfield, 1968, p. 10), technology is fundamental to the production process. A new method, technique, or machine can alter literally overnight the production possibilities open to a business. It may fundamentally alter the choices of factor mix that are made and may therefore substantially alter the cost effectiveness of a given locational choice.

The most comprehensive early treatment of technical knowledge as an input factor was made by Joseph Schumpeter (1939). He saw the accumulation of knowledge as a function of two discrete but correlated processes, *invention* and *innovation*. He envisaged invention as the introduction of new production processes and techniques to the existing stock of knowledge and innovation as the adoption of those processes and their translation into actual production processes. Schumpeter further distinguished *autonomous* from *induced* invention. Autonomous invention is the long-term, spontaneous, and apparently random contribution of occasional geniuses who invent things. They extend the existing stock of technical knowledge by the application of intuitive thought to the existing body of technology. Induced invention, by contrast, is the fruit of a deliberate expenditure of time, effort, and resources for the purposes of generating new knowledge. Today we would see this element of invention as being generated by research and development activity (R&D).

Critics of Schumpeter's classification see one of the main problems as trying to divide the indivisible. If invention is the first practical application of some abstract idea, how does one distinguish whether "making it work" was defined as part of the invention itself or of the innovation process? Does innovation differ from invention only in scale of application? The first automated-spindle cotton picker was made to work in the United States in 1889. Was this its invention? Full commercial use had to wait until 1948 for the development of the invention by the inventor using the capital of the firm that took it up. Invention and innovation together? Clearly, semantic hairsplitting is unhelpful to our present purpose, but it should encourage you, while accepting the Schumpeter typology for the moment, to give careful thought to the gray areas of its application. For Schumpeter, innovation is seen in terms of commercial exploitation at a significant scale, and this is important to the present discussion because it implies the widespread application of capital and labor inputs whose spatial availability is, as we have seen, a constraining influence on location. Similarly, induced invention, or the application of R&D, demands significant inputs of other production factors. Both innovations and the sources of R&D activities may therefore have a different probability of occurring over a region or nation than the more esoteric autonomous invention. If this is the case, the availability of technology as a factor of production for its potential users may also vary significantly over space in the same way as other factors of production.

Variations in the Spatial Distribution of Technical Knowledge

Technical knowledge is simply a subset of total knowledge, and the processes of communication and interaction are fundamental determinants. Following from this, it might be argued that locations in space with the highest probability for the generation of new technical knowledge of a spontaneous (autonomous) nature will be the points of greatest human interaction.

From what we know so far of the space economy, these points would be identified as the foci of the communications and interaction network, in particular, as the higher-order central places. Historical evidence for the United States supports this view. Table 4.1 shows the dominance of existing and emerging higher-order centers in the late nineteenth century in terms of their shares of the number of patents granted. The dominance of cities in the traditional industrial heartland of the Northeast is very apparent. By the 1950s,

Table 4.1 PERCENTAGE OF PATENTS
GRANTED IN SELECTED U.S.
CITIES, 1860–1900

	1860	1880	1900
Total	100.00	100.00	100.00
New York	14.75	13.19	10.27
Chicago	0.90	3.87	6.92
Philadelphia	4.88	4.28	3.98
St. Louis	0.81	1.37	1.62
Boston	2.91	3.32	2.02
Baltimore	0.95	1.54	1.11
Pittsburgh	1.13	1.61	2.05
Cleveland	0.95	1.05	1.58
San Francisco	0.60	1.32	1.08
Detroit	0.21	0.76	0.87
Los Angeles	0.00	0.02	0.49

Source: Data from Pred (1966). tables 3.1–3.3.

as Figure 4.9 shows, some broader geographic spread was evident, but the spatial bias of past developments was still there.

The sources of much of the autonomous invention that contributes to the stock of technical knowledge may thus be identified as localized in their distribution. Central places in general, but in particular those of higher order with greater levels of functional interaction and wide hinterlands, dominate and polarize the availability of invention.

A similar tendency toward metropolitan spatial concentration exists in the case of induced invention. Malecki (1980) has carefully documented the geographic structure of R&D in the United States. As he points out, "Research and development has some remarkable geographical patterns. . . . In all countries for which data are available, R and D has been found to be consistently more concentrated geographically than either population or industrial activity" (p. 3). Figure 4.10 maps the geographic pattern of two aspects of R&D in the United States: R&D employees and R&D laboratories. Malecki calculated that 87.9 percent of industrial R&D laboratories in 1975 were located in 177 metropolitan areas that contained a much smaller share (67.2 percent) of the nation's population. The spatial distribution of R&D employees was rather less highly concentrated than that of laboratories largely because general research laboratories "are often located away from large urban areas" (p. 3). Even so, the metropolitan spatial bias is very apparent.

In spatial terms, the processes of induced invention (the conscious pro-

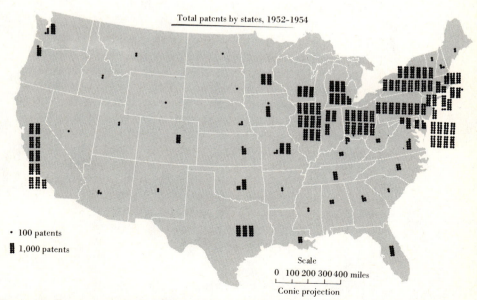

Total patents by states, 1952–1954

• 100 patents

▌ 1,000 patents

Scale

0 100 200 300 400 miles

Conic projection

Figure 4.9 One measure of the distribution of technical knowledge in the United States: the distribution of patents, 1952–1954. *Source:* E. L. Ullman (1958), "Regional Development and the Geography of Concentration," *Papers and Proceedings of the Regional Science Association* 4, map 4. Reproduced by permission.

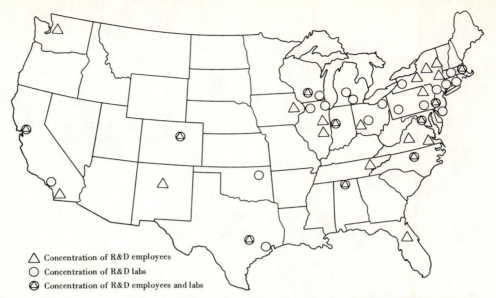

Figure 4.10 Urban complexes of research and development in the United States. *Source:* E. J. Malecki (1980), "Dimensions of R&D Location in the United States," *Research Policy* 9: 2–22, fig. 1.

motion of new technical knowledge) and innovation (the conversion of ideas into practice and process) have much in common. Both depend on the application of substantial investment inputs and the willingness of entrepreneurs to use their capital resources in this way. They therefore tend to exhibit a space preference for "capital-rich" locations with ready availability of investment funds and with, in Schumpeter's definition, a favorable "entrepreneurial climate."

Thus in effect, all forms of invention and innovation have a tendency toward spatially localized patterns of evolution (Siebert, 1969). Autonomous invention tends to take place with greatest probability at the points of greatest interaction and information exchange in space. Induced invention and innovation have a stronger association with the foci of interaction toward the apex of the urban hierarchy that are economically the most successful—the *centers of control.*

The Mobility of Technical Knowledge

Given that new technical knowledge does not originate in all regions at the same rate, its availability at different locations depends, like other factors, on its mobility. As before, with complete and free mobility, spatial disparities in its initial origin would have no significance, since the factor would move to attract high returns at favorable locations.

Technical knowledge is not perfectly mobile any more than capital is, though once again some writers have assumed it to be so.* Like all information, its movement is attenuated by the effects of distance, and its precise flow patterns are determined by a variety of complex factors. Suffice it to say at this stage that the exchange of technical knowledge over space tends to be sharply affected by the particular communications network in existence and by the relative spatial locations of senders and receivers. There is a strong conservative element in information flow. Existing structures tend to maintain and reinforce themselves because, to a considerable degree, the existing spatial distribution of economic activities determines the potential application of new knowledge and information.

The Availability of Technical Knowledge over Space

Technical knowledge may therefore be considered a spatially localized factor input for most economic activities. Its localization, as we have seen, tends to be oriented toward the larger and more successful existing concentrations of production and the foci of the geographic network of communications. In terms of mobility, it behaves like the other nonland production factors: it is sensitive to movement over space; it tends to be attenuated by distance, diffusing only slowly from its origin; and it is channeled along existing lines of movement and interaction.

In view of this, it tends in locational terms to be a strong polarizing agent in the evolution of economic activity. In particular, its polarizing function—that is, its capacity to draw development to its location—is especially powerful when other factor inputs such as labor and capital tend to be relatively mobile: "Process, product and organizational innovations have a strong polarizing incidence. The more mobile capital and labor [are], the stronger are the polarizing effects which technical knowledge induces with respect to these determinants" (Siebert, 1969, p. 40).

For industries that may be described as "technology-intensive," the costs involved in relation to the technical knowledge factor are essentially the opportunity costs a business would incur by being "in the wrong place at the wrong time." They would be notional costs that would add up to the total savings that might have been made if they were better placed—the costs of "forgone opportunities." The quantification of such costs in such a way as to compare them with money costs of labor or capital is a matter of fine judgment. For industries like electronics, the relative importance of "new ideas" as an input is likely to be so great as to encourage a location close to the main foci of technical knowledge in this field, even if the cost of other factor inputs at that point is high.

One of the problems associated with the evaluation of technology as a factor of location as well as a factor of production is, as we pointed out earlier,

---

* See, for example, Borts and Stein (1964, p. 81).

measurement. It is not difficult to gain some idea of the way in which technology-producing activities are located in space. When it comes to measuring technology-consuming activities, say, those in aircraft and missile manufacture and the making of electronic and scientific equipment and chemicals, it is equally clear that they are to be found, together with most other industries, in precisely the same regions. Indeed, the technology-producing activities and those that consume technology as a major input are very largely one and the same. But who is to say that the one attracted the other? Both grew up more or less in parallel. As the aircraft and missile industry grew up in the West, so did the research and development expenditure that kept it ahead of the world. Similarly, as the vehicles industry evolved in the Midwest, so did vehicle technology and the research activity that goes with it. If there is a direct, perceptible link between them, it is their twin association with success. Successful industries these days remain successful by investing in invention and innovation. Embedded within a regional structure of industries with heavy research and development expenditure are identifiable geographic clusters of research-producing activities. The Silicon Valley complex in California, the cluster of electronics industries along Route 128 near Boston, the Cape Canaveral complex, the science-based industrial complex along the Queen Elizabeth expressway near Toronto, and the cluster along the M4 west of London are well known. The complexity of the issues surrounding the production and consumption of technical knowledge in modern society and the sheer difficulty of measuring any aspect of the whole phenomenon make generalization unprofitable in this particular case.

At best, we can argue that technology is critical to economic growth in the advanced industrial nations. It draws a high proportion of the venture capital generated by both industry and government. As a factor, we have suggested that technology is not as perfectly mobile as many economic models would have us accept. There are places where it is more available than others, and deliberate expenditure to create it has a localized pattern of allocation. Even the transmission of technical knowledge through the channels of communication tends to be between such centers of specialization and thus reinforces its relative spatial concentration. As to its role as a location factor, we can do no more than suggest that there is a geographic association between research and development expenditure and successful industries. For a new enterprise, whether it simply searches out success or the technological innovation that tends to accompany it, the opportunity costs of being "in the wrong place at the wrong time" could be critical in locational terms.

THE LOCATIONAL IMPACT OF SPATIAL VARIATIONS IN FACTOR COSTS

Now that we have seen something of the "real-world" variety of the structural and spatial conditions affecting the input of production factors to economic activities, let us make some summary generalizations on the impact of this variety on the location of economic activity in space.

The tradition of analysis in location theory has been to take a single factor of production and to show how it would "distort" locational patterns made with reference to other criteria. We shall begin this discussion with a brief review of this kind of analytic approach, concluding with the general locational issues that arise where, under more realistic conditions, production factors are combined together in a given production function.

Under the greatly simplified conditions of his model, Weber (1909) devised a technique for the evaluation of the impact of localized sources of relatively cheap labor on the location of an industry, the initial site for which had been determined by reference to transport costs alone. He begins with the proposition that "when labor costs are varied an industry deviates from its transport locations in proportion to the size of its labor coefficient" (p. xxv). In other words, it was established that the locational pull of an additional factor on a location previously set by other factors would depend on the relative importance (labor coefficient) of that factor in relation to all other factor inputs for the industry.

Weber went on to determine the precise spatial extent of the locational shift under the conditions of his model by the use of isodapanes (see Chapter 2). He introduced the notion of the *critical isodapane*, which has particular relevance here. This is the isodapane whose value represents the additional transportation costs (above those of the minimum-transportation-cost point) that are equivalent to the savings that might be made through reductions in other production costs at alternative locations. The method is illustrated in Figure 4.11. Total production costs, which, for the moment, we assume to be simply transportation plus labor, are $5 per unit of output higher at *A*, the minimum-transportation point, than at *B*, *C*, or *D*. This means that some differential in nontransportation production costs exists between them. For instance, labor may be $5 per unit of output higher at *A*, the minimum-

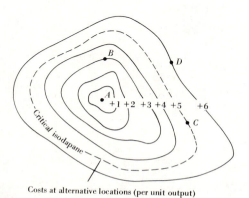

Costs at alternative locations (per unit output)

	A	*B*	*C*	*D*
Transport costs	$10	$13	$15	$16
Production costs	$20	$15	$15	$15
Total costs	$30	$28	$30	$31

Figure 4.11 Deviation of location from the minimum-transportation-cost point to the point of lowest production costs.

transportation-cost point, than at *B, C,* or *D*; alternatively, it might be capital, perhaps as a function of development cost subsidies or taxes.

The critical isodapane, therefore, has a value of $5. It can be seen that of the three alternative locations, *C* lies exactly on the critical isodapane, so in the absence of other factors, a producer would be indifferent as to location at either *A* or *C*. Location *D* incurs excessive transportation costs, while *B* appears as the location at which total costs of production including transportation are minimized. As a general rule, therefore, if the location with lower production costs lies inside the critical isodapane, it is worthwhile for the producer to transfer to that point from the minimum-transportation-cost location. Strictly speaking, a separate critical isodapane exists for each alternative location, though in Figure 4.11, the production cost savings are the same at each location other than *A*, so only a single one need be used.

Weber thus established that the evaluation of least-cost location for an economic activity using diverse factor inputs is essentially one of *spatial substitution*. In Weber's special case, the producer substitutes transportation outlays for labor outlays. His isodapane technique provided a simple method for solving this two-factor substitution problem in spatial terms within the constraints of his own model.

Applying Isard's substitution technique to the same problem, it can be shown that while the theoretical implications are the same, the method is perhaps a little simpler. Figure 4.12 depicts a similar situation by means of an outlay substitution line (*FJLMNR*). The substitution in this case is again between transportation and labor, though it is possible to apply the method to all other input costs. Under the conditions assumed in Figure 4.12, labor costs are equal at locations *F, G, E,* and *H* ($20 per unit of output) but lower at *J, L, M, N,* and *R*. Point *F* is the minimum-transportation-cost point, but *J*, with only slightly higher transportation costs, has labor costs that are $5 cheaper. *J* therefore lies on a lower iso-outlay line ($50) and thus represents the optimal solution. It is worth noting that the solution might be affected by a shift in the slope of the iso-outlay lines, that is, in the relative cost importance of labor and

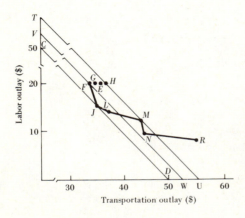

Figure 4.12 Substitution relationships between labor outlays and transportation outlays. *Source:* W. Isard (1956), *Location and Space Economy* (Cambridge, Mass. MIT Press), fig. 25. Copyright © 1956 by the Massachusetts Institute of Technology. Reprinted with permission.

transportation in the total situation. This is logically the parallel of Weber's solution by labor coefficient.

Not only is the substitution method simpler to apply, but it is also considerably more flexible in its application. Substitutions over a wide range of input factors can be examined (providing that they are expressed in two-dimensional form). It is also possible, as was pointed out in Chapter 3, to take into account the curved and stepped nature of realistic transportation rates.

In general terms, although we began with the land factor of production and the role of transportation costs, each factor with its specific localization pattern and mobility characteristics exerts a more or less powerful influence on location in space. The particular factor or combination of factors that has the dominant role varies from sector to sector and from industry to industry. For some, like agriculture, a fundamental tie to a totally fixed factor is the dominant force (though, of course, land is "mobile" in the sense that the use to which it is put can change); for others, it is a pull to sites of bulky raw materials, pockets of specialized labor, or "seedbeds" of new technical knowledge.

For most economic activities, however, an identifiable pull to one or other production factor is difficult to isolate. In effect, as we pointed out at the beginning of this chapter, the production function offers a producer choices. Where labor is a demanded factor in short supply, it may be possible to substitute capital. Alternatively, some new form of technology may be adopted or developed that raises the same productivity from the same labor force. Similarly, where capital is in short supply, perhaps in remote regions or in inner-city locations where finance capitalists view investment as risky, labor can be applied as a substitute. Since in the modern world the kaleidoscope of possible factor substitutions is changed almost from day to day by new developments in technology, generalization about production factors as location factors is made extremely difficult.

The simplicity of the analytic approach where one factor is isolated and its locational pull is estimated by making the classical "other things being equal" assumption ignores a salient feature of modern economic life. At the extremes, where activities are tied to particular factor needs, there are imperatives, but over the entire range between, there are factor mix choices. Perhaps in the nineteenth and early twentieth centuries, the imperatives tended to hold sway. But certainly in more recent years, there appear to be more choices and fewer imperatives in the locational decision. One imperative that has been drastically reduced during the past 150 years has been that imposed by the "tyranny of distance." Both through the transportation revolution and the accompanying revolution in the communications media, flows of people, goods, credit, and new ideas have all been made easier. As we have suggested, mobility in these phenomena is still not as perfect as the economists' models frequently assume. Production factors are still to some degree localized and exert localizing forces on economic activities that demand them. But in the second half of the twentieth century, the nature of these localizing forces is more subtle and complex. Perhaps the nineteenth century was the heyday of the influence of key factor inputs (generalized in the literature as location factors) as a positive determinate force underlying economic geography. We shall return to the factors of production in Part Two.

In this chapter, we have introduced some of the complexity associated with the choices open to a business in selecting a mix of production factors. One of the choices, however, has been constrained still—choosing the scale of operations. To evaluate this, the producer needs to know something of the demand that production is to satisfy. It is to these questions of demand, scale, and the associated spatial form of agglomeration that we turn our attention in the next chapter.

FURTHER READING

Davis, L. (1966). "The Capital Markets and Industrial Concentration," *Economic History Review* 19: 255–272.

Estall, R. C. (1972). "Some Observations on the Internal Mobility of Investment Capital," *Area* 4: 193–198.

Gertler, M. S. (1984). "Regional Capital Theory," *Progress in Human Geography* 8: 50–81.

Isard, W. (1956). *Location and Space Economy*. Cambridge, Mass.: MIT Press, chap. 6.

Malecki, E. J. (1980). "Dimensions of R&D Location in the United States," *Research Policy* 9: 2–22.

Walker, R., and M. Storper (1981). "Capital and Industrial Location," *Progress in Human Geography* 5: 473–509.

Chapter
5

Demand, Scale, and Agglomeration

In Chapter 4, we looked at spatial differences in production costs. We showed how each would-be producer might decide on the amounts and proportions of the factor inputs needed and how variations over space in the availability of production factors would influence the attitude to location. In effect, the producer is faced with a *space-cost surface* for each factor and with a composite cost surface for all inputs added together, depending on which combinations are chosen. The space-cost curve, introduced in Chapter 2, then, is simply a two-dimensional vertical section cut through such a three-dimensional cost surface.

Highly localized activities will tend to respond to steep cost variations, whereas "footloose" ones (those not tied to specific locations) will respond to plateaulike conditions. So far we have proceeded on the assumption that the locational problem is the same in every case. Find the lowest point on the production cost surface, and that is the point of maximum profit and best location. In Figure 5.1a, this would be within the shaded area bounded by the spatial margins to profitability, a situation we met in Chapter 2 (see Figure 2.6). But this solution applies only where demand (or revenue received by the firm) is spatially constant, as shown by the horizontal line in Figure 5.1a. When we

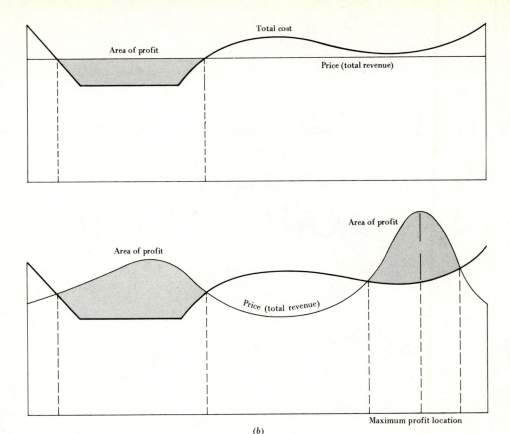

Figure 5.1 The impact of variations in revenue on the spatial pattern of profitability: (*a*) Spatial margins to profitability with total revenue constant over space; (*b*) Spatial margins to profitability with total revenue variable over space.

allow for the fact that demand inevitably varies from place to place, the locational problem, and its solution, is very different. In the hypothetical situation of Figure 5.1*b*, for example, the area of profit is not now a single contiguous zone. It appears over space as two separate regions. Second, the material source, though it provides the least-cost location, no longer provides the maximum-profit one.

To this point we have proceeded tacitly on the assumption that the only real consideration in locational choice was the minimization of costs in serving a given market, but we recognized the importance of spatial variations in demand (through variations in population distribution) as early as Chapter 1. We are in good company in adopting this approach. In common with many location theorists, we have chosen to ignore the subject of demand variations and their impact until we could get the notion of the least-cost location across in simple terms. Now we must set the record straight.

SPATIAL VARIATIONS IN DEMAND

The Spatial Distribution of Demand

In looking for some approximation of the spatial distribution of household consumer demand, it is clear that it will vary to some degree, with the pattern of incomes. In general terms, the higher the level of household income, the higher the anticipated level of demand. However, bearing in mind our preliminary discussion of demand in Chapter 1 and what we know from our own daily experiences about consumer choice, there are at least three other important determinants of the level of demand for a good or service:

1. The market price of the good or service
2. The relative prices of all other goods and services in the consumer's bundle of demands
3. The weightings given by the consumer to each element in the bundle of tastes and preferences

We know, even for the constrained conditions of the simplified model in Chapter 1, that the price of a single good and the relative prices of a bundle of goods will vary over space. We can further assume, given the nature of the species, that the way people weight their tastes and preferences varies with age, sex, income level, culture, and most of the other variables attributable to humanity. Under these circumstances, we are certainly never going to get very close to an accurate overall measure of aggregate consumer demand. Even for a single product with a clearly defined consumer group, the measurement of spatial patterns of demand is a slippery task outside the controlled world of the simplified model. Various attempts have been made, however, to measure the intensity of aggregate demand over space, and we can make use of them here, tempering our criticism of the methods by the sheer difficulty of the task. The fact that these studies were carried out in the 1950s and are factually very dated need not worry us unduly. We are concerned here with their conceptual basis, not their empirical accuracy.

One possible way of describing spatial variations in the intensity of demand is to use the concept of *potential,* an idea closely related to the gravity model that we looked at in Chapter 1. Recall that the gravity model tells us that interaction between two areas will be related to the product of their masses divided by the distance separating them. We can extend this idea to encompass the level of interaction between the one area (i) and all other areas (n) by simply adding together the interaction between i and every other area. For example, interaction between area i and area 1 is calculated as for the gravity model:

$$I_{i1} = \frac{P_i P_1}{d_{i1}^b}$$

Similarly, interaction between area i and area 2 is calculated as

$$I_{i2} = \frac{P_i P_2}{d_{i2}^b}$$

This process is repeated for all other areas to give the *population potential* at area i (we can use the symbol PP_i to denote this). Because the process involves merely adding together a whole series of pairs of transactions, we can express the population potential in general terms as

$$PP_i = \sum_{j=1}^{n} \frac{P_j}{d_{ij}^b}$$

For place i, therefore, its population potential gives some indication of the "intensity of the possibility of interaction" of place i with all other places in the system. In a sense, it is a measure of the "nearness" of place i to the population of the entire system. If we carry out the same calculation for every place in the system, the resulting values can be mapped using isopleths to give a *surface of population potential*.

If, as we assumed initially in Chapter 1, every individual in the population did have an identical income, spent the same proportion of that income on purchasing goods, and had similar tastes, a population potential surface would represent a reasonable approximation of a household sector demand surface. But since these assumptions can no longer be retained, one possible solution would be to weight population by some measure of income as in Figure 5.2.

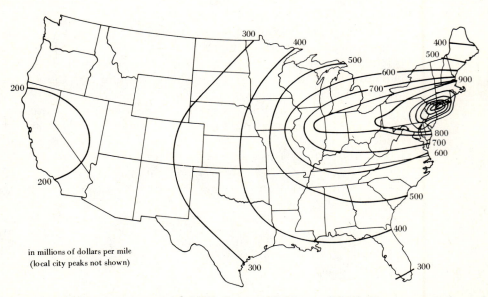

in millions of dollars per mile
(local city peaks not shown)

Figure 5.2 U.S. income potential, 1956. *Source:* W. Warntz (1956), *Macrogeography and Income Fronts* (Philadelphia: Regional Science Research Institute), fig. 1.

An alternative measure, used by Harris (1954) in his classic study, was the value of retail sales per county, from which he derived a *market potential,*

$$MP_i = \sum_{j=1}^{n} \frac{M_j}{d_{ij}}$$

where MP_i = market potential at i
   M_j = size of market at j measured in terms of retail sales per county
   d_{ij} = distance between i and j based on an estimate of the transportation cost, allowing for long-haul economies

Market potential at point i, therefore, is obtained by dividing the retail sales of each jth point (group of counties) by the distance between i and j and summing the result. The process is carried out for all points to yield values on which an isopleth map of the market potential surface can be based.

Figure 5.3*a* indicates that market potential in the United States in the 1950s was consistently high in a belt between Illinois and the East Coast (an area roughly coterminus with the traditional manufacturing belt), reaching a peak at New York City. Away from this ridge of maximum potential, there was a consistent decline to the north, south, and west, with a subsidiary peak in southern California. If market potential is a valid measure of the spatial pattern of demand, it follows that location at points of high potential should offer closer proximity to the market, a better assessment of current and changing demands, faster delivery, and other locational advantages (Isard, 1960).

Market potential provides an indication of the general proximity of a location in relation to total demand, the peak giving an approximation of the maximum sales location. It gives no indication, however, of the cost involved in transporting goods to that market. According to Harris, the minimum-transport-cost point in relation to the national market may be derived as follows:

$$TC_i = \sum_{j=1}^{n} M_j d_{ij}$$

where TC_i = total transportation costs at i and M_j and d_{ij} are as defined in the formula for market potential. The calculation of transportation costs at i involves multiplying the retail sales of each jth area by its distance from i. The transportation cost surface is then obtained by repeating the calculation for each point and constructing cost isopleths. Figure 5.3*b*, the transportation cost surface, reveals a rather different pattern from the market potential surface. The location at which total transportation costs to the entire U.S. market were minimized in the 1950s was Fort Wayne, Indiana.

So far we have been concerned with spatial variations in the pattern of aggregate demand as this is reflected by such indicators of household consumption as income or retail sales potential. However, not all sectors of economic activity are concerned directly with this market. Manufacturing is to a great extent its own market. Roughly 80 percent of all U.S. industries make use of materials previously processed by other industries. To estimate the spatial

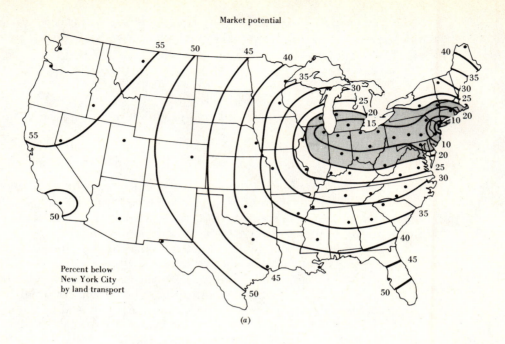

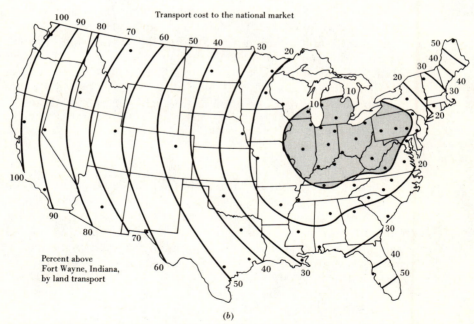

Figure 5.3 (*a*) Market potential and (*b*) transport cost surfaces in the United States. *Source:* C. D. Harris (1954), "The Market as a Factor in the Localization of Industry in the U.S.," *Annals of the Association of American Geographers* 44: 315–348, figs. 4 and 7. Reprinted with permission.

pattern of this market for intermediate and capital goods, Harris substituted employment for manufacturing in the potential and transportation cost formulas. Figure 5.4 shows the surfaces of manufacturing potential and transportation cost to the U.S. "manufacturing market." In the former case, the optimal location was New York City; in the latter, Cleveland.

A problem therefore exists: which of these locations, New York or Fort Wayne in the case of the total market, New York or Cleveland in the case of the manufacturing market, offered the most advantageous market location? The differences between the two surfaces arise because in the market potential case, the contribution of a given-sized market declines with increasing distance, whereas in the transportation cost case, the contribution of the same market increases with distance. Harris gives the example of the impact of the Pacific Coast market on the values for Chicago. Pacific Coast retail sales represented roughly 11 percent of total U.S. sales, yet they contributed less than 5 percent to Chicago's market potential because of the magnitude of the distance between the coast and Chicago. Because of this distance, however, they accounted for 22 percent of the transportation cost of serving a nationwide market from Chicago.

Despite the many problems with the market potential approach—not the least being the exclusion of Canada from Harris's calculations—it does offer a useful perspective. It gives us a rough impression of potential spatial variations in demand, an impression that could be made more graphic by converting the contours into a three-dimensional surface. Of course, the actual demand for a product will vary geographically for a host of additional reasons including market prices, individual tastes and preferences, and the availability of acceptable substitutes.

Nevertheless, the point we are trying to make is simply that demand—like all the other factors we have considered that influence the location of economic activities—is extremely variable over space. It becomes another complex input to the substitution problem that faces every potential producer. The point of lowest production cost is only in rare circumstances likely to coincide exactly with the point of maximum profit. A producer must look "forward" to demand as well as "backward" to cost in making a judgment of the best location. However, another important component in the producer's substitution problem in a sense "closes the circle" by connecting the size of a producer's market with the costs of production. This is the question of the appropriate volume or scale of output to be produced. This factor, though not obviously spatial in itself, has the most far-reaching locational implications. In choosing the *scale of production*, the producer makes a key decision that both influences and is influenced by costs and the level of demand.

ECONOMIES OF SCALE

Let us for the moment go back to the producer we encountered at the beginning of Chapter 4. The problem at that stage was primarily getting the right factor mix for the product to be produced. We ignored the decision of how

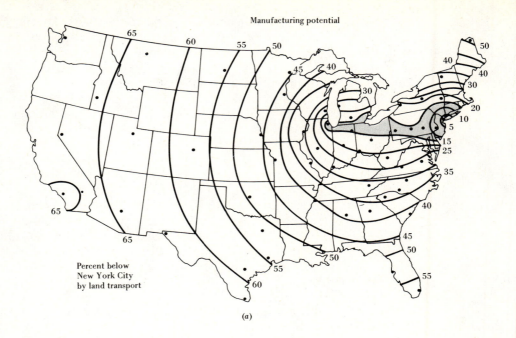

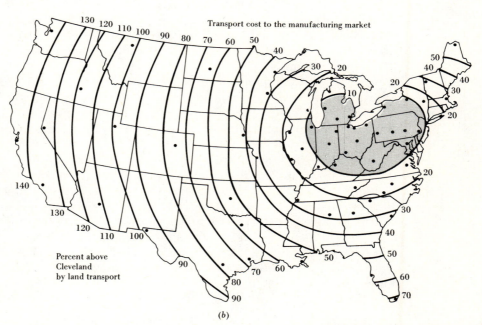

Figure 5.4 (*a*) Manufacturing potential and (*b*) transport costs to the national manufacturing market in the United States. *Source:* C. D. Harris (1954), "The Market as a Factor in the Localization of Industry in the U.S.," *Annals of the Association of American Geographers* 44: 315–348, figs. 28 and 29.

much to produce. In fact, one of the first decisions, and in some ways a difficult one to get right, is how many units of output to produce. Hinged on this, of course, is the question of the amount of each production factor that will be needed to meet the chosen level of output, and we demonstrated in Figure 4.3 how these amounts could be determined for a given level of output. Perhaps the simple answer to the question of how many units is, "As many as can be sold at a profit." This clearly links scale with levels of demand and, since lower prices will generally produce greater sales, also with price. Some effort in market research will need to precede the scale decision.

But there are also important implications for costs in the choice of scale of operation. To this point, we have followed the example of most of the early location theorists and assumed, once again for the sake of simplicity, that the average cost of producing each additional unit of output remains the same no matter how many units are produced. We have in fact assumed away the cost implications of scale. This situation is shown graphically in Figure 5.5*a* with average costs as a horizontal straight line. We have further assumed that there is no lower limt to the scale of production. Not only can a producer (if he has a buyer) produce one unit or 10,000, but in production terms alone, there will be no difference in the cost per unit.

In reality, there is a minimum scale of production necessary for competition. For the most part, this is determined by prevailing technology. As far as the progression of costs with scale is concerned, this is more likely to describe a U-shape or L-shape like those shown in Figure 5.5*b, c,* and *d.*

Above the initial threshold, average costs tend at first to fall as the number of units of production begins to rise. They may then begin to rise beyond a certain size of production or to level out. For some activities, this "straight and level" situation appears to be maintained, while for others, costs rise at the

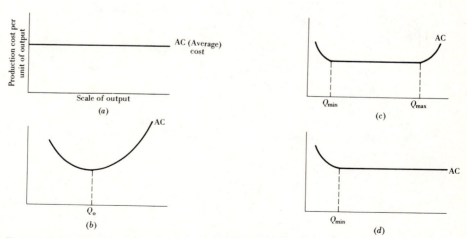

Figure 5.5 Various possible cost-scale curves. *Source:* J. S. Bain (1968), *Industrial Organization,* 2d ed. (New York: Wiley), figs. 2–5. Reprinted with permission.

higher levels of output to give an extended U form. We shall return to the implications of these cost considerations in more detail later, but for the moment, let us briefly explore the question of *returns to scale* in production to see how they come about.

The Basis of Scale Economies

To begin, let us assume that our producer has not yet decided on a fixed scale of operations. The market for the product seems unlimited, and the producer sets out to keep on increasing the size of the operation. As before, let us assume for the sake of simplicity that only two factors are used, labor and capital. Suppose also that the capital input is held constant so that the short-run effect on production of varying amounts of labor can be seen. It might well be found that under these circumstances, the progression of output with variable inputs of the labor factor follows the trend shown in Table 5.1.

At first, the effect of hiring an additional worker will increase dramatically the level of output achieved. It is not difficult to see why. If our production unit is a farm of fixed size and capital equipment, a second pair of hands on the farm will be enormously valuable. The new worker in the hypothetical example triples the total output, adding 1,000 output units to the total. This is the *marginal productivity* (the amount added to production) by two units, as opposed to one unit, of labor. Average productivity (the total output now divided by 2) is lower than marginal productivity at 750. Hiring a third pushes up marginal productivity by another 1,250 units, and average productivity rises to 917. At this stage we can talk of *increasing returns to scale*. As the variable factor is added, both marginal and average returns in output also increase.

On hiring the fourth worker, the increase in marginal returns is for the first time lower than before, but average returns are still rising. By the time the fifth worker is hired, marginal returns have slipped back to the level that the second worker's inclusion brought about, but average returns have increased to their

Table 5.1   RETURNS TO SCALE

Labor units added (no. of workers)	Total output	Marginal productivity (change in output)	Average productivity (total output per worker, rounded)
1	500	+500	500
2	1,500	+1,000	750
3	2,750	+1,250	917
4	3,900	+1,150	975
5	4,900	+1,000	980
6	5,800	+900	967
7	6,550	+750	936
8	7,150	+600	894

peak at 980 production units. From here on, as extra units of labor are added, both marginal and average returns decrease with the scale of the input.

The graph in Figure 5.6 shows the basic trends. It is worth noting that marginal output falls while average output is still rising. This is because the fifth worker added, although producing less than the fourth, is still producing more than the average output of all earlier workers. Output clearly varies with scale, as our producer would have discovered. In this case, the scale is the scale of input of a single factor with all others held constant. Had our producer chosen to vary capital inputs while holding land and labor constant or to vary land inputs with labor and capital constant, a similar progression would have resulted. First, there is a stage of *increasing* (average) *returns*. This is followed by *diminishing returns*, when average output falls as more units of the variable factor are added. Finally come *negative returns*, when the marginal return of the next unit of the factor is less than that of the first unit to be added.

In judging the scale of operations, our producer should be aware, then, that in hiring variable factors of production, physical or technical constraints may make certain factor combinations at first more and then increasingly less productive. Just as there will be some minimum threshold size, so also will there be a point higher up the scale of production where diminishing returns are likely to set in as the supply of certain factors tends to become fixed relative to others.

Let us look at the situation from the cost viewpoint. Hiring factors costs money, and our producer should be looking for the point at which most units of output for the money outlay are received (always assuming that such quantities can actually be sold). We could attempt to calculate the producer's total *variable* costs per unit of output. (These must be distinguished from fixed costs, which we will come to in a moment.) Let us suppose, to keep the arithmetic simple, that the going wage rate is $10,000 per year per worker. The unit cost calculations for variable factor inputs would then come out as in Table 5.2. It is obvious that beyond the fifth worker hired, labor costs per unit will start to rise, although there have been substantial *scale economies* in the cost of labor up to this point. If we graph the situation as in Figure 5.7, we can begin to see where the U-shaped cost-scale curves encountered at the beginning of the chapter have come from. Taking our simplified model using only two factors as repre-

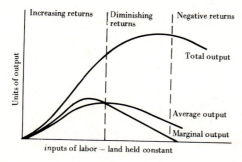

Figure 5.6 Returns to scale with one variable factor input.

Table 5.2   VARIABLE COSTS AND SCALE OF OUTPUT

Labor units added (no. of workers at $10,000 per year)	Total output (units per year)	Total labor cost per year ($)	Average labor cost per unit of output ($)
1	500	10,000	20.00
2	1,500	20,000	13.33
3	2,750	30,000	10.91
4	3,900	40,000	10.26
5	4,900	50,000	10.20
6	5,800	60,000	10.34
7	6,550	70,000	10.69
8	7,150	80,000	11.19

sentative of the more general situation where factors are used in combination, we can see how variable costs of production change with increasing output. If there were only variable costs to be considered and demand conditions were met, it would clearly be best for our business owner to produce at the scale of 4,900 units a year at a cost of $10.20 per unit in factor outlays.

We are almost there but not quite. There is the additional question of *fixed costs*. Our simple case study assumed that labor was a variable cost, and we have looked at its progression as production levels rise. But we assumed land, capital, and technology to be fixed as labor varied. Fixed or not, they must still be paid for, and though they will not vary as production rises, they must still be added to the total cost per unit of output. In effect, under the chosen strategy, they will be once-and-for-all payments made when the business is set up. (If it is decided to change them, of course, they will take on the characteristics of variable costs.)

How will fixed costs vary over units of production? Suppose that there is a single initial outlay on these fixed "overheads" of $20,000. For one unit of output, the effective cost will be all of this at $20,000 per unit. But for two units,

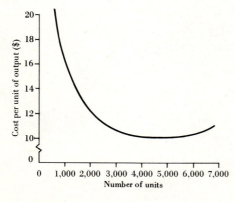

Figure 5.7 Variable costs per unit of output: labor variable, capital constant.

the cost per unit will be $10,000; for three, $666.66; and so on. The curve produced will be of the type shown in Figure 5.8. Starting at 1,000 units of output, fixed costs stand at $20.00 per unit, but by the time 8,000 units have been produced, they are down to $2.50. Beyond this point, they will fall still further until in the long run they become infinitesimal (if they can remain fixed long enough).

All that remains is to add the fixed-cost curve and the variable-cost curve to obtain the *total-cost curve* for our business owner's enterprise (Figure 5.9). The effect of the sharp improvement in returns to scale due to the spreading of fixed costs over more units has been to push the minimum unit cost point from the level associated with the five-worker firm to the level of the seven-worker firm (Table 5.3).

Our business owner, then, in attempting to decide the scale at which to operate, will now be aware that over a certain range of scale in production, economies of scale will occur and the unit cost will tend to fall as each new unit of output is added. At first this fall in unit costs will be sharp as fixed costs are paid off over more units of output. But as the size of the output increases, restrictions on the physical or technical availability of factors begin to activate the law of diminishing returns, and *diseconomies of scale* begin to appear. But for long-run conditions, this is by no means invariably the case. If the entrepreneur is lucky, there may be a definable scale point determining exactly at what scale unit costs are lowest—the bottom of the U-shaped cost-scale curve. This is known as the *minimum optimal scale* (MOS).

We have talked in analytic terms about the basis of scale economies; let us now investigate just how such economies may arise.

Specialization of Manpower and Equipment

In looking more generally at the factors underlying scale economies in production, the logical place to begin is with Adam Smith's classic principle, the *division of labor*.

In 1776, in the first chapter of *The Wealth of Nations*, Adam Smith described at great length the advantages in production brought about by the division of labor. His description of a pinmaker, which follows, has possibly never been bettered as a way of making the principle clear.

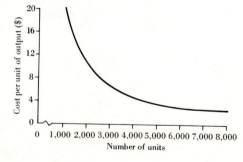

Figure 5.8 Fixed costs per unit of output for capital.

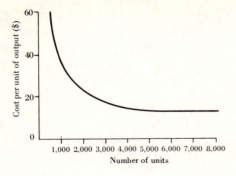

Figure 5.9 Total costs per unit of output, fixed and variable.

A workman not educated to his business nor acquainted with the use of the machinery employed in it could scarce, perhaps, with his utmost industry, make one pin in a day, and certainly could not make twenty. But in the way in which this business is now carried on, not only the whole work is a peculiar trade, but it is divided into a number of branches, of which greater part are peculiar trades. One man draws out the wire, another straights it, a third cuts it, a fourth points it, a fifth grinds it at the top for receiving the head; to make the head requires two or three distinct operations, to put it on is a peculiar business, to whiten the pins is another; it is even a trade by itself to put them into the paper; and the important business of making a pin is in this manner, divided into about eighteen distinct operations, which, in some manufactories are all performed by distinct hands, though in others the same men will perform two or three of them.

I have seen a small manufactory of this kind where ten men only were employed and where some of them consequently performed two or three distinct operations. . . . Each person . . . making a tenth part of forty-eight thousand pins might be considered as making four thousand eight hundred pins in a day. But if they had all wrought separately and independently and without any of them being educated to this particular business, they could certainly not each of them have made twenty, perhaps not one pin in a day, that is, certainly, not the two hundred and fortieth, not the four thousand eight-hundredth part of what they are

Table 5.3   TOTAL COST PER UNIT OF OUTPUT

No. of workers	Total output (units per year)	Total cost ($20,000 fixed costs plus $10,000 per worker)	Average cost per unit of output ($)
1	500	30,000	60.00
2	1,500	40,000	26.66
3	2,750	50,000	18.18
4	3,900	60,000	15.38
5	4,900	70,000	14.29
6	5,800	80,000	13.79
7	6,550	90,000	13.74
8	7,150	100,000	13.98

at present capable of performing, in consequence of a proper division and combination of their different operations.

It is in itself instructive to consider that all of these operations are now more than likely performed by a single machine under the control of a single person and that their combined output would render Smith's then staggering output figures quite trivial.

In analyzing the key features of cost saving through producing so many units of output per worker, three basic elements can be identified:

1. *Increase of dexterity.* A person who works at the same task for some time will acquire a degree of basic skill and rhythm, giving an economy of motion and yielding a higher rate of output for a longer period. People with natural gifts can apply them to tasks for which they are best suited.

2. *Saving of time lost in passing from one task to another.* Where one or a small number of workers produce complex articles like automobiles in a "one-off" mode or where a single farmer handles a variety of jobs around the farm, a great deal of labor time is used up in moving from task to task. To add labor and assign specialized continuous production tasks to each individual in a complementary system immediately produces a rise in output for the same factor cost.

3. *Capability of equipping specialist workers with equally specialized machines.* Simplifying work tasks enhances the possibility of inventing machinery to facilitate production. In this way, one person can do the work of many. Humanity would have had to wait a long time for a machine to be invented that would make a whole automobile the way a small group of workers would. But by breaking the exercise down into enough simple tasks, the automated production line assembly plant became a possibility shortly after the turn of the twentieth century.

The pinmaker might have been the classic example of division of labor in the late eighteenth century, but there is no doubt that the automobile manufacturer led the field throughout much of the twentieth. It was Henry Ford who learned the most from Adam Smith's three principles of division of labor. In particular, by bringing the product to the worker—the assembly line principle—the most complex tasks on even large items of equipment could be performed simply, efficiently, and with minimum loss of labor time. (Ford was not in fact the inventor of the assembly line, but he developed it most extensively.)

Division of labor up to the level permitted by the size of the market is one of the means by which an increase in the input of labor as a production factor produces lower unit costs in output. As we have shown, however, there is a limit to its application. Overstaffing—too many workers for the capital equipment and space available—pushes the scale-cost curve upward as the law of diminishing returns takes its toll.

Just as specialization in the application of one factor, labor, brings about scale economies, so too will specialization in the use of capital or land.

Let us turn to the case of capital equipment. We have seen that the specialization of tasks facilitates the more efficient use of labor and that this permits greater opportunities for the invention of machines. Once available, machines can further enhance the productivity of the system, this time through the combined application of capital and labor. In accordance with the principles we established for the human worker, a machine performing a simple repetitive task has certain important cost advantages. It is likely to be simple in design, using fewer complex parts and performing operations each of which contributes directly to generating the product. A highly complex machine, besides having more parts to go wrong, may have some parts waiting unproductive while other parts do their task. Specialization in the use of machines contributes to production efficiency. True, some complex machines cannot meet the requirement that every part contribute output at all times, but part of the drive behind technological improvements in design is to achieve more output per hour of machine time.

In strict cost terms, a machine is bought with a *single outlay* to perform a productive task. It is not, like a worker, paid by the hour and is not subject to human contraints like getting tired, bored, or hungry. Humanitarian organizations do not spring up to protect machines from exploitation, and machines do not renegotiate the terms on which they produce if the cost of living changes. Thus the simple economic principle for getting the best return from the use of a machine is this: having bought it and installed it, make use of it to produce output every feasible second of its working life. As a fixed outlay (per machine), capital equipment has the cost characteristics we discussed earlier. The more units of output it produces, the more units of output bear the initial cost and the cheaper the unit cost of each marginal unit. The use of machines, then, contributes heavily to the economies of scale achieved by a business as its output rises. Remember, though, that other factors are needed to make the machine function. By adding more machines—each one paying off its cost as it produces more units—and therefore making the input of capital a variable, albeit "lumpy," cost factor, the other factors also need to be expanded. Too many machines in too small a space (land fixed-capital variable) will give ultimate diseconomies from diminishing returns. So will too many machines and too few workers to operate and maintain them (labor fixed-capital variable).

We need to look more closely at this "lumpiness" in capital inputs since it has profound effects on scale. For the most part, this constraint depends on the state of technology at a given time. An economist would call this the problem of *indivisibility*. In its simplest sense, this means that machines tend to come in single, whole units with a given capacity for output. Take the case of the sewing machine. For the handicraft worker using a needle and thread to make shirts at a rate of seven a week, the sewing machine has significant indivisibility problems. First, the purchase of the machine costs money, a substantial sum by comparison with the needle, which has a much lower indivisibility threshold. To justify the outlay on a sewing machine, the worker must earn enough from using the machine to repay the fixed cost of buying it. Remember, the first shirt made will effectively bear the whole cost of the machine plus materials and labor. It will be a long way along the production run before the worker is able to produce a shirt at the same unit cost as by needle-and-thread methods. Let us say that the shirtmaker needs to produce 30 shirts a week to achieve this; he or she must then be able to make 23 more shirts a week to justify the purchase of one sewing machine. What if no more than 15 can be sold? The shirtmaker would be sensible to stick with needle and thread. There is therefore an *indivisibility threshold* of 23 shirts to be overcome before beginning to mechanize. Supposing that the shirtmaker could meet the indivisibility threshold of the first sewing machine, consider the ramifications of buying a second one. Like sewing machines, metal presses, drop forges, blast furnaces, power stations, and all other machines tend to present indivisibility problems. To proceed from zero to one or from one to two requires that certain increased demand conditions be met in addition to a step-up in the availability of the other factors. Staffing the second blast furnace or finding space for it, for instance, demands an equivalent jump in the provision of labor and land.

When we talked earlier of the lower threshold limit to production imposed by technical factors, this sort of consideration prompted it. The scale-cost curve for a given form of business does not begin at zero and follow a smooth progression. Where significant inputs of capital are involved, there is an indivisibility threshold, like that for our shirt manfuacturer, below which production is impossible unless the method of operation is different.

The Economy of the Large Machine

A further point on the subject of the scale economies associated with the use of capital equipment is worth making. The cost of producing many types of capital equipment increases less rapidly than the capacity of such equipment. The laws of mathematics more than those of economics determine the nature of the relationship. Consider the cost of raising the capacity of an oil storage tank in relation to its capacity to store oil. The first (cost) increases regularly in proportion to the surface area of the tank—that is, the *square* of its dimensions. The second (capacity) increases with the inside volume, the *cube* of its dimensions.

Take another example from the field of transport. A ship is essentially a container, so it has the technical advantages in capacity of the oil storage tank. In addition, however, it must move over a viscous medium—water—and energy is used up in overcoming the resistance both of this and, to a lesser extent, the air through which it moves. The "interface" between ship and water is the area over which the surface of the ship is in contact with the water. Increasing the size of a ship increases this water-ship interface by the square of the ship's dimensions below the water line. Its capacity is increased by the cube of its overall dimensions. Up to a technically determined point, therefore, a large ship will require less horsepower per unit of carrying capacity than a small one.

The advent of the wide-bodied jet airliner, which had to await technical developments in the thrust capacity of jet engines, exploits the same technical principle. Doubling all the dimensions of an aircraft gives an eightfold increase in cubic capacity for a fourfold increase in air resistance. The passenger-carrying capacity of the jumbo series of jets has therefore increased dramatically for a less than proportionate increase in engine thrust and wing lift. Once again, of course, such increases have indivisibility connotations, as the international airlines have found to their cost. The large machine must generate sufficient output to reap its scale benefits, and, we must constantly emphasize, the market must enable this to take place. Further restrictions on the economies of the large machine come from the problems of making the technically correct amounts of other factors available. The need to add the land factor in extending airport runways to parallel jumbo jet developments is one example. In the same context, restrictions in the amounts of the labor factor to meet the huge capacity expansions following the introduction of the wide-bodied jet have brought diminishing returns. Many travelers have experienced the time diseconomies of the big jet brought about by a less than proportional increase in the supply of baggage handlers, immigration officers, and customs officials as passenger capacity has expanded.

Economies of Massed Reserves

Another of the advantages of size in the scale of a business's operations becomes clear when contingency plans have to be made against the effects of machine failures or breakdowns. For any business, regardless of size, the breakdown of a machine is a costly event. As we said earlier, machines pay their way most efficiently by round-the-clock usage on activities that produce returns. Should a machine fail, its downtime costs money. This is a contingency against which all businesses need to take out some insurance. How does the small operation with a single machine proceed? The quick answer might be to have a second machine always ready in reserve. This would be cost-foolish, however, because every unit of output produced by the one machine would have the cost of two machines in its overhead. More sensibly, it might carry reserve stocks of the moving or most vulnerable parts, but this too goes directly into the overhead costs chargeable against every item of output. Here is another of the entrepreneur's basic trade-off problems: risk not carrying spares, bearing the cost of breakdowns as they come, or go for more security by carrying basic spares even though holding them unused adds cost to each unit of output? Either way, expenses raise the unit costs of output.

Bigger operations, however, have a significant cost advantage over smaller ones in one key respect. Where there are a large number of machines in operation, the loss of one makes for a smaller proportion of production loss than where there is only a single machine. Equally, though in this case there are more machines to fail, there is a low probability that more than one will fail *at the same time* by virtue of the breakdown of the same part. Thus supplies of parts held in a bigger plant service the potential needs of more units of output.

Crisis demands of any kind can more easily be met by a larger plant than by a smaller one. Sudden increases in demand at one point may take up slack capacity at another, making overall factor usage more efficient.

For the smaller operation, sudden shifts in the demand for resources or the allocation of capacity in emergency can only be met by carrying reserve stocks. These use resources without producing returns. The electricity grid network provides a good example of the massed reserves principle. Separate power generators not connected together would each need to carry contingency reserves against emergency or breakdown. Connecting them all in a grid system means that a far smaller allocation of total resources needs to go into reserve stocks. Failures at one point can be compensated by allocation of slack capacity from another, and the entire system needs only a minimal backup in terms of spare parts. The same part is most unlikely to fail everywhere at the same time.

Economies of Large-Scale Purchasing

Everything said so far in this discussion contributes to the notion that larger production units are (to a point) more cost-efficient per unit of output. To achieve these economies, each producer must seek to find a market of sufficient size to absorb the extra units of output that take the operation nearer the scale optimum.

Take the owner of a small plant producing high-quality paper for the publishing industry. At the current scale of output, unit costs are relatively high. The plant is somewhere on the low side of the output needed to achieve the minimum optimal scale. Costs are high, but given sufficient demand to permit the purchase of the latest high-capacity polishing and cutting machine, unit costs could be brought down sharply. At this stage, the publisher of a new glossy magazine turns up to discuss a possible long-term order for large quantities of high-quality paper. From the paper producer's viewpoint, capturing the order would take the firm over the indivisibility threshold for the new machine. It could then produce all its output at a lower unit cost, including that to existing customers and any new customers who might be attracted by the lower prices it could then offer. What should the plant owner do? Any sensible business person would obviously try to gain new business by offering at least a major discount to the publisher. The owner would, in effect, offer to share some of the gains from the return made on becoming more scale-efficient. The new business would in any case lower unit costs, and being more price-competitive, the firm might well attract even more trade and become even more scale-efficient.

What does the publisher gain from all this? It gains high-quality paper at a discount price. Why? Because its demand has made the papermaker more efficient by enabling it to profit from scale economies. Take the situation one step further back. The papermaker, once the deal has been made, can go to its suppliers and in turn offer them an increase in demand for the long term. If they were not already into diminishing returns and had either slack capacity or, like the papermaker, an indivisibility threshold, it would be sensible to demand a hefty discount too. So it goes on. Large-scale purchasing may well make suppliers more efficient and attract discount rates to the buyer.

There is another area in which economies are achieved for large-scale purchases. We saw in Chapter 3 that freight rates on large consignments such as full carloads are a good deal more favorable than those on smaller quantities. We can set this situation clearly within the context of scale economies. A railway company invests capital in a freight car. That car is only at maximum efficiency where all of it is used all of the time to produce returns on the initial capital investment. That way the unit costs of the items carried in it can be reduced to a minimum. With a half-carload, 50 percent of the capital equipment paid for out of scarce resources returns no income to the operator. To compensate, higher rates are charged where the market will bear it.

Big may not necessarily be beautiful, but it certainly tends to be cost-efficient—other things, of course, being equal. However, the cost efficiency of size tends to disappear beyond a certain scale of operations. Also realize that bigness and efficiency are very different for different industries. A big, efficient manufacturer of diamond rings may employ fewer than 20 people in a second-floor workshop, whereas the smallest of the major American automobile producers may employ upward of 30,000 on acres of land and still be below the notional optimal scale. There are some advantages of scale and size that we have chosen not to discuss here. These are concerned with the sizes of firms and organizations rather than plants—a story taken up in more detail in Part Two. In general terms, then, it can be asserted that over certain ranges of scale, there are significant cost advantages to be obtained from concentrating production into larger-scale processing units.

Under conditions of free entry and perfect competition, plants, firms, farms, even towns and cities—in fact all identifiable production systems—would strive toward an equilibrium condition, their numbers, sizes, and, in part, location depending on their minimum optimal scales of production and the share of the total market that they could expect to achieve at such scales. From such a concept as the *minimum efficient scale* (MES), once again given conditions of free entry and perfect competition, much of the structure of the space economy could theoretically be determined: the equilibrium number of plants, firms, multiplant firms, and so on in a given industry at a given point in time.

However, under more realistic conditions, such simple solutions to complex questions tend to disappear, although, like the hexagon and the von Thünen ring, the minimum optimal scale may be considered a latent force molding the form of economic structures. The precise nature of its influence, although relatively simple under the conditions of the constrained model of Chapter 1, is, however, extremely complex once such constraints are removed. Perhaps the most potent force for variability in a realistic situation is the fact that the MES itself is constantly changing and that there is no uniquely identifiable scale to which any economic enterprise can strive in an open, constantly developing system. In particular, technological change may make it possible to produce efficiently and competitively at quite small volumes in some cases. This point will be developed more fully in Chapter 10.

ECONOMIES OF SCALE IN REALITY

While it is relatively simple to put forward logical reasons for the existence of scale economies in theory, it is quite another matter to prove their existence in practice. Businesses can survive in an environment with less than perfect competition whether their operations are at the technically efficient scale or not. There is great variation in the minimum optimal scale both between industries and, perhaps less obviously, within industries. Table 5.4 reflects one of the very few attempts to make detailed technical esimates of the MES. The figures are not up-to-date, but this is not of vital importance. The point of the table is to illustrate the range of scale differences in the real world. The industries are ranked in order of the percentage of total U.K. capacity necessary to achieve the minimum optimal scale of production. The variation is enormous, from 80 percent in the case of the integrated steel plant to 1 percent for a bakery. The table also shows the "cost penalty" of producing at half the minimum efficient scale. Again, there is substantial variation between different industries. For example, while bakery plants needed only 1 percent of total capacity to produce at the most efficient scale, total costs would be 15 percent higher if the bakery were half the most efficient size. The penalty was as high as 30 percent when the effects of material costs were removed and just the actual costs of transforming the materials into finished products were considered ("value added").

For the United States, comparable detail in technical estimates is harder to come by. Commenting on technical considerations of scale for a sample of 20 industries in *Industrial Organization*, Bain made the following observations:

1. Two industries (automobiles and typewriters) had "very important" scale economies. In each, the minimum optimal (efficient) plant scale exceeded 10 percent of total market capacity. Unit cost penalities at half the efficient scale would be on the order of 5 percent.
2. Five industries (cement, farm machinery, tractors, rayon, and steel) had "moderately important" plant scale economies, with efficient plants operating at between 4 and 6 percent of total market capacity. Again, unit cost penalties at half the efficient scale were on the order of 5 percent.
3. Nine industries (cigarettes, petroleum products, rubber tires, flour, canned goods, liquor, soap, shoes, and meat) had unimportant scale economies, with either small efficient plant sizes in general or "flat" scale curves.

Economies of scale at the plant level are extremely sensitive to technologi-

Table 5.4   ESTIMATES OF THE MINIMUM EFFICIENT SCALE FOR NEW PLANTS

	Capacity or output	Percent of U.K. capacity or output	Percent increase in costs at 50% of MES	
			Total costs per unit (including materials)	Value added per unit
Steel: integrated with strip mill	4 million tons per year	80	8	13
Motor cars: range of models	1 million cars per year	50	6	13
Steel: Blast and LD furnace	9 million tons per year	33	5–10	12–17
Synthetic fibers: polymer plant	80,000 tons per year	33	5	23
Soaps and detergents	70,000 tons per year	20	2.5	20
Domestic appliances: range of ten	500,000 appliances per year	20	8	12
Chemicals: sulfuric acid plant	1 million tons per year	30	1	19
Chemicals: ethylene plant	300,000 tons per year	25	9	30
Motor cars: one model and variants	500,000 cars per year	25	6	10
Cement: portland works	2 million tons per year	10	9	17
General-purpose oil refinery	10 million tons per year	10	5	27
Beer: brewery	1 million barrels per year	3	9	55
Bread: baking plant	30 sacks flour per hour	1	15	30

Source: After C. F. Pratten (1971), "Economies of Scale in Manufacturing Industry," *Department of Applied Economics Occasional Papers,* No. 28, University of Cambridge, table 30.1.

cal change. In some periods, as in the nineteenth century with the introduction of steam power, for example, new technologies increased the significance of producing in larger and larger quantities. The present-day effects of some kinds of technology point in the opposite direction. In particular, the introduction of highly automated and flexible manufacturing systems, which allow machines and assembly lines to be reprogrammed very quickly, is beginning to reduce the significance of very large production runs in a whole range of industries. One of the major results of the new electronic and computer-aided production technology is that it permits rapid switching from one part of a process to another and may even allow the tailoring of production to the requirements of individual customers. Flexible automation thus offers the possibility of manufacturing goods cheaply in small volumes. The effects are beginning to be felt, for example, in the automobile industry, in which the minimum optimal scale at the plant level seems to be falling.

All of this does not necessarily imply that economies of scale are no longer important. But they are probably more variable and complex than is often believed to be the case. Scale factors, certainly in the past and less certainly in the present, have exerted a significant impact on business production costs. This makes life for the producer choosing a location even more difficult. First, a choice has to be made of the right factor mix. Second, the location has to be found where distance-related costs are lowest in collecting factors and material together. On top of all this, an appropriate scale of production has to be chosen. These things, of course, trade off against each other. Opening a technically efficient scale of plant in a prime position to dominate a large market may be a good solution even though more factors and materials may have to be moved over greater distances. The requirements of a large, possibly more technically efficient plant will differ from a small one not only in quantity but in quality as well. Indeed, the entire locational problem may look significantly different with each successive estimate of the scale of the production unit. In making ready-to-wear clothing, for instance, there may well be scope for plants to operate successfully over a wide range of plant sizes. For a small operation, the producer might look for cheap premises, accessibility to downtown fashion stores, nearness to cloth wholesalers, and so on. For a large one, the criteria change. Now maybe it will be a new, custom-built factory on a green-field site with good access to freeways, near a pool of reliable and adaptable labor.

Looked at this way, scale has an important influence on location even though, of itself, it is a nonspatial, technical consideration. A trap into which we may sometimes fall is to believe that only geography may affect geography. In this case, without a knowledge of the technical and economic importance of scale, we cannot comprehend fully the importance of one of the critical factors taken into account in locating a factory. Not that this makes the problem simpler; quite the reverse. As we have suggested, the minimum efficient scale is essentially a theoretical abstraction. It varies over space, through time, and between and within industries. Moreover, in some industries, technological changes come so fast that plants are obsolescent before they are completed and the production process is always at least one innovation behind. Small wonder, then, that the question of scale generates much dispute and proves difficult to handle in any model whether designed by geographers or economists. We cannot, however, choose to ignore it, and next we shall look more closely at the locational impact of scale.

SCALE, DEMAND, AND LOCATION

At this stage in our analysis, the substitution problem of choosing a location has three main dimensions:

1. The costs incurred in assembling input factors over space

2. Technical considerations of the minimum efficient scale of production
3. Estimation of demand conditions and gaining access to the market

All three are, of course, inextricably tied together. Factor costs and scale affect price; price affects the level of demand; the level of demand affects scale; and to complete the circle, scale affects factor costs. There are several ways of going around this particular circle, but the important consideration for us at this stage is to realize that the process is circular. It begins nowhere and ends nowhere. We choose to cut into it where we like, but the act of cutting in by itself destroys something of the reality of the situation.

One of the acknowledged weaknesses of early location theory was its tendency to neglect scale. When scale was taken into account, it was treated as a secondary factor, helpful in determining the size of the market area for a given location. The prior assumption was made that the plant had already been located on the basis of other criteria, usually traditional ones based on minimization of input factor costs. By treating the problem in this way, the vital interaction of costs, scale, and demand in choosing the location in the first place is lost.

Hoover (1937), for instance, showed the way in which reductions in unit costs with increasing scale reduced the delivered price of goods for an established plant. This extended the market area over a wider range than would be the case with the assumption of constant returns to scale. To show this, Hoover used the *margin line*. This showed that delivered prices at the edge of a producer's market varied as the spatial extent of the market itself varied. The underlying idea is relatively simple. Suppose that an already located firm serving a spatially dispersed market finds a way of expanding its territory. At least two things will happen. Total transport costs will rise as the product is shipped to more customers who are farther away. Compensating for this, if the firm is operating below its minimum efficient scale, a rise in the volume of output will lower the unit costs of production. There will be a circular effect. If the fall in costs due to the additional sales volume exceeds the additional shipment charges, lower delivered prices can be offered at the margin of the market, and the firm can carry on expanding. Where the trade-off between scale economies and extra shipment charges equalizes itself, further expansion would mean higher delivered prices beyond this margin.

Isard (1956) showed how this works for two competing producers, and the situation he describes is shown in Figure 5.10. Assume that potential consumers of the good produced at A are located geographically along the line AB. If only consumers located at A want to buy the good, the producer's marginal cost (and in this special case, delivered price also) will be equivalent to the amount AK. But if other consumers perhaps as far away as L are willing to buy the good, marginal production costs could fall. This is because (assuming previous output levels to be somewhere below the MES) scale economies in producing the greater volume of output reduce unit costs of production. However, in tapping the new market to achieve these scale economies, there will be additional transportation charges to be paid. The slope of the line JG

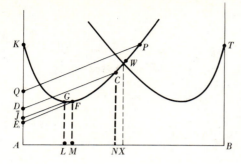

Figure 5.10 Relationship between scale of output and spatial extent of the market. *Source:* W. Isard (1956), *Location and Space Economy* (Cambridge, Mass.: MIT Press), fig. 27. Copyright © 1956 by the Massachusetts Institute of Technology. Reprinted with permission.

shows a linear uniform freight rate, and adding the appropriate shipment cost to the basic production cost, the new delivered price to the customer at *L* becomes *LG*. Further expansion of the geographic extent of the market, say, to the site of *M*, permits still further production economies (down to *AE*), but adding the additional costs of transportation incurred (the slope *EF*) gives a delivered price of *MF*.

This, however, represents the optimal combination for this case of market area served and scale of production. Any further increase in demand will incur higher unit production costs as well as the extra shipment charges for covering more distance. Delivered prices to customers rise to, say, *NC* or *XW*. The curve connecting points *KGFCP* is called a margin line. Any point along this line, as we have seen, expresses the delivered price at the edge of the market area. Before going on to look at the second producer, consider for a moment what would happen if transportation rates rose steeply with distance (that is, if the slope of lines like *JG* or *DC* were steeper). The benefits of production scale economies would be bought at a very high price in extra shipment costs as markets farther afield were tapped and would quickly disappear. Whatever the technical feasibility of scale economies at the plant, steeply rising space costs would preclude the achievement of scale economies. Thus not only are economies of scale limited by the size of the market in a nonspatial sense, but they may also be limited to the spatial extent of the market and the character of transportation rates.

Figure 5.10 also shows a second producer of the same good, located at *B* with identical initial costs *BT*. This producer's margin line has the same form as that of the producer at *A*, so that the market is shared equally at *X*, the point where the two margin lines intersect. If, however, scale economies are more readily available to one firm than to the other, it is possible, as in Figure 5.11, for the former to charge a lower delivered price over a wider area. Our discussion of spatial pricing policies in Chapter 3 is useful at this point. In Figure 3.22, for example, one reason why firm X could lower its delivered price might well be that it was able to produce at a larger scale than producer Y. In Figure 5.11, then, *A*'s market area is increased from *X* to *Y* at the expense of *B* because *A* can achieve greater scale economies in production. Figure 5.11 shows the

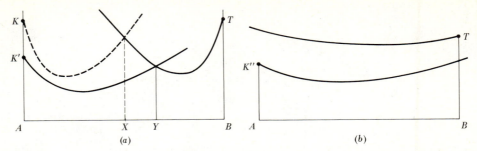

Figure 5.11 Achievement of economies of scale permits the extension of a firm's market by reducing the delivered price. *Source:* W. Isard (1956), *Location and Space Economy* (Cambridge, Mass.: MIT Press), fig. 30. Copyright © 1956 by the Massachusetts Institute of Technology. Reprinted with permission.

extreme case in which *A*'s delivered price is below that of *B* at all points within the market area, thus excluding *B* completely.

In this way, both Hoover and Isard showed that scale of production is inextricably linked with the level of demand available to a producer. As we have pointed out, however, this kind of analysis neglects the fact that the optimal location itself varies according to scale of output. Not that Isard himself is not aware of this:

> In any location decision, the scale of output is one of several basic, interdependent variables. As scale varies, so may the substitution points between any part of transport outlays, between any two sets of outlays, between outlays and revenues and so forth. (Isard, 1956, p. 175)

Let us look at another piece of the puzzle that has been filled in, this time by Leon Moses (1958). His analysis involved what Isard identified as scale affecting any two sets of outlays. To follow Moses's methodology, it is necessary to recall the techniques for the analysis of substitution previously outlined in Chapters 2 and 4. Essentially, Moses set out to demonstrate that the optimal combination of inputs and the optimal location of the plant may change as the scale of output changes.

Consider two locations where the production of a good can possibly take place. One is at the material site, the other at the market. All that is needed in production here is a raw material and some labor.

First, take the production possibilities for a given total outlay at the material site. Suppose to start with that the producer estimates total expenditure at *X* dollars per year. For location at the material site, materials are relatively cheap, while labor is dearer in the ratio of 3 : 2. To add a touch of reality, one might consider the site to be a booming mining town where labor, still relatively scarce, is at a premium. As Figure 5.12*a* shows, our producer could draw up the budget or iso-outlay line against these constraints. At the limits, either 60 units of materials or 40 units of labor could be obtained for the expenditure of a total of *X* dollars. The line drawn between these two extremes gives the

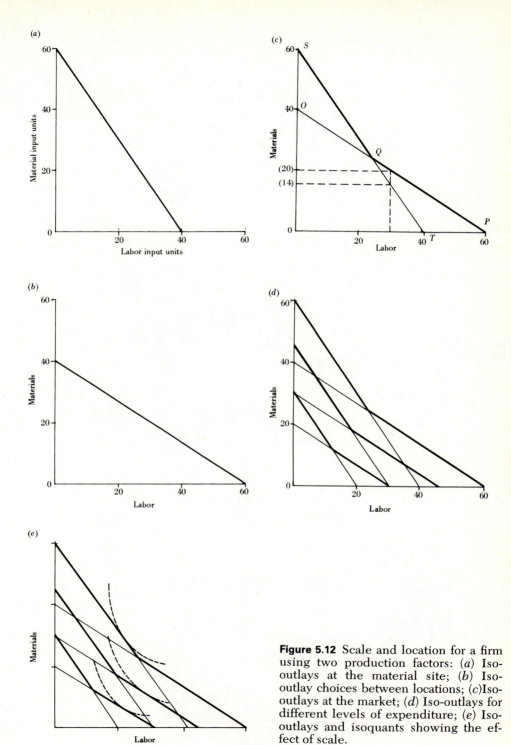

Figure 5.12 Scale and location for a firm using two production factors: (*a*) Iso-outlays at the material site; (*b*) Iso-outlay choices between locations; (*c*)Iso-outlays at the market; (*d*) Iso-outlays for different levels of expenditure; (*e*) Iso-outlays and isoquants showing the effect of scale.

possible combinations of both labor and materials for the same total expenditure. It is steeply sloping and reflects the fact that three units of materials trade off against two units of labor—the latter being the more expensive.

Let us shift our attention now to the market site where the iso-outlay situation is shown in Figure 5.12b. This time, three units of labor get two units of materials—the relative trading positions are reversed. Here labor is cheaper while materials are dearer since they now have a transportation item in their cost. The iso-outlay line slopes the other way. For the same total expenditure of X dollars, the producer can buy either 60 units of labor or 40 units of materials.

If we superimpose the two iso-outlay lines as in Figure 5.12c, the choices in outlays between the two locations are clearly laid out. Suppose the producer decides that the best production system functions with 30 units of labor. If the location at the material site were to be chosen, it would be necessary to consult the lower half of the line ST. This would show that with this choice of location at 30 units of labor input, 14 units of materials could be purchased within the balance of the X-dollar expenditure limit.

Before becoming committed, it is to be hoped that the producer bothers to have a look at the same sum for the alternative location at the market. This time, 30 units of labor would be looked for on the lower half of the line OP. Following it across to see how many units of material the firm could have for the balance of its expenditure, this time it would find that the answer is 20. It would be able to have more material units from the balance of expenditure with a constant input of labor if the right location were chosen. For the amount of inputs at the intersection of the solid curves, it would not matter where the factory was located. But for every other combination of inputs, there are advantages in location either at the market or at the source of materials, depending on the amounts of each it is proposed to use in production.

As Figure 5.12d shows, the iso-outlay line for X dollars, which we used in the preceding example, is only one of many possibilities (in fact, an infinite number of them). For every budget figure the producer could name, there would be a particular substitution possibility or iso-outlay line for materials and labor.

Now what about scale? One thing we have learned about scale is that at different levels, there are different factor mixes. A small operation may need far more labor in relation to materials, while a bigger one may demand more materials and less labor. These depend on the particular production functions of the establishment and the technical possibilities available for producing the good. Remember that in Chapter 4 we showed the production function as a curved line indicating the possible factor combinations for a given level of output. Clearly, the technical possibilities in production will vary with the scale of the output to be produced. We have, then, superimposed these production possibility curves for different levels of output or isoquants on the set of iso-outlay lines used in the earlier example. These showed that for a small plant, in this hypothetical case, the best location would be the market. For a large one, the point at which the isoquant just touches the lowest possible iso-outlay line is above Q, and therefore the material source would be the best location. Thus as the scale of output changes, so does the optimum location.

Most location theory has adopted a rather one-sided approach to the problem of defining the optimal location. Whereas the least-cost school concentrates on spatial variations in costs with little attention to demand as a primary locational force, the market area approach overemphasizes demand at the expense of cost variations. Both tend to neglect the influence of scale of output. But the major conclusion arising out of the preceding discussion is that the locational problem is tripartite. Variations in costs, revenues, and scale must be considered simultaneously. Change in any one of these will alter the optimal location that is the point of maximum profits.

Perhaps one of the reasons for the failure of location theorists to treat scale of production adequately has been the problem of isolating its specific locational effect. Scale interacts with all the elements of the locational problem. By influencing the nature of the factor inputs in the production function, it affects the whole character of an industry. As a result, the particular importance assigned to production factors as location factors changes with scale. Similarly, there is an interaction with demand and the market. As we have seen, economies of scale can be achieved up to an extent permitted by the size of the effective market. But by affecting the basic costs of producers and, consequently, the delivered prices they can offer, scale economies also contribute to the increases in the size and spatial extent of markets. In all this, the impact of space itself is pervasive. The ability of a plant to achieve scale economies may, as we have seen, depend on the relationship between the production economies derived from serving a wider market and the transportation costs that must be overcome to obtain access to such a market. Scale is a difficult subject to treat analytically. It is equally difficult to treat empirically. Problems are encountered particularly in attempting to translate the assumed theoretical benefits of scale economies into measurable quantities amenable to empirical testing. This is especially true with scale economies assumed to be derived from the geographic clustering of economic activities, and it is to these that we now turn our attention.

AGGLOMERATION: LOCALIZED EXTERNAL ECONOMIES OF SCALE

One of the characteristic features of economic activities is their marked tendency to occur in spatial clusters. Regular spatial groupings of specialized activities were shown to exist even on the homogeneous landscape of Chapter 1. The chapters that followed demonstrated still more the propensity for economic units to cluster at material sites, terminals and transshipment points, cheap labor locations, and the focal points of major markets as businesses moved in to exploit the comparative advantage offered by these points in space. But the process of clustering itself offers further economies of a particular kind to those who take part in it.

The scale economies we considered earlier in this chapter were internal economies of scale. They arose from the savings to be made within an individual plant from producing at larger volumes. In contrast, external economies of scale are savings that a plant or firm gains from its connection with other plants or firms. One source of such external economies is explicitly spatial. By clustering in close spatial proximity to other activities, it is believed, firms will benefit from a particular kind of external economy of scale that Alfred Weber called *economies of agglomeration.*

In this case, the external economies experienced by a production unit derive from its particular locational association with a larger-scale spatial cluster of economic activities. Scale economies internal to the major cluster— for example, the economies achieved by a city during the early phases of its growth—are passed on as external economies to the individual production units that make it up. Without necessarily raising their own scale of production, their spatial association and functional linkage with the larger agglomeration permit them to derive cost economies at second hand from scale factors operating outside themselves. Under these conditions, agglomeration economies become yet another element contributing to the producer's decision to locate and therefore have a significant impact on the location of economic activity in space. Weber visualized agglomeration economies as exerting a deviational force on the minimum-transport-cost location in a similar way to that exerted by cheap labor locations (see Chapter 4). In both cases, the existence of a locational deviation was seen to depend on the location of the attracting force in relation to the critical isodapane. For agglomeration to occur, the critical isodapanes of the firms in question must intersect as in Figure 5.13*b*; where intersection does not occur, as in Figure 5.13*a*, agglomeration will not take place.

Although Weber's analysis is consistent with the rest of his location theory, it is perhaps one of its weaker aspects, for two reasons. First, he failed to probe sufficiently deeply into the nature of agglomeration economies and, second, he adopted a highly simplistic view of the mechanism of relocation. Agglomeration economies differ from other locational economies, such as cheap labor or materials, in that they depend on the coincident decisions of a number of firms.

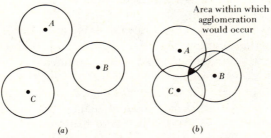

Figure 5.13 Weber's concept of agglomeration: (*a*) The critical isodapanes do not intersect: there is no agglomeration; (*b*) The critical isodapanes do intersect: agglomeration may occur.

An agglomeration point is merely a place to which a number of persons engaged in industry decide to resort. Without the decision it does not exist; after the decision it is there. Looked at from another point of view, a point of agglomeration is not one to which it is to the advantage of any single producer to transfer his plant. While it may be to the advantage of two producers to come together, neither will gain unless the other also acts. (Daggett, 1955, p. 450)

One perceived advantage of such spatial agglomeration may be the reduction of uncertainty.* For example, if a producer observes that other firms in the industry are clustered in a particular location, it may be inclined also to locate there on the grounds that if they are surviving there, conditions must be satisfactory. It may also be sensible for firms selling to a particular market to locate close to each other in order to minimize uncertainty. This particular problem can be illustrated in the following way.

Two sellers of a good are competing for the best location from which to serve their market. This example is the classic problem analyzed many years ago by the economist Harold Hotelling. Here we use the game theory approach of Stevens (1961). Game theory is a method of analyzing conflict situations. it aims at providing the best solution to problems in which the outcome for one player depends not only on his or her own actions but also on the actions of an opponent. The rules of the game are as follows:

1. Imagine a limited stretch of highway along which potential consumers of hot dogs are evenly spread.
2. Each customer is prepared to buy one hot dog no matter what the price may be (in technical language, their demand is infinitely inelastic), but each will buy from the seller with the lowest total price.
3. Each hot dog costs $1 at the point of sale, to which must be added the cost of the customer's travel to the point of sale. In terms of our discussion of spatial pricing policies in Chapter 3, we are dealing here with an FOB pricing system.
4. The two hot dog sellers are spatially mobile—they can move without cost—but they are confined to the five equally spaced locations *a* to *e* shown in Figure 5.14.

Obviously, given these rules, the success of each hot dog seller (measured in terms of one's net advantage over the other) depends not only on its own location but also on the location of the other seller. In other words, their locations are interdependent, and they are competing for the location that will give them the largest volume of sales. Let us see how this competitive struggle may be resolved, using Figure 5.15 as a guide.

Suppose that seller I is quicker off the mark than seller II and enters the market first. In the time period T_1, it can, by locating at *a*, serve the entire market and make sales of 40 hot dogs. When seller II eventually enters, its

---

* Webber (1972) has provided a detailed discussion of the impact of uncertainty on location.

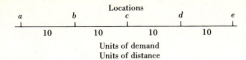

Figure 5.14 Possible locations for two competitors in an evenly spaced linear market.

location decision is fundamentally influenced by the fact that seller I is already in business. Seller II assumes that seller I's location is fixed, so by locating at b, while seller I is at a, Seller II gains a very considerable advantage. As Figure 5.15 shows, it captures the entire market between b and e, gaining a net advantage over seller I of $30 (made up of the $30 of sales between b and e plus the $5 for the market between b and a less the $5 sales that seller I achieves). This is obviously a severe financial blow to seller I, so it retaliates spatially by moving to location c. In time period T_3, therefore, the market is divided in such a way that seller I has converted its net disadvantage of $30 to a net advantage of $10 ($25 − $15). This, of course, forces seller II to reconsider its locational strategy. By seller II's relocating at c, immediately adjacent to seller I, neither gains an advantage over the other, but each gains exactly the same volume of sales. In the period T_4, seller II supplies all the hot dog needs of the consumers between a and c and seller I supplies all the customers between c and e.

We can summarize this competitive situation as a payoff matrix as in Table 5.5, in which the values represent the net advantage of one seller over the other. Note that although there are nine cells in which a zero appears, only one of these—the one corresponding to the situation where both sellers are located at c—represents a stable or equilibrium situation for the sellers. If they both locate at c, neither can gain an advantage from moving elsewhere. The reverse is true of the other eight zero positions, where it will always be advantageous

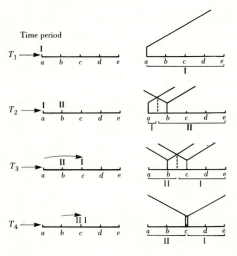

Figure 5.15 Competition and locational change in the two-seller situation.

Table 5.5 PAYOFF MATRIX FOR THE GAME THEORY SOLUTION OF THE TWO-SELLER PROBLEM

		Locational strategies of seller II					
		a	b	c	d	e	Row minima
Locational strategies of seller I	a	0	−30	−20	−10	0	−30
	b	+30	0	−10	0	+10	−10
	c	+20	+10	0	+10	+20	0
	d	+10	0	−10	0	+30	−10
	e	0	−10	−20	−30	0	−30
Column maxima		+30	+10	0	+10	+30	

FOB price per unit is $1. Plus signs represent payoffs to Seller I, whose strategies are listed in the left-hand column. Minus signs represent payoffs to Seller II, whose strategies are listed across the top of the table.

Source: B.H. Stevens (1961), "An Application of Game Theory to a Problem in Location Strategy," *Papers and Proceedings of the Regional Science Association* 7:143–57, table 1. Reproduced by permission.

for one of the sellers to relocate. Thus if seller I is at *b* and seller II at *d*, neither gains a net advantage; however, if either one of them moved to *c*, it would immediately achieve a net advantage of $10. If both were located together at the extremes of the market (that is, both at *e* or both at *a*), a move by one seller to the next nearest location would produce a large net advantage but also an unstable situation, because the other seller would then retaliate by relocating. Clearly, then, the best position for both sellers, given potential retaliation from each, is together at the center of the market—that is, to *agglomerate.*

An important basis of agglomeration economies is the connections or linkages between economic activities within a relatively restricted geographic area. In the final analysis, of course, any firm is but one part of a complex chain of production—a series of transactions—held together by direct or indirect linkages between a series of firms. Through such linkages, external economies are transmitted to the individual production unit via its network of interconnections with other elements in a system. These linkages are of three main types: production linkages, service linkages, and marketing linkages. Figure 5.16 illustrates these backward and forward linkages between firms.

An agglomeration economy may exist, therefore, where some or all of these linkages are present within a relatively small geographic area, thus either lowering a firm's costs or increasing its revenue (or both). In addition, other economies may be derived, as it were, by association. Complementary or similar industries, by recruiting and training a labor force, for instance, provide a localized cluster of particular labor skills. These skill pools add to the attractiveness of such areas for particular specialized industries.

Commonly, a distinction is made between two types of agglomeration economy: localization economies and urbanization economies. *Localization economies* are gained by firms in a single industry (or a set of closely related industries) at a single location and accrue to the individual production units

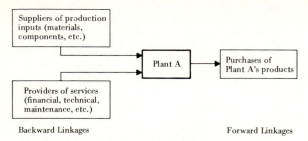

Figure 5.16 Major functional linkages of a firm.

through the overall enlarged output of the industry as a whole at that location. *Urbanization economies,* by contrast, apply to all firms in all industries at a single location and reflect external economies passed to enterprises as a result of savings from the large-scale operation of the agglomeration as a whole. The economies of scale to the higher-order systems such as the localized industry or the city as an economic unit are essentially the same as those discussed previously for the individual producer—specialization, economies of massed reserves, and economies of large-scale purchasing. Indeed, a real distinction between localization and urbanization economies is often difficult to establish.

Although all firms have linkages of the type shown in Figure 5.16, the literature dealing with economies of agglomeration in general has tended to focus primarily on those at the smaller end of the plant size spectrum. Here, highly specialized trades under separate ownership exploit a complex constellation of linkages to each other, often replicating in this way the characteristics found under single ownership in the multiplant firm. At this scale, it seems easy to identify one of the keys to agglomeration economies as the minimization of the distance between each linked firm and its trading partners. Under these circumstances, the economies achieved would not be fundamentally different from those outlined by Adam Smith in his description of the pinmaker, except that in this case each subprocess is under separate ownership. Each small firm can specialize in its particular trade to a degree that promotes a high level of efficiency in the operations of both workers and machines. Where the firms cluster together in the same street, block, or quarter, the movement of goods along the production line, instead of being by conveyor belt, can be by handcart along the street or by truck. In the handcart example, there are obvious immediate savings to each producer since the "conveyor belt" (the street) is paid for and maintained not by any individual firm but by the community at large.

Apart from the apparently obvious transfer cost benefits of close spatial juxtaposition and the "production line in the street," certain other advantages associated with plants keep their linkages within a closely confined area. Hoover (1948) points to the particular scale economies that come from specialization of function:

> Certain operations and services that a firm in a smaller place would do for itself

can in the city be farmed out to separate enterprises specializing in these operations and operating at a scale large enough to do the job more cheaply. (p. 120)

The economies of scale from serving the large city market are reputed to make
the small specialist cost-efficient since its level of output can move closer to the
minimum efficient scale. These internal economies are then passed on as
external economies to the other firms that use their goods or services.

In the same way, the economies of massed reserves that lowered unit costs
for the single large firm can also be seen to operate at the level of the urban-
industrial agglomeration. This may apply to the supply of materials, labor, or
perhaps floor space. For instance, an isolated firm may have to tie up its funds
by carrying a considerable inventory of materials and components to cover it
for possible delays in delivery or temporary shortages. The firm that is part of
an agglomeration of activities, however, may well be able to call on supplies at
very short notice. This is because the high level of aggregate demand, as we
have seen, permits the operation of a wide range of specialist factors or merchants. As a result, far less capital needs to be immobilized in inventories. In
the case of the printing industries of New York City, for example, it was found
that firms needed only 9.7 percent of their total assets in inventories, compared
with 19 percent for printing firms outside New York (Hall, 1959). The benefits
of access to a large labor pool for the individual firm were discussed in Chapter
4. Large agglomerations offer a wide range of skilled and unskilled labor that
can be drawn on relatively easily to meet sudden shifts in production activity.
Similarly, industrial and commercial floor space of all kinds is widely available
in most major agglomerations and can be taken up or discarded to meet short-
term shifts in production needs.

Economies of large-scale purchasing are also available as external economies to the firm in an urban-industrial complex. The advantage of the single
large firm, obtaining favorable rates on the bulk purchase of supplies and
services, may also be available to the cluster of firms of all sizes in agglomerations. In the case of transportation, for example, small firms that individually
ship small quantities of their goods can share the services of freight forwarders.
Dealing with a large number of spatially concentrated firms, these can combine shipments to make up full-carload lots and theoretically pass on some of
the resulting economies to the customer. Another of the acknowledged benefits of agglomeration for certain industries is the rapidity with which communication can take place between customer and supplier. This is particularly
important where direct and frequent contact is essential.

For all these reasons, therefore, the economies of agglomeration have been
accorded a key position in studies of the location of economic activities. But
what of the empirical evidence to verify their existence? A number of classic
studies of urban-industrial areas seemed to provide substantial support for the
significance of local linkages between firms. Many cities were found to contain
highly specialized concentrations of particular industries in specific quarters.
Figure 5.17 shows a map of the jewelry quarter in Birmingham, England, in the
1940s. The extremely fine division of labor between the myriad (mostly very

Key

■ Goldsmiths and manufacturing jewelers
□ Silversmiths
⊡ Electroplaters
□ Medalists
⊟ Gilt and imitation jewelery

▼ Factors and merchants
▽ Dealers in bullion and precious stones
▼ Jewelers' material suppliers

◆ Gem setting
◇ Stamping and piercing
◇ Engraving, polishing, and enameling
⊖ Die sinkers
△ Jewelery repairer
▲ Refiners
◇ General outwork

○ Manufacturers of optical goods
⊕ Manufacturers of fancy leather goods
⊖ Watchmakers
⊙ Miscellaneous manufacturers

Figure 5.17 The Birmingham, England, jewelry quarter, 1948. *Source:* M. J. Wise (1949), "On the Evolution of Jewelry and Gun Quarters in Birmingham," *Transactions of the Institute of British Geographers* 15: 57–72, fig. 1.

small) firms is abundantly clear. A similar pattern was identified by Peter Hall in his study of the industries of London and by Max Hall in his work on the manufacturing industries of New York City.

It seemed reasonable to infer from these observations that a major reason for such tight spatial clusters was the simple one of spatial proximity between firms linked together in a functional sense. This was almost certainly true under particular organizational and technological circumstances. When most firms were small, single plants and the costs (in terms of both money and time) of moving materials and semifinished products were high, it made a lot of sense for such firms to locate as close as possible to one another. But both the organization of economic activities and the nature of technology have changed. In addition, most urban areas have undergone massive physical transformation, and most of the old industrial quarters have disappeared.

Organizationally, the trend has been one of a geographic spreading out of interfirm linkages. During the late 1960s and early 1970s, a number of studies that actually measured the linkages between firms found that only a relatively small proportion could be described as local. For example, only 37 percent of the inputs to a large sample of manufacturers in Philadelphia actually came from Philadelphia itself (Karaska, 1969). A similar finding emerged from Gilmour's (1974) study of firms in Montreal, where only 31 percent of the inputs bought by almost 200 manufacturers originated within Montreal and only 27 percent of these firms' output was sold to other Montreal firms. Scott's more recent and very detailed studies (1983a, 1983b, 1984; Scott and Angel, 1987) of the printed circuit industry and the women's dress industry in Los Angeles and of semiconductor production in Silicon Valley showed that the economies of spatial clustering remain extremely important, particularly for small firms in which there is a high degree of organizational interdependence between plants. Figure 5.18 shows the intensity of spatial clustering of semiconductor manufacturing establishments in Silicon Valley.

Thus it is dangerous to generalize too far about the existence of localization economies based on short-distance linkages. Though such linkages may remain very important in some industries and for some firms, depending on their particular scale and technology, for others the spatial horizon of interfirm linkages has become the world itself. This is an issue to which we shall return again in Part Two.

Our attention so far has been devoted to the economies of agglomeration, both for firms within closely related industries (localization economies) and for all firms in all kinds of industries (urbanization economies). But agglomeration may also generate *diseconomies*: beyond a certain scale, disadvantages of spatial clustering may appear. There may be a point at which an expanding urban agglomeration becomes incapable of maintaining its efficiency. Problems such as congestion and clogged transportation arteries, soaring land prices, pollution, and administrative overload begin to transform urbanization economies into diseconomies, generating what Weber called *deglomerative* tendencies.

One of the general economies of agglomeration afforded to the constituent

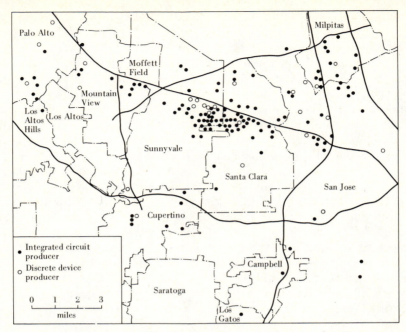

Figure 5.18 Geographic distribution of semiconductor manufacturing establishments in Silicon Valley, California. *Source:* A. J. Scott and D. P. Angel (1987), "The U.S. Semiconductor Industry: A Locational Analysis," *Environment and Planning* 19: 875–912, fig. 5.

firms of a major urban-industrial complex is the provision of wide-ranging public utility and welfare services. As the size of such complexes increases, the unit costs of providing utility services such as power generation, urban transportation, water supply, and sewage disposal should fall. Scale economies should, in theory, continue to increase until a point is reached that is equivalent to the minimum efficient scale of the individual plant. Up to this point, lower costs should be passed on to the individual user in the agglomeration as external economies of scale. In theory, therefore, there should be optimal sizes for agglomerations just as there are for individual plants, sizes at which public utilities and services are provided at optimum levels of efficiency, maximizing the external economies that they pass on to constituent firms.

It should be possible to calculate a minimum optimal scale of agglomeration just as one can measure the MOS of a factory. But the problem is infinitely more complex in the case of an agglomeration. Isard (1956) attempted to show how this might be done by means of his net economy curves (economies less diseconomies) for utilities such as power generation, urban transportation, water supply, and sewage disposal. Figure 5.19 shows a set of hypothetical net economy curves for agglomerations of varying size, with the individual curves summed to give a total economy curve. Unfortunately, as Isard himself points out, such a procedure involves numerous problems, in-

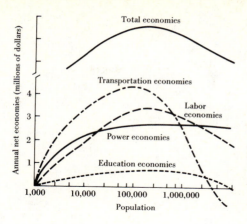

Figure 5.19 Hypothetical net economy curves. *Source:* W. Isard (1956), *Location and Space Economy* (Cambridge, Mass.: MIT Press), fig. 35. Copyright © 1956 by the Massachusetts Institute of Technology. Reprinted with permission.

cluding the question of weighting each curve according to relative importance and the fact that there is a degree of interdependence between, for example, power economies and transportation economies.

Until a few years ago, there seemed to be no noticeable slackening in the process of urban growth in the Western economies. Although the central cities of most of the older metropolitan areas had been in decline, this had been offset by growth in the suburbs, but within the same broad area. Indeed, every U.S. census from 1900 to 1970 showed that the metropolitan population was increasing at a faster rate than the population as a whole. By the 1970s, however, the position seemed to have changed. Some observers began to comment on the onset of counterurbanization, the reversal of the long-standing tendency toward spatial concentration of human activity. Growth of population and of economic activity began to shift toward what had been termed the intermetropolitan periphery and to smaller towns and cities. The reasons underlying such a major reorientation—if that is what it turns out to be—are extremely complex, but they may in part be a reflection of the diseconomies of agglomeration.

SUMMARY

Relaxing our initial assumption about the spatial uniformity of demand and giving more explicit attention to the question of scale economies in production has introduced a substantial degree of sophistication into our analysis. It has also, however, promoted a sharp increase in the complexity of the subject matter. By restricting our early investigation to cost considerations alone, we were able to deal with a single source of variation in the conditions facing the firm. By opening up questions of variable demand and scale, however, not only have new sources of complexity been introduced, but, more important, we have been able to show the existence of key *interaction effects* among costs, scale, and demand. Each influences the other in an unbroken circle of cause and effect.

Lösch (1954), reviewing Weber's least-cost approach to location, made the following observation:

> The fundamental error consists in seeking the place of least cost. This is as absurd as to consider the point of largest sales as the proper location. Every such one-sided orientation is wrong. Only search for the place of greatest profit is right. (pp. 28–29)

Lösch is, of course, correct, but the problem is that what he recommends is far easier said than done. The place of greatest profit will not depend only on choices related to material and factor inputs and their costs of assembly. It will also depend on the chosen scale of production, the size and geographic disposition of the market, and the existence of external economies of scale. As we have emphasized, all are closely interrelated, and the real choice of the location of greatest profit involves a multiway problem of substitution in which there are many complex trade-offs.

FURTHER READING

Bain, J. S. (1968). *Industrial Organization,* 2d ed. New York: Wiley, chaps. 4 and 6.

Gilmour, J. M. (1974). "External Economies of Scale, Inter-industrial Linkages and Decision Making in Manufacturing," in F. E. I. Hamilton (ed.), *Spatial Perspectives on Industrial Organization and Decision Making.* London: Wiley, pp. 363–393.

Greenhut, M. L. (1956). *Plant Location in Theory and Practice.* Chapel Hill: University of North Carolina Press, chaps. 2 and 6.

Harris, C. D. (1954). "The Market as a Factor in the Localization of Industry in the U.S.," *Annals of the Association of American Geographers* 44: 315–348.

Hoare, A. G. (1985). "Industrial Linkage Studies," in M. Pacione (ed.), *Progress in Industrial Geography.* London: Croom Helm, chap. 2.

Isard, W. (1956). *Location and Space Economy.* Cambridge, Mass: MIT Press, chap. 8.

Moses, L. N. (1958). "Location and the Theory of Production," *Quarterly Journal of Economics* 72: 259–272.

Scott, A. J. (1983a). "Industrial Organization and the Logic of Intrametropolitan Location: I. Theoretical Considerations," *Economic Geography* 59: 233–249.

Scott, A. J. (1983b). "Industrial Organization and the Logic of Intrametropolitan Location: II. A Case Study of the Printed Circuit Industry in the Greater Los Angeles Region," *Economic Geography* 59: 343–367.

Scott, A. J. (1984). "Industrial Organization and the Logic of Intrametropolitan Location: III. A Case Study of the Women's Dress Industry in the Greater Los Angeles Region," *Economic Geography* 60: 3–27.

Weber, A. (1909). *Theory of the Location of Industries.* Chicago: University of Chicago Press, chap. 5.

Chapter
6

The Time-Space Dimension: Cumulative Economic Development

In Chapters 1 through 5, we adopted a very specific viewpoint to understanding the geography of economic activity. We focused specifically on the spatial dimension and progressively introduced individual variables or location factors to make our initial simplified economic landscape more realistic. In each of Chapters 2 through 5, we introduced just a little dynamism by describing how the nature or operations of each individual factor may change through time. But we have not looked systematically at change over time. Neither have we tried to fit the individual factors together to make an interconnected and coherent whole. In this final chapter of Part One, we aim to do both of these things by looking at how spatial economic change may occur. We shall have much more to say about this topic in later chapters and from alternative perspectives. But in this chapter, it is important to be aware that we are still operating within the specific conceptual framework adopted throughout Part One.

How, in a general sense, does an economic landscape evolve or change over time? What processes are involved? Particular economic landscapes evolve in particular ways. But can we make any general observations that apply to most circumstances? One starting point is to separate out two aspects of the development process:

1. The initial formation of a cluster of economic activity
2. The subsequent growth and development of that cluster

INITIAL TRIGGERS TO DEVELOPMENT

It is almost impossible to generalize about why economic development begins in some places and not in others. We can point to certain kinds of favorable circumstance, such as a good natural harbor, an easy crossing point of a river, or a valuable energy or mineral resource, and argue that such circumstances explain development. But although favorable circumstances like these may well be important, they do not provide a full explanation. They may be necessary elements in an explanation, but they are not sufficient on their own. Each needs to be set within the specific economic, political, social, cultural, and technological circumstances of the time period in question.

The Swedish writer Gunnar Myrdal (1957) took a broad cut through the problem of initial triggers to economic development:

> Within broad limits the power of attraction today of a center has its origin mainly in the historical accident that something once started there, and not in a number of other places where it could equally well or better have started, and that the start met with success. (p. 26)

Such "historical accidents" are well documented in the literature of economic and business history, most notably in the idea that many successful businesses have grown in the hometowns of their founders—the entrepreneur's place of birth or the town in which he or she eventually settled. On a much longer time scale, the historical accident of a particular place taking on a key political role as a national or regional capital or administrative center is another initial trigger commonly identified in the literature.

In areas of totally new settlement, the initial stimulus for development of a particular location is likely to be exogenous, that is, to come from outside. The geographer James Vance (1970) has shown how the initial growth and development of the U.S. urban system can be interpreted in terms of such exogenous forces rather than as the result of the kind of central place theory we discussed in Chapter 1.

Figure 6.1 is Vance's summary comparison of the central place model and what he terms the mercantile model based on externally generated trade. As we saw in Chapter 1, the development of a central place in Christaller's theory is related to the demand for goods and services generated by the local population (of individual households and businesses). Vance argues that although this may be a good description of urban development in "old lands," it is less appropriate in the context of undeveloped "new lands." Vance identified the key trigger to urban development in the latter case as the external trade links created by merchants and wholesalers. These formed "points of attachment" (Figure 6.1) in the new lands, which, as they took hold and grew, themselves became central places serving the now emerging hinterland. The point of

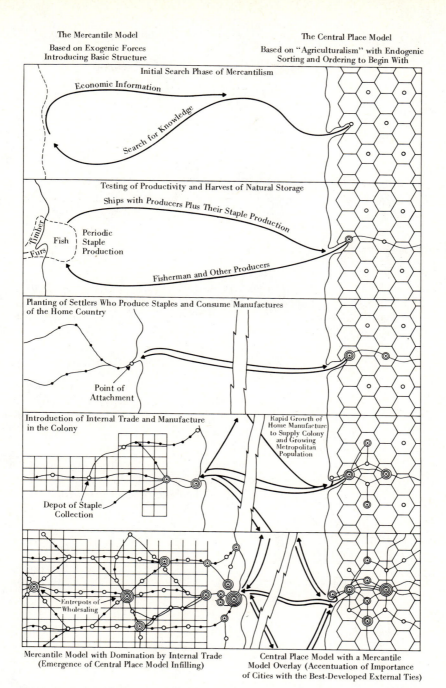

Figure 6.1 Initial growth of a region as the result of external or internal forces. *Source:* J. E. Vance, Jr. (1970), *The Merchant's World: The Geography of Wholesaling* (Englewood Cliffs, N.J.: Prentice-Hall), fig. 18.

relevance is that the initial trigger for development was induced externally rather than internally.

SUBSEQUENT GROWTH AND DEVELOPMENT: A CUMULATIVE PROCESS

Of course, not every location at which economic activity is initiated necessarily grows into a major center. History is littered with the remains of deserted settlements that flowered briefly and then withered. But beyond a certain threshold level, once triggered by some initial motivating force, economic development tends to be a cumulative process, at least for a certain period of time. How does such a cumulative process operate?

The Multiplier Mechanism and Cumulative Development

The concept of the *multiplier* is fundamental to an understanding of the cumulative process of development. To make the explanation as simple as possible, let us begin by assuming that the initial trigger for development at a particular location in a previously little developed area is generated from outside the area. In other words, development is driven by *export demand*. *Export base theory* is one way of understanding the operation of the multiplier. It is based on the idea that any region, be it town, city, state, or nation, has two mutually exclusive sectors to its economy:

1. The *export base sector* (E_{ex}). This includes all activities for which the effective demand is external to the region itself; that is, their level is set by forces outside the region. This sector consists of the region's export activities.
2. The *residentiary sector* (E_{res}). This includes all activities for which the effective demand is internal to the region itself—production systems supplying the day-to-day needs of the resident population, such as the retail trade and local services.

In shorthand terms, therefore, the total economy of a region E_t is

$$E_t = E_{ex} + E_{res}$$

The export base sector is considered to play the primary part in the promotion of economic development.

Export base theory is based on the idea that at least in the short term, the level of residentiary activity depends entirely on the level of export base activity. In other words,

$$E_{res} = f(E_{ex})$$

It is also assumed that the relationship between the two over time remains the same (expressed by the constant k):

$$E_{res} = k(E_{ex})$$

On the assumption that the relative proportions of export base and residentiary activity remain the same, it is possible to calculate the impact of a change in the export base sector on the residentiary sector and therefore on the regional economy as a whole. Suppose that our hypothetical region has a total labor force to begin with of 1,000. The export base sector accounts for 400 of these, the residentiary sector for 600. Thus the existing ratio between residentiary and export base activity measured in employment terms of $1 : 1.5$ ($k = 1.5$). In other words, for every two jobs in the export base sector there are three jobs in the residentiary sector.

On the basis of a constant relationship, we can calculate the total impact of an increase in export base activity such as the opening of a new manufacturing plant employing, say, a labor force of 2,000. The residentiary sector will gain

$$E_{res} = 1.5(2,000) = 3,000 \text{ new jobs}$$

Therefore, where the export base ratio is $1 : 1.5$, an increase of 2,000 in export sector employment will yield an increase of 5,000 in total regional employment. Assuming that the ratio remains constant over time, it may be interpreted as a simple type of economic multiplier known as the *export base multiplier*. Written in a form more clearly identified with multiplier notation, it may be expressed as

$$E_t = \frac{1}{1 - a} (E_{ex})$$

where E_t and E_{ex} are, respectively, total and export base employment (or income), as before, and a is the parameter that determines the constant relationship between export base and residential activity. (In our example, where 60 percent of total employment is in the residentiary sector, a would have a value of 0.6). The term $1/(1 - a)$ is the multiplier, which in this case has a value of 2.5.

Direct and Indirect Multiplier Effects: The Income Multiplier

Our simple example shows only the direct impact on employment of an increase in export activities. The point about the multiplier, however, is that the direct effects are followed by a series of indirect effects in a chainlike sequence as expansion induced in one sector has repercussions on other sectors, though the effect becomes less and less pronounced as "distance" from the original stimulus increases (just as the ripples created by throwing a stone into a pond decline in intensity outward). Although we have visualized the relationship in terms of an employment multiplier, the true basis for the changes that take place "down the line" of the multiplier is income. This is derived from the increased demand for goods and services, which in turn pulls in the full range of production factors. Labor is only one of these factors, but it is often the only one that can be readily quantified.

One of the most useful explanations of the income multiplier in a slightly expanded export base framework was provided by Tiebout (1962). Tiebout was concerned with it as a tool for predicting short-term economic changes, whereas we are more interested in the multiplier as a conceptual device. However, his explanation of the way the multiplier works is so clear that we cannot do better than to base our discussion on his work. Tiebout made a basic distinction between short-term and long-term perspectives. In the short term, let us assume that an area's economy consists of three sectors: exports, local investment, and local consumption (in other words, we subdivide the residentiary sector into two categories). The population of the area derives income from each of these. In the short term, income from both export and local investment sectors depends directly on external forces, while income from the local consumption sector is determined by local spending from income generated in the other two sectors. Income from local consumption is based on two steps:

1. Residents spend some of their income on local goods and services. This creates what we can call *local sales dollars*. Many variable factors influence how much income is spent on local goods and services: one of the most important is, of course, the level of income. Let us suppose that, on average, local residents tend to spend 50 percent of their income on local goods and services. In technical terms, we can say that the *propensity to consume locally* is 0.5. (The other half of the residents' income may go into savings, be paid in taxes, or be spent outside the area). If we imagine that local income from export activity increases by $1.00, we might suppose that 50 cents of this would remain as local income through spending in the local consumption sector. But this is not so, because we have to take the second step into account.

2. Only part of the local sales dollar remains within the local area to become local income. Some of the 50 cents of the extra dollar is used by local suppliers to pay for inputs that originate outside—imported goods and wages to nonresidents, for example. But part of the 50 cents will certainly remain as local income—local wages, profits, payments to local businesses, and so on. Again, assume that we can measure the average proportion that remains locally and that it is 40 percent. Using the same terminology as before, we can say that the *income propensity* of the local sales dollar is 0.4.

Combining these two steps, we can say that for every dollar of local income, 50 cents will be spent locally and 40 percent of this 50 cents—that is, 20 cents—will remain as local income. But this is only the *direct* effect. Part of the additional 20 cents will be spent locally (half, or 10 cents), and 0.4 of this will remain as local income—4 cents. Again, half of this is spent locally and 0.4 of this remains as local income. Thus as the multiplier process proceeds, each step contributes less and less. We can use the kind of multiplier formula introduced earlier to express this chain reaction:

Total income increase
= increase in (export plus
    local investment) income $\times \dfrac{1}{1 - \text{(propensity to consume locally} \times \text{income created per \$ of local consumption sales)}}$

Inserting the values from our discussion, we can calculate the total effect of increasing income by \$1.00:

$$\text{Total income increase} = \$1.00 \times \frac{1}{1 - (0.5 \times 0.4)}$$

$$= \$1.00 \times \frac{1}{1 - 0.2} = \$1.25$$

Thus the multiplier effect of increasing income by \$1.00 is to increase total income to \$1.25 through spending in the local consumption sector.

As we observed earlier, this is a short-run view. But what if we take, say, a 10- or 20-year view? In such a longer-term analysis, we would expect that *local investment* income will be less dependent on external forces and influenced more by local income. For example, increases in income from export and local consumption sectors will stimulate the need for housing, plant and equipment, and so on. So in addition to the propensity to consume locally, we must consider the *propensity to invest in local capital goods*. Again for simplicity, assume that we can calculate such a propensity and that it is 0.2 (that is, 20 cents of every dollar of local income is spent on local investment) and that the income created per dollar of local investment sales is 0.5 (that is, half of the expenditure on local investment remains in the local area).

Our multiplier formula can be adjusted to take account of these changes. In its revised form, local investment combines with local consumption. If, again, export income increases by \$1.00 we can trace the total multiplier effect of this increase:

Total income increase

= increase in export income $\times \dfrac{1}{1 - [\text{(propensity to consume locally} \times \text{income created per \$ of local consumption sales)} + \text{(propensity to invest locally} \times \text{income created per \$ of local investment sales)}]}$

Inserting our hypothetical values, we obtain

$$\text{Total income increase} = \$1.00 \times \frac{1}{1 - [(0.5 \times 0.4) + (0.2 \times 0.5)]}$$

$$= \$1.00 \times \frac{1}{0.7} = \$1.43$$

Thus when both the local consumption and local investment sectors contribute to local income, the multiplier value is increased. Using larger values, an increase in export-generated income of $100,000 would create an additional $43,000 through spending on local consumption and local investment.

It should now be clear that if exports are the primary determinant of local growth, regions or cities with a strong export orientation and high multipliers will be much more sensitive to the impact of any initial kick, positive or negative, than those with low multiplier relations. It is worth noting in passing that the multiplier relation may apply in the reverse sense. For instance, in the employment example used earlier, should redundancies or factory closures take place, for every two jobs lost in the export base sector, three would be lost in the residentiary sector.

An employment multiplier of 2.5 is pretty high. Very few regions today would expect to generate three jobs in the residentiary sector for every two in the export sector. In fact, the strength of the multiplier depends on something we discussed in detail in Chapter 5: the nature of the linkages in the regional economy. For some regions, the new income from the expanding export base would leak away almost instantly as its recipients send it back home to the family, buy goods made outside the area, or invest more safely elsewhere. This is, perhaps, the classic profile of the little developed region in the present day, where it is all too easy to mail money home, import fancy goods, or invest with a telephone call to a big-city stockbroker. In less accessible places or at a time, say, in the prerailway era when most places were less accessible to each other, factors of production, goods, and services would have to be provided locally, and consequently, much more of the newly won income would remain within the region to be recycled locally, generating more wealth and greater multiplier effects.

Although export base theory has provided valuable insights into the operation of the growth process and has been very widely used, it has been subjected to considerable criticism. Among the many technical problems is identifying precisely which activities form the export base since many enterprises serve both local and export markets. A further difficulty concerns regional delimitation, especially since with increasing size, the export base–residentiary ratio changes because larger regions tend to be more self-sufficient.

For these and other reasons, including the recognition that activities other than exports exert a multiplier effect, more recent work on multiplier theory has shifted toward the notion that regardless of its export or residentiary classification, all new investment in the production system will have some multiplier effects on a regional or urban economy. This is important nowadays because of the observed shift in emphasis from manufacturing and secondary activities in general to the service sectors.

Detail of the Multiplier Process: Input-Output Analysis

Tracing the actual impact of new or expanded activities through the complexities of a local economic system requires some knowledge of the interrelationships between sectors in the economy. The technique of *input-output analysis** allows us to do just that: it provides a detailed description of the interrelationships of all sectors of the economy (though in practice the level of detail is restricted by difficulties in obtaining the right kinds of data). As its name implies, input-output analysis shows what inputs of one sector are the outputs of another sector.

Table 6.1 is a hypothetical input-output table for a six-sector economy (A to F) plus the household sector (H). We can regard the household sector as the origin of labor inputs into the other sectors. Note that the order in which the sectors appear is the same in the rows as in the columns. Each entry in the table is an *input coefficient* that shows the amount of inputs from each industry needed to produce one dollar's worth of the output of any given industry. Thus for industry B to produce an output worth $1.00, it needs to purchase 26 cents of inputs from industry A, 7 cents of inputs from other firms in industry B, 4 cents of inputs from C, 2 cents from D, 11 cents from F, and 32 cents from H (these represent wages paid to labor). If we assume, for simplicity, that these coefficients remain the same over time, we can easily calculate the effects of an increase or decrease in the output of one industry on all other sectors.

The direct impact of an increase of $1.00 worth of output by sector B can be seen by reading down the B column. B is clearly a labor-intensive industry because the largest impact would be on the household sector, which is the one supplying labor. Sector A is most closely linked to B, so the effect on A would be considerable; at the other extreme, the direct impact on sector E of expansion by B would be nil. But this does not mean that E is not affected at all. As we have seen, expansion is induced in sector A, but sector A uses 6 cents' worth of inputs from sector E to produce $1.00 of output, so E would benefit *indirectly* from expansion in sector B even though there is no direct input link from E to B. In theory, therefore, the more detailed the input-output tables, the more closely we can trace the total multiplier effect (direct and indirect) of an increase in activity in one sector.

Thus the impact of increased investment not only affects the general economy of the area in question but also has a varying impact on the individual sectors of the economy. The export base and related approaches are concerned essentially with the aggregate effects of an increase in investment, income, or employment, whereas newer approaches such as input-output analysis tend to examine the detailed effect of multiplier relations on individual sectors.

Figure 6.2 describes the multiplier mechanism and input-output processes in a simplified diagrammatic manner. It depicts the entry of a new industrial

---

* This is a relatively complex technique, the technical details of which need not concern us here. Sound and very readable introductions are provided by Hewings (1985) and Miernyk (1965).

Table 6.1   HYPOTHETICAL INPUT COEFFICIENT TABLE (DIRECT PURCHASES PER DOLLAR OF OUTPUT)

Industries producing	Industries purchasing (in cents)						
	A	B	C	D	E	F	H[a]
A	16	26	3	5	13	13	19
B	8	7	18	3	8	18	24
C	11	4	21	3	13	7	7
D	17	2	5	21	16	9	6
E	6	0	3	36	8	4	12
F	3	11	18	15	5	13	11
H	25	32	18	13	18	20	1

[a] H represents the household sector (i.e., labor inputs).

Source: W. H. Miernyk (1965), *The Elements of Input-Output Analysis* (New York: Random House), table 3.4. Copyright © 1965 by Random House, Inc. Reprinted with permission.

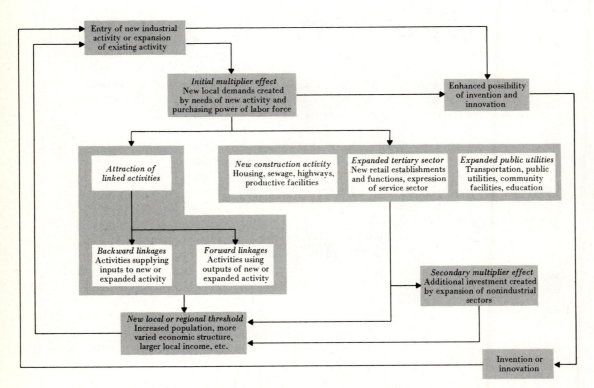

Figure 6.2 The multiplier mechanism and the process of circular and cumulative growth. *Source:* After A. Pred (1966), *The Spatial Dynamics of U.S. Urban-industrial Growth* (Cambridge, Mass.: MIT Press), fig. 2.1. Copyright © 1966 by the Massachusetts Institute of Technology. Reprinted with permission.

activity into an area and shows how this generates a set of new local demands. These may be derived from the requirements for local products by the factory itself, or they may be generated by the purchasing power associated with the arrival and settlement of the additional labor force. In this way, the successful growth area draws in a whole spectrum of new businesses. An early boom in the construction industries is followed by new developments in the service, trade, and transportation sectors. Industries supplying needed inputs to the initial developers may then follow (*backward linkages*), and these, too, create more local demand through their own needs for inputs and the increments to the area's income generated by the possibility of using the outputs of the new or expanded activity (*forward linkages*). The entire process is cumulative, each new development generating additional multiplier effects to draw in new enterprises. This trend is reinforced as the growth of the area increases (in terms of population, income, and range of economic activities). In effect, new industrial and commercial thresholds are achieved, permitting the influx of larger-scale activities, each generating its own set of multipliers. As Figure 6.2 shows, the expansion of nonindustrial activities generated by the initial multiplier itself creates a secondary multiplier, which adds to overall growth and contributes toward the attainment of even higher thresholds.

As we have seen, however, not all the incomes earned locally are spent locally as new incomes are leaked away. Furthermore, as we have shown, different production systems generate different linkages within a region. The precise nature of the local multiplier effect will therefore depend on the particular characteristics of the new industry's production function and the proportion of the total induced income spent locally compared to that leaked through the purchase of goods produced elsewhere and imported.

The complexity of the multiplier and input-output mechanisms makes their empirical study extremely difficult. A particularly bold attempt to do so was Isard and Kuenne's classic 1953 study of the impact on employment in the greater New York–Philadelphia region of a proposed large-scale increase in steel production in the region. Table 6.2 summarizes their results and is included here because it illustrates a number of interesting features relevant to our discussion. Column 1 gives estimates of the inputs required from each of the 45 sectors by the new steel and steel-fabricating activities. Obviously, inputs from the iron and steel industry itself are particularly important, and the new installation requires $121,170,500 of iron and steel inputs (13.4 percent of all inputs), but many other sectors are also significantly affected. However, not all the input requirements will be produced locally, as column 2 shows. Most services (rows 34–44) will be supplied locally as the high percentage values in column 2 reveal, but the extent to which other inputs are supplied locally depends on the area's existing economic structure. Column 3 shows how much each local sector must expand to keep the new steel production in operation. But this expansion itself generates further input requirements, and a second round of expansion is generated (column 4).

Column 6 shows the cumulative result of six rounds of expansion. The overall impact of installing new steel capacity was calculated as

Table 6.2 DIRECT AND INDIRECT REPERCUSSIONS OF THE INSTALLATION OF NEW STEEL CAPACITY, GREATER NEW YORK–PHILADELPHIA REGION

	Input requirements of initial steel and steel fabricating activities (in $ thousand) (1)	Minimum percentage of input requirements to be produced in area (2)	First-round expansions in area (in $ thousand) (3)	Second-round expansions in area (in $ thousand) (4)	Third-round expansions in area (in $ thousand) (5)	Sum of round expansions in area (in $ thousand) (6)	Total new employees corresponding to round expansions (7)	Total new employees in initial steel and steel fabricating activities (8)	Overall total of new employees (9)
1. Agriculture and fisheries	50.0	0	0.0	0.	0.	0.	0		0
2. Food and kindred products	294.6	60	176.8	17,660.	8,249.	42,492.	1,833		1,833
3. Tobacco manufacturers	0.0	0	0.0	0.	0.	0.	0		0
4. Textile mill products	3,864.7	10	386.5	406.	39.	1,280.	142		142
5. Apparel	1,285.6	75	964.2	10,124.	3,461.	21,155.	2,302		2,302
6. Lumber and wood products	5,610.7	5	280.5	93.	36.	450.	64		64
7. Furniture and fixtures	1,753.4	33	578.6	802.	198.	2,000.	234		234
8. Paper and allied products	4,818.7	40	1,927.5	1,674.	1,297.	6,574.	426		426
9. Printing and publishing	425.5	90	383.0	5,929.	3,014.	14,617.	1,667		1,667
10. Chemicals	10,626.4	45	4,781.9	3,599.	1,630.	12,077.	601		601
11. Products of petroleum and coal	10,936.6	25	2,734.2	2,547.	1,118.	7,634.	228		228
12. Rubber products	8,381.5	15	1,257.2	355.	102.	1,879.	169		169

#	Industry								
13.	Leather and leather products	647.7	20	129.5	679.	194.	1,371.	150	150
14.	Stone, clay, and glass products	9,031.7	15	1,354.8	441.	139.	2,083.	268	268
15.	Iron and steel	121,170.5	50	60,585.3	13,566.	2,965.	78,335.	6,093	17,759
16.	Nonferrous metals	33,997.4	20	6,799.5	1,667.	381.	9,063.	505	505
17.	Plumbing and heating supplies	3,192.4	25	798.1	248.	50.	1,189.	118	3,758
18.	Fabricated structural metal products	3,480.7	40	1,392.3	312.	33.	1,809.	151	1,571
19.	Other fabricated metal products	31,770.9	40	12,708.4	2,146.	561.	16,121.	1,537	11,597
20.	Agricultural, mining, and construction machinery	3,651.3	5	182.6	46.	11.	251.	22	729
21.	Metal-working machinery	7,389.1	25	1,847.3	270.	43.	2,210.	289	2,705 / 2,994
22.	Other machinery (except electric)	28,463.6	40	11,385.4	2,675.	551.	15,384.	1,486	28,607 / 30,093
23.	Motors and generators	11,265.9	20	2,253.2	226.	42.	2,560.	301	
24.	Radios	4,562.2	30	1,368.7	428.	101.	2,026.	192 ⎱	10,392 ⎱ 12,312
25.	Other electrical machinery	21,773.9	50	10,887.0	2,011.	432.	13,903.	1,427 ⎰	
26.	Motor vehicles	50,530.8	10	5,053.1	742.	260.	6,421.	389	8,770 / 9,159

(continued)

Table 6.2 DIRECT AND INDIRECT REPERCUSSIONS OF THE INSTALLATION OF NEW STEEL CAPACITY, GREATER NEW YORK–PHILADELPHIA REGION (*continued*)

	Input requirements of initial steel and steel fabricating activities (in $ thousand) (1)	Minimum percentage of input requirements to be produced in area (2)	First-round expansions in area (in $ thousand) (3)	Second-round expansions in area (in $ thousand) (4)	Third-round expansions in area (in $ thousand) (5)	Sum of round expansions in area (in $ thousand) (6)	Total new employees corresponding to round expansions (7)	Total new employees in initial steel and steel fabricating activities (8)	Overall total of new employees (9)
27. Other transportation equipment	2,605.5	20	521.1	276.	69.	958.	117	4,605	4,722
28. Professional and scientific equipment	3,221.4	50	1,610.7	801.	287.	3,123.	416		416
29. Miscellaneous manufacturing	5,116.8	60	3,070.1	2,888.	982.	8,418.	845	6,108	6,953
30. Coal, gas, and electric power	7,767.0	50	3,883.5	1,843.	2,693.	11,079.	1,100		1,100
31. Railroad transportation	13,575.8	75	10,181.9	6,010.	2,390.	21,532.	3,308		3,308
32. Ocean transportation	457.3	75	343.0	331.	170.	1,021.	110		110
33. Other transportation	4,179.4	95	3,970.4	8,422.	2,836.	19,694.	2,394		2,394

No.	Industry									
34.	Trade	13,969.8	95	13,271.3	36,585.	11,855.	83,642.	13,874	13,874	
35.	Communications	1,790.7	90	1,611.6	2,409.	1,283.	7,305.	1,191	1,191	
36.	Finance and insurance	3,086.2	90	2,777.6	9,472.	5,062.	25,252.	2,329	2,329	
37.	Rental	3,018.8	95	2,867.9	26,222.	9,603.	55,680.	909	909	
38.	Business services	5,338.5	95	5,071.6	2,385.	2,406.	13,384.	1,305	1,305	
39.	Personal and repair services	396.9	95	377.1	14,399.	5,688.	24,212.	4,443	4,443	
40.	Medical, educational and nonprofit organizations	000.0	90	0.0	9,811.	2,160.	17,271.	4,370	4,370	
41.	Amusements	000.0	90	0.0	3,677.	1,066.	6,591.			
42.	Scrap and miscellaneous industries	8,388.2	50	4,194.1	2,054.	727.	7,411.	1,100	771	
43.	Undistributed	103,638.6	50	51,819.3	5,875.	6,019.	69,236.			
44.	Eating and drinking places	000.0	95	0.0	16,916.	3,903.	29,551.	7,208	3,705	
45.	Households	348,281.0	82	285,590.4	63,002.	80,894.	509,578.			
	Total	903,807.7		521,377.2	282,024.	164,400.	1,177,822.	70,089	88,680	158,769

Source: W. Isard and R. Kuenne (1953). "The Impact of Steel on the Greater New York–Philadelphia Region," *Review of Economics and Statistics* 35:297, table 5. Reproduced with permission.

$1,177,822,000. The impact in employment terms is as shown in columns 7, 8, and 9. The total impact is an increase of 158,769 employees. Of these, 88,680 are the direct result of the steel expansion itself, but a further 70,089 jobs are created by the induced expansion of the other sectors. Figure 6.3 places this new steel installation in the context of our cumulative growth model of Figure 6.2. Each box in the diagram is drawn at a scale proportional to the increased employment in each sector that results from the new steel installation.

Propulsive Industry: Key Triggers to Cumulative Growth

Some economic activities exert a more powerful effect on development in an economic system than others. But we need to be careful in distinguishing between the impact of new or expanded activity on the economy and its impact on the area in which the expanded activity is located. The two should not be confused, though they often are. If we examine the development of an economy, we can observe that certain sectors are especially important (though their precise identity is likely to change over time). The French economist François Perroux (1955) identified these sectors as poles of growth.*

Writing of economic growth in general, Perroux (1955) claimed, "Growth does not appear everywhere at the same time; it becomes manifest at points or poles of growth, with variable intensity; it spreads through different channels with variable terminal effects on the whole economy" (p. 93). In Perroux's terms, then, *poles* are industries or firms, not geographic locations. We can perhaps best envisage them as sectors in an input-output matrix of the kind discussed earlier. Polarization in this sense depends on the development of a *propulsive industry or firm* that seems to have certain important characteristics. First, the industry or firm should be relatively large if it is to generate sufficient direct and indirect effects, though size alone is not sufficient. Second, it should be relatively fast-growing. Third, it should have a high intensity of input-output relationships (or linkages) with other industries or firms in order for the effect of its growth to be transmitted. Fourth, it should be innovative.

Industries or firms with these characteristics are likely to be the leaders, the poles around which the economy clusters. But it does not necessarily follow that geographic clustering will occur in a direct one-to-one relationship. The geographic impact of an economic growth pole will depend on a number of factors, particularly the intensity of the local linkages generated and the importance of agglomeration economies to the activity in question. Overall generalization is difficult on the precise spatial impact of the multiplier

---

* The term *growth pole* has been widely used and abused. Though completely nongeographic in Perroux's original conception, it became transformed into a geographic concept, particularly by regional planners. It is better to use the term *growth center* for geographic applications, reserving the term *growth pole* for its original meaning. For a comprehensive discussion of growth poles, see Darwent (1969), Hansen (1967), Lasuen (1969), Hermansen (1972), and Higgins and Savoie (1988).

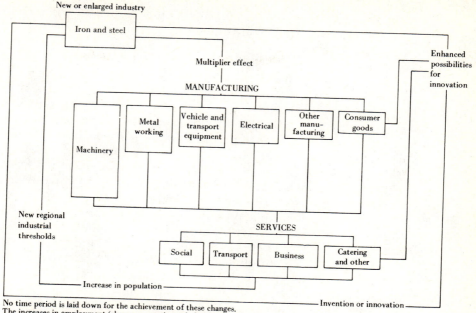

Figure 6.3 The multiplier effects of the location of an integrated steel plant in the New York–Philadelphia area. *Source:* F. E. I. Hamilton (1967), "Models of Industrial Location," in R. J. Chorley and P. Haggett (eds.), *Models in Geography* (London: Methuen), fig. 10.15.

process, especially since it forms only a part of the complex cumulative growth mechanism. However, there do appear to be significant spatial polarizing influences present in the working of the multiplier, influences that are reinforced by other considerations, in particular the operation of scale factors and the geographic clustering of innovations. Let us look briefly at each of these in turn.

Scale Factors and Cumulative Growth

Most writers looking at the way in which growth and development tend to become concentrated in the same companies, industries, regions, or nations sooner or later turn to economies of scale as a basic determinant. The old adage expressed variously as "much makes more" or "them as has gits" seems to imply that the bigger and more powerful the object of the comment, the more likely it is that it will grow bigger and still more powerful. In short, they are commenting on the cumulative properties of growth.

In earlier chapters, we tried to show how locations of particular initial advantage drew people, money, and machines. These, in their turn, attracted more investment through the demands they exerted and others have come in to satisfy. Each successive phase of development in a growing region creates an expanding source of demand for both capital and consumer goods. Bearing this in mind, let us now apply what we learned in Chapter 5 about internal and external economies of scale to see how these demand increases influence regional growth. What we shall discover is another force for cumulative growth.

Take a typical firm in an undeveloped but developing region. It has assembled factor inputs to reap the anticipated profits. Assuming a U-shaped scale-cost curve at the plant (see Figure 3.7b), it is reasonable to assume that at this early stage of development, scale economies will be forthcoming as output builds up. As this happens, unit costs will fall sharply as the fixed costs of the initial startup are spread over more units of output. With any luck, this will enable the firm to set lower delivered prices to its customers. These may stimulate still more demand and raise output levels further to approach the minimum efficient scale for the particular plant under current technology. For a booming export sector plant with perhaps a temporary monopoly of the market (perhaps it controls some new breakthrough in technology), the rise in output may be sharp enough to take it rapidly through the threshold levels needed to support more advanced and even more cost-efficient machinery. At the level of the single producing unit, then, *internal scale economies* will develop as output rises. But as we saw in Chapter 5 in the context of external economies of scale, other industries will benefit from the internal economies of the growing plant. Depending on the type and strength of its linkages with other plants and businesses in the region, the growth impulses derived from one plant will be passed on to others.

If the plant in question is one of the region's export leaders, attracting large income from demand outside the region but still dependent on strong local linkages, it would exert a propulsive force on the regional economy. Let us follow the situation through. Suppose that a mining concern subcontracts work to local truckers to carry materials from the mine to the processing plant. Other subcontractors might provide maintenance services, security patrols, workers' canteen facilities, accountancy services, and so on. In addition, power, gas, and water might be drawn from local utility companies, and a contribution to local taxes would pay for public service of various kinds. Suppose that output rises rapidly in the mine in response to external demand, which responds well to the firm's competitive prices. What happens to the subcontractors and linked services? Demands on them clearly increase. Perhaps for the trucking company it presents the opportunity to buy for the first time a giant mineral transporter with ten times the capacity of its existing equipment. Ten potentially costly trucks can be written off, and the wages of nine drivers can be saved—the trucking company moves to a new, vastly more efficient scale consistent with the new level of output demanded from it. For the security subcontractor, physical expansion permits them to buy an all-purpose jeep covering a wider area more efficiently. Here, too, one or two workers may be dispensed with as the more mechanized operation comes into being. The point is certainly clear: increased demand by the propulsive sector industry raises the level of output in the first round for its linked suppliers, giving them effective external economies of scale and making them more cost-efficient. They, too, pass on further scale economies by exerting extra demand in the second round to still another group of enterprises. The impulse from the growth sector then works its way through that part of the regional economy to which it has income links, offering others the opportunity to benefit from external economies of scale. Benefits are passed on to utility companies—perhaps it is worth building a new, more efficient gas plant or power station. The demand for public-sector services like highways, sanitation works, and schools for the children of the expanding work force at the mine is similarly increased, and the region's tax base (ideally) increases with it. Thus through the general impact of external economies of scale, the rising income derived from more efficient production in one sector is passed through the economy. It passes through by means of the multiplier effect as gains in one sector rebound on others. What the scale mechanism does then is to give an extra twist to the income benefits of the multiplier.

As Weber pointed out, scale economies also offer a stimulus toward spatial agglomeration when it comes to deciding a location for new investment. More than this, however, each new investment in an area of agglomeration tends to enhance cost efficiency (it generates economies of scale internal to the region) and, through the multiplier-accelerator-scale mechanism, strengthens its pull on future investment decisions. This continues until some point of diseconomy is reached and deglomerative forces set in.

At least in the early stages of growth, then, substantial cumulative, self-multiplying forces are at work in the impact of scale economies on economic development in space. The early progression tends to be exponential, with the rate of growth a positive function of the size already attained. But at a later stage, growth rates fall as scale diseconomies begin to appear. In this context, we can again return to the concept of the *threshold*, which we introduced early in Chapter 1. Recall that the term was defined as the minimum demand necessary for a firm to function. In terms of the economic development of an area, as its size increases, it provides increasingly higher thresholds for businesses to operate. More specifically, however, some writers argue that we can envisage a certain threshold—a critical minimum size—that is necessary for economic growth to become self-sustaining. In the case of urban areas, for example, Wilbur Thompson (1965) describes such a threshold as one

> short of which growth is not inevitable and even the very existence of the place is not assured, but beyond which absolute contraction is highly unlikely, even though the growth rate may slacken, at times even to zero. In sum, at a certain range of scale . . . some growth mechanism, similar to a ratchet, comes into being, locking in past growth and preventing contraction. (p. 22)

Thompson goes on to identify a number of possible reasons for the survival of the very large urban center. Large centers tend to be more economically diversified than small centers, thus cushioning themselves against decline associated with obsolescent activities. Large centers tend to be politically more powerful—they represent more votes—and as such they may be able to acquire government aid in times of adversity. (Thompson was, of course, writing long before the economic crises of the mid-1970s and the resulting urban and regional problems, about which we shall have much more to say in Part Two of this book.) Whether this is true or not, the massive fixed capital of large cities makes total contraction unlikely. As Thompson observed, "No nation is so affluent that it can afford to throw away a major city" (p. 23). The very large urban center is also self-justifying as a product market, with its concentration of both economic activities and consumer population.

Although there is a good deal of apparent logic in the argument for the existence of regional thresholds, there has been singularly little success in attempts to define them more closely. Nevertheless, it is clear that scale, once achieved, is a powerful influence on subsequent growth. But such influence is more than simply economic; it also has behavioral connotations, particularly in terms of the concentration of innovative activity.

INNOVATION AND CUMULATIVE GROWTH

Technological change plays an extremely important—indeed vital—part in the process of economic development. It is the volatile dynamic factor that permits a constant reevaluation of the production possibilities of various factor combinations, offsetting the appearance of diminishing returns. In short, technological improvements allow for increasing productivity and are a key part of the mechanism for increasing returns to scale.

As we saw in Chapter 4, invention and innovation constitute the ingredients of technological progress. Invention may be autonomous or induced—in the first case largely a random process, in the second the result of deliberate expenditure of resources and effort. Innovation represents the adoption of inventions and their application to the actual production process and is heavily dependent on the availability of investment funds and the right entrepreneurial climate.

Both induced invention and innovation are often closely associated with the existence of successful, expanding economic systems possessing the appropriate resources and entrepreneurial attitudes. Even autonomous invention has a greater probability of occurrence under conditions favoring high levels of human interaction and in situations where inventiveness is socially acceptable—that is, in growth situations.

Thus a growing, successful economic system, say, an industrial city, draws to it the ideas of spatially dispersed inventors searching for sponsorship, pulls in the skills of migrants, invests its own funds in the search for invention, and uses its accumulating capital and labor to apply this flood of new technology. Pred (1966) described this situation as it applied to urban-industrial centers in the United States toward the end of the nineteenth century and concluded that

> new enlarged urban industries and their multiplier effects created the employment opportunities that successively attracted "active" and "passive" migrants to

the infant metropolises, and eventually led to additional manufacturing growth by directly or indirectly enhancing the possibility of invention and innovation. (p. 39)

This effect is shown in Figure 6.2, where the multiplier effects are further strengthened by the innovative process. This, then, is another of the powerful forces for cumulative growth and for the spatial concentration of growth at focal points in the network of human interaction.

Wilbur Thompson (1968) in fact suggested that, increasingly, the major advantage of the large metropolitan center is not so much its economic base in the traditional sense but rather its innovative strength as reflected in its university and research institutions and other bodies whose concern is creativity and change. He emphasizes particularly the key role of entrepreneurial skill in regional growth:

> The large urban area would seem to have a great advantage in the critical functions of invention, innovation, promotion, and rationalization of the new. The stabilization and even institutionalization of entrepreneurship may be the principal strength of the large urban area. . . . A population of 50,000 that gives birth to, say, only one commercial or industrial genius every decade might get caught between geniuses at a time of great economic trial such as the loss of a large employer, but in a population cluster of 5 million, with an average flow of ten per year, a serious and prolonged crisis in local economic leadership seems highly improbable. (p. 53)

Once again, there appears to be a positive relationship between the rate of growth experienced and the scale of development already attained. Success tends to breed success in this respect, and as Schumpeter (1939) pointed out, an important ingredient in this relationship is the particular set of behavioral attributes of the society operating the economic system. The spirit of adventure, commercial optimism, and what Hansen (1967) called "frontiermindedness" all contribute to stimulate invention and, even more important, encourage risk taking in innovation. Where such optimism meets with success, a powerful behavioral mechanism for self-cycling growth is brought into being.

TRANSMISSION OF GROWTH IMPULSES: CENTER-PERIPHERY RELATIONSHIPS

The result of the operation of a circular and cumulative process of economic development is that the geographic pattern of development is uneven. Such unevenness is structured in a particular manner into two major components:

1. A dominant *center* or core
2. A subdominant *periphery*

This twofold conceptualization is, of course, an oversimplification of the complexity of the real world, but it is a useful simplification because it helps us to focus on some of the basic processes involved in the geography of economic development.

Center and periphery are functionally related to each other in a dynamic sense. According to John Friedmann (1973):

> Core regions are defined as territorially organized subsystems of society which have a high capacity for generating and absorbing innovative change; peripheral regions are subsystems whose development path is determined chiefly by core region institutions with respect to which they stand in a relation of substantial dependency. Core and periphery together constitute a complete spatial system. (p. 67)

There is general agreement that at any single point in time, the center tends to dominate the periphery. But there is basic disagreement about the nature of the center-periphery relationship over time. Theorists who regard the market mechanism as operating to restore a situation of equilibrium or balance argue that over time, the differences between center and periphery will be reduced and perhaps even disappear. Figure 6.4 suggests how this might occur. First, the demand generated by the center for goods and services will enable peripheral areas to grow richer. This may occur simply by the outflow of money in payment for materials, goods, or services supplied by the periphery to the center. Second, the movement of labor to the center in response to new employment opportunities will in the long run create a shortage of labor in the periphery and result in a rise of wages and incomes there. Third, the need for inputs from the periphery will promote a compensatory movement of the capital accumulated in the center, seeking out more profitable investment opportunities. In other words, new production systems may be established in the periphery—possibly to process natural resources—that will themselves act as triggers to development and set in motion cumulative growth

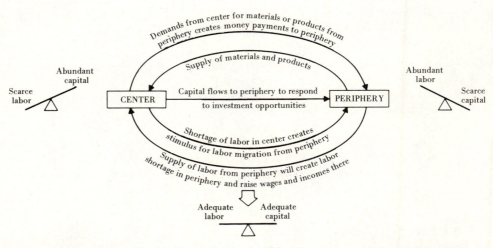

Figure 6.4 Center-periphery relationships as a balancing or an evening-out process.

processes at the production location. In effect, the equilibrium view is based on the idea that both capital and labor will tend to flow freely from areas of low return (interest, wages) to areas of high return and that in doing so, the resulting shortages of supply of capital and labor in the periphery will raise their price and lead to an eventual equalization between center and periphery. (We saw in Chapter 4, however, that infinite factor mobility is a feature of the economists' models but not necessarily of real life.)

Spread Effects

Myrdal (1957) collectively termed these kinds of processes *spread effects*. He argued that although spread effects may certainly exist, the market mechanism does not inevitably produce them in such a way that geographic imbalances in economic development are wiped out. Within developed economies, which are highly integrated economic and political systems, spread effects do operate but in a very uneven geographic manner.

Geographers have identified two important constraints on the way in which spread effects operate: the effect of geographic distance and the role of the urban hierarchy.

The Attenuating Effect of Geographic Distance   A recurring theme throughout Part One of this book is the attenuating effect of distance on the intensity of human activity. Insofar as the spread effect is simply a blanket term for a number of phenomena that have a spatial expression, we should not be surprised to find that the spread of growth impulses tends to fall off very rapidly with distance from their source. In other words, the spread effects exerted by the development of a growth node are most effective in areas close to the node itself.

We have already referred to some of these short-distance spread effects at various points in this book. The desire to minimize transportation costs has led to the exploitation of the nearest resources first, whether these are industrial raw materials or agricultural products. We saw in Chapter 3 how the increased demand generated by the expanding urban-industrial core of northwestern Europe in the nineteenth century led to a progressive outward shift of agricultural supply zones (see Figure 3.31). Similarly, industries such as iron and steel manufacture in Britain and North America began by using local ores (usually associated with coal measures) and gradually substituted long-distance sources of iron ore. As Norcliffe (1975) points out, therefore, "The tendency is to exhaust raw material supplies in the heartland first; this exhaustion process concentrates material-oriented activity in the periphery" (p. 38).

A further impetus toward spread at the edges of growth centers is the need for increasing inputs of land. As an immobile factor, land draws uses to itself. Space for residential development, for "distance-sensitive" agricultural practices like intensive horticulture, and for recreational needs is a primary requirement. For the most part, the land required to produce them is appropriated by the physical expansion of the growth node. This results in a movement outward of the bid rent and land use surfaces as the factor is drawn into use. At the margins of urban areas, rural land is converted to urban uses. As we pointed out in Chapter 1, one likely repercussion of actual or anticipated urban expansion is the inversion of the agricultural land use pattern in the innermost zones. Thus Sinclair's adaptation of von Thünen is based on the short-distance spread effects of expanding urban areas.

A similar distance-decay influence characterizes the more general spatial changes that have been occurring in urban and metropolitan centers. The large-scale exodus of both people and manufacturing activities from the center of urban areas has been predominantly directed toward the edges of such centers rather than toward new centers in the periphery, although there are certain exceptions, one of which is the demand for recreation and amenity. Where this is not the case, it is frequently a response to the efforts of government to control the evolution of the urban system rather than a reflection of the normal working of market forces. The issue of suburbanization is too wide to be taken up in detail here, although two key features can be identified. First, spatial changes in the growth and location of manufacturing industry have led to major shifts at the intraurban scale. In the period since the Second World War in particular, industry has shown a marked tendency to move out to the suburbs. The blight, congestion, and security problems of operating in the traditional inner urban areas have combined with the effects of planning policy to push industry outward. The pull of the suburban industrial park with all its amenities has drawn new and expanding manufacturing plants in all activities but those still tied to the inner city. The second feature of suburbanization of people and activities is reflected in the extension of commuting fields, indicating the enormous attractive power of the center on daily journey-to-work patterns.

In both cases, the spread is relatively limited spatially; there is a strong distance-decay component whereby the economic momentum of the center loses its impact with increasing distance. For other input needs as well as land, the sensitivity to distance of factor movements in one direction and investment flows in the other ensures that the nearer opportunities are the first to be absorbed. Of course, technological developments in the "space-adjusting" technologies of transportation and communication extend the geographic reach of spread effects. But they do so in a highly selective manner, favoring the parts of geographic space best connected into the transportation and communications network. In this sense, *relative* geographic distance, rather than absolute distance, is the crucial factor.

The Role of the Urban Hierarchy   The second way in which growth impulses are transmitted in a spatial system with a center-periphery structure is from town to town through the urban hierarchy. Again, we have already discussed the basis of interurban interaction at several points in earlier chapters. The basis of such interaction is the degree of complementarity between places as modified by the existence of intervening opportunity and distance (transferability). We observed as early as Chapter 1 that the direction of flow depends very much on the kind of urban hierarchy considered.

In the particular case of the diffusion of innovations, Pred (1973) observes that

> regardless of where a growth-inducing innovation originates or initially enters a system of cities, it is very likely to quickly appear in some or all of the system's largest units. That is, even if an innovation did not originate or first enter a system of cities at one of a few nationally dominant metropolitan areas, those places would tend to be early to adopt because of their high contact probabilities with a number of other places. . . . It is also clear that, within this probabilistic contact-field framework, large cities would benefit additionally from indirect contacts. More precisely, because of their high frequency of direct contacts with one another, large cities have the potential to quickly snap up innovations originating or entering at smaller centers with which they have no direct contact, but which lie within each other's regional sphere of influence. (p. 36)

Insofar as the less rigid hierarchical structure devised by Lösch is more flexible than that of Christaller, we would not expect the transmission of growth impulses necessarily to follow a rigid progression from high- to low-order centers (that is, passed down from big cities to towns and villages). On the contrary, the observed tendency is for both information and innovation activity to be concentrated and recycled in larger, higher-order urban centers and for interaction to be greatest between large centers. Once again, this means that the large, already existing centers have a pronounced growth advantage that they tend to conserve among themselves. In essence, therefore, the economic system is structured on the urban hierarchy and on the links and flows within it. The hierarchy articulates economic activity at any given point in time and also channels growth over time, though in more complex and conservative ways than a direct high-order–to–low-order progression would suggest.

Backwash or Polarizing Effects

Myrdal's view was that in most cases, *backwash effects* of the center on the periphery are more important and more extensive than spread effects. In other words, growth in the center tends to produce not a parallel growth in the periphery but a counterpoised decline, stagnation, or, at best, lower rates of growth. Figure 6.5 gives a simple comparison between spread and backwash effects. In Figure 6.5*a*, spread effects dominate over backwash effects: the benefits of economic development extend from the center into the periphery. In Figure 6.5*b*, backwash effects dominate, and development becomes even more concentrated at the center.

We noted three major ways in which classical equilibrium theory sees growth being transmitted to the periphery. We can take each of these and argue that each frequently operates in exactly the opposite way—as a backwash

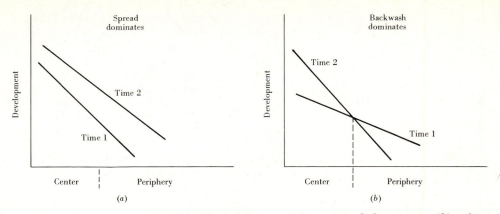

Figure 6.5 Myrdal's spread-backwash model: (*a*) When spread dominates; (*b*) When backwash dominates. *Source:* After P. Haggett (1979), *Geography: A Modern Synthesis,* 3d ed. (New York: Harper & Row), fig. 21.10.

effect instead of a spread effect—and thus increases, rather than decreases, the center-periphery disparity.

1. Purchase of Goods by the Center from the Periphery   There is no doubt that this occurs. Most of the primary resources of agriculture, minerals, and other raw materials for industry that are consumed by center businesses and population do, of course, originate in the periphery. Consequently, there is a flow of money from center to periphery in payment for such goods. Many peripheral areas also tend to have considerable tourist potential, and this, though seasonal, also promotes an income flow from the center to the periphery. The problem is that, with the exception of tourism, the demand for many of the periphery's primary goods tends to be highly inelastic. In other words, it does not change in proportion to the changing income of the center: as income levels rise in the center, its demands for primary resources do not grow at the same rate. Not only are primary goods, in general, relatively low earners, but also, as we have shown, the income earned by the periphery tends to be leaked away by being spent on goods that the periphery itself cannot provide. Such urban centers as exist in the periphery tend to be lower-order centers, which we know have a restricted variety of goods to offer. The production of high-order consumer goods—cars, refrigerators, and television sets, for example—tends to be largely in the control of business in the core centers. Thus much of the income generated by the center tends to return to the center in payment for goods and services. The scale economies existing in the center and the accumulated competitive benefits of its early start inhibit the production of such goods and services in the periphery.

2. Migration of Labor from Periphery to Center   Proponents of equilibrium argue that as labor flows from the low-wage periphery to high-wage urban centers, certain equalizing forces come into play. First, within the periphery, there is assumed to be a situation of diminishing returns to land in agricultural production. (Some degree of rural overpopulation is usually postulated.) Under these conditions, a loss of population has economic benefit. It takes the periphery from a less efficient to a more efficient combination of land and labor in its factor mix. Second, the falling supply of labor in the periphery relative to demand will produce a rise in wages. This increase in wage levels will enable the periphery to "close up" on the center, where the opposite supply-demand situation for labor will prevail.

There can be no doubt that massive migrations have taken place from rural to urban areas, particularly since the industrial revolution. Our concern here is not with the mechanism of migration itself but with its effects. There is no clear evidence that the income levels of the periphery have converged with those of the center as a result of such migration. (A glance back at Figure 5.5 will confirm this.) However, there is no unequivocal evidence that it is the selectivity of migration that makes this equalizing process fail. Many writers, including Myrdal, have argued that it is the young, the educated, the skilled, and the adventurous who tend to form an excessive proportion of migrants. As Parr (1966) wrote:

> One adverse feature of outmigration is its selective character. Generally the migrants represent the best workers (who may not even be unemployed), the younger elements (good trainee material) and would-be local administrators and entrepreneurs: in other words, the area is sapped of its vital and most needed elements. Also the age distribution of the population may well become skewed in favor of the older groups. (pp. 152–153)

Although the selectivity argument may be less clear than such statements suggest (in that it oversimplifies the migration mechanism), any loss of active population by migration will result in some loss of local income. The local tax base may fall, local spending may be lower, and consequently, local thresholds may decline, thus either inhibiting the entry of new businesses or leading to the failure of those already there.

3. Flows of Capital Between Center and Periphery Classical theory pointed to the role of capital flows as a balancing factor inducing investment and income formation in the periphery. Capital was seen to flow out in search of the high interest rates and greater marginal productivity in resource-rich but capital-poor regions. However, in reality, the greater part of the returns to investment, the profits from new production systems, and the multiplier effect from production linkages accrues in the growth center to the finance houses and corporations sponsoring and providing equipment for new development. Indeed, there is evidence of a net capital transfer from lagging to growing regions.

It is true that local suppliers may receive a boost from new projects in the periphery and local incomes and purchasing power may rise. However, in general, a good deal of the final demand impact of new development is lost from the local multiplier mechanism and finds its way back to the growth center in exchange for the consumer goods that the new levels of income bring within the range of the population. As Myrdal (1957) observes, "Trade operates with the same fundamental bias [as capital movement] in favor of the richer and progressive regions against the other regions" (p. 28). Thus where compensatory factors (such as government measures) are not in operation, migration losses and net capital transfers promote trigger effects through falling local demand, and these generate cumulative multiplier effects in the reverse direction to those that we have previously discussed. Such a mechanism may be triggered not only by the effects of migration and capital outflow but also by increasing unemployment or any force that reduces aggregate regional demand over the long run (Figure 6.6). This, in essence, is the motivating force for cumulative negative development, which Myrdal envisaged in his notion of backwash.

CUMULATIVE GROWTH AND TIME-SPACE CONVERGENCE: A SUMMARY MODEL

The primary focus of the chapters in Part One of this book has been on the spatial dimension of economic activity and on the influence of geographic space on mostly economic variables to create an economic-geographic

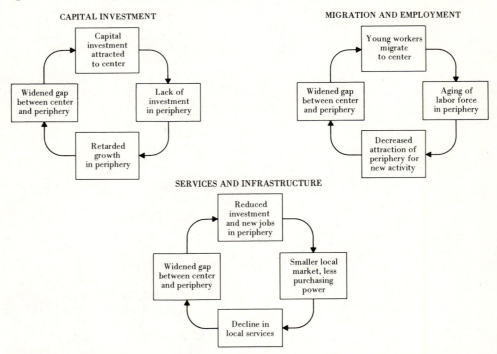

Figure 6.6 Some examples of backwash effects in the Myrdal model. *Source:* After P. Haggett (1979), *Geography: A Modern Synthesis,* 3d ed. (New York: Harper & Row), fig. 21.11.

landscape. Understanding the location of economic activities in that landscape—their *location in space*—has been based on a method of adding increasing complexity to an initially greatly simplified situation. Within such an avowedly spatial approach, the role of the space-adjusting technologies of transportation and communication is of central importance. In concluding Part One, therefore, it is useful to combine the time-space convergence model introduced in Chapter 3 with the cumulative causation model discussed in the present chapter to produce a composite model of spatial development. The emphasis in this composite model is on the interdependence between development at particular points in geographic space and the evolving transportation and communication links between these points. Figure 6.7 shows how the two separate models of time-space convergence and cumulative causation may be linked.

Two processes are in operation, both of which are cumulative and circular in their own right but are also mutually reinforcing: changes in the one influence and are influenced by changes in the other. Let us set out the stages involved, remembering that the process is circular and that any starting point is bound to be arbitrary:

1. Expansion of activity leads to a demand for better communication or accessibility—to suppliers of materials, to customers, to other functionally linked activities.
2. This leads by way of technological change and its diffusion to the adoption of transport innovations and in turn to improved communication between places. Time-space or cost-space convergence is made possible.
3. The combination of stages 1 and 2 enhances the potential for specialization in the developing economic system as the effective market is expanded. Greater spatial concentration and centralization of control are facilitated.
4. The accumulated result of the changes so far permits still greater demand for accessibility, which, when satisfied, produces even higher levels of interaction. The attractiveness of the pivotal centers is further enhanced, and new investment flows in. Inmigration is stimulated, and new thresholds of consumer demand are attained. A new cycle of mutually reinforcing changes is begun.

Recycling growth of this kind, however, produces its problems, and, in particular, resource limitations appear. In the long run, the need for additional space becomes one such major problem. Space has to be adapted to accommodate new or growing activities. As we suggested earlier, this will tend to occur on the edges of the growth center. Steps 9 and 10 of Figure 6.7 reflect this and also the possibility of expansion vertically through skyscraper construction.

The key point in this general spatial reorganization is this:

The process of spatial reorganization in the form of centralization and specialization will accelerate most rapidly at those places which stand to benefit most from increasing accessibility. In other words, transport innovations are most

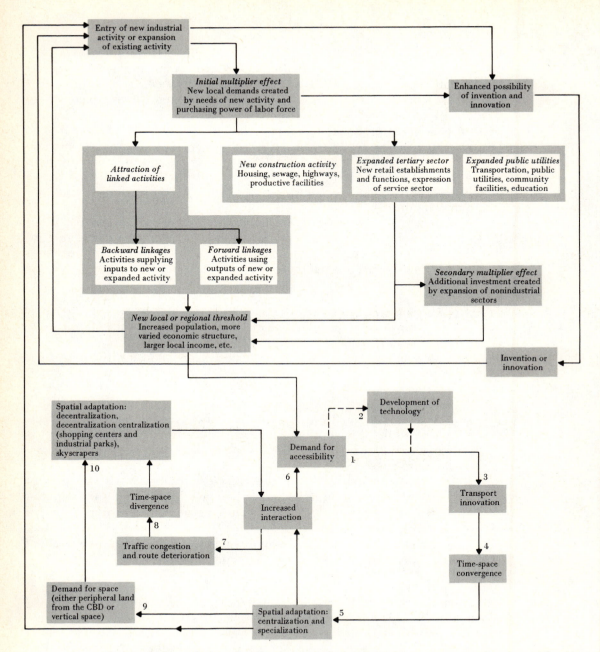

Figure 6.7 Cumulative development and time-space convergence. *Sources:* After A. Pred (1966), *The Spatial Dynamics of U.S. Urban-industrial Growth* (Cambridge, Mass.: MIT Press), fig. 2.1. Copyright © 1966 by the Massachusetts Institute of Technology. Reprinted with permission; and D. G. Janelle (1969), "Spatial Reorganization: A Model and a Concept," *Annals of the Association of American Geographers* 59: 348–364, fig. 3. Reproduced with permission.

likely between those places which will benefit most from a lessening in the expenditure of time (cost or effort) to attain needed and desirable goods and services." (Janelle, 1969, p. 357)

Once again it can be shown that places that gain some initial advantage are likely to retain this advantage, benefiting from the cumulative processes of growth and transportation change. Places that are already large and economically powerful will be the most likely beneficiaries of innovation, which will further enhance their accessibility to other important places. In this way, the already high levels of interaction between high-order, key centers are likely to be increased even further while smaller, less significant places may well be bypassed by the development of high-order transport linkages.

Again, in the interests of integration, we can usefully see Taaffe, Morrill, and Gould's classic (1963) model of transportation development as a broad summary of this twofold process of urban growth and change in accessibility. Figure 6.8 shows that as the network evolves, certain routes develop to a greater extent than others in association with the increased importance of particular urban centers.

The "ideal-typical sequence" begins (stage 1) as a pattern of scattered, poorly connected small ports along the coastline. In stage 2, two major routes penetrate inland, perhaps to tap a valuable mineral resource or to facilitate the movement of desired agricultural produce from an area in the interior to the

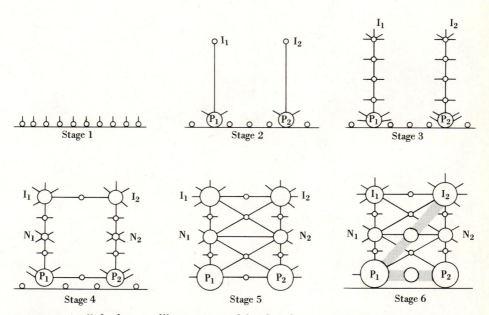

Figure 6.8 An "ideal-typical" sequence of the development of a transportation network. *Source:* E. J. Taaffe, R. L. Morrill, and P. R. Gould (1963), "Transport Expansion in Underdeveloped Countries," *Geographical Review* 53: 503–529, fig. 1. Copyrighted by the American Geographical Society of New York.

coastal market. The result of the increased interaction is to increase the size of the two terminal ports and (stage 3) to initiate new urban growth at strategic points on the routes. Stage 4 shows how, as a consequence of lateral route development, the competitive position of the two major centers and those at the interior terminal points is further enhanced. Subsequently (stage 5), almost complete interconnection may be achieved, but because certain centers are already more powerful than others, the further development of high-priority linkages (stage 6) is likely.

The overall result of these spatial changes is a constantly evolving and changing space economy in which forces of polarization tend to outweigh the forces of spread. But as Friedmann has shown, the degree of imbalance between center and periphery is very much related to a country's overall level of economic, social, and political development. He identifies four major stages in the sequence of spatial organization (Figure 6.9), stages which have certain broad parallels with the transportation mode of Figure 6.8. The diagram is self-explanatory, though we should observe that the final stage (4) implies a deliberate planning policy aimed at the total integration of the urban-economic system.

Thus, in general, we can identify in the space economy powerful forces that serve to amplify the impulses set up by initial triggers to geographic differentiation. In spatial terms, given that there are no serious blockages in factor inputs or in the willingness of society to consume the products of its leading economic subsystems, the end product of the cumulative process in a free market system tends to be the creation of localized growth centers. A characteristic feature of such centers is that beyond a certain stage of development—the threshold—they appear to exert a considerable influence on the subsequent development of the space economy while acquiring strong self-perpetuating momentum through derived advantages of their early growth. Away from the localized growth centers, the intensity and rate of development have a marked tendency to decline with distance, although their expansionary momentum results in spread effects that raise levels of development in their immediate proximity. As a result of time lags in development, the periphery often remains in nearly permanent subordination to the center, cumulative growth in the one being associated with cumulative decline in the other. Geographic space, and the constraints that it imposes, has a particularly important impact in molding the forms produced by the processes of economic development and change.

However, the notion that economic development is a cumulative process, though a powerful and useful concept, does pose some problems. It suggests that economic development is relatively smooth and progressive for the center and relatively smooth and regressive for the periphery. An alternative view, recently put forward in the geographic literature by Clark, Gertler, and Whiteman (1986), suggests that regional economic development is "a highly dynamic, rapidly fluctuating investment process in time" (p. 4). This suggests that we need to look beyond the time-space models of cumulative causation to

1 *Independent local centers, no hierarchy.* Typical preindustrial structure; each city lies at the center of a small regional enclave; growth possibilities are soon exhausted; the economy tends to stagnate.

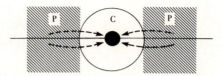

2 *A single strong center.* Structure is typical for the period of incipient industrialization; a periphery (P) emerges; local economies are undermined in consequence of a mass movement of would-be entrepreneurs, intellectuals, and labor to the center (C); the national economy is virtually reduced to a single metropolitan region, with only limited growth possibilities; continued stagnation of the periphery may lead to social and political unrest.

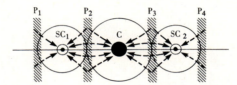

3 *A single national center, strong peripheral subcenters.* The first stage toward a solution during the period of industrial maturation; strategic subcenters (SC$_n$) are developed, thereby reducing the periphery on a national scale to smaller, more manageable inter-metropolitan peripheries (P$_n$); hypertrophy of national center is avoided while important resources from the periphery are brought into the productive cycle of the national economy; growth potential for the nation is enhanced, but problems of poverty and cultural backwardness persist in intermetropolitan peripheries.

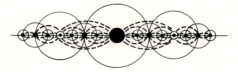

4 *A functionally interdependent system of cities.* Organized complexity is the final solution to be aimed for during the period of industrial maturation, but it will subsequently give place to other configurations; major goals of spatial organization are fulfilled: national integration, efficiency in location, maximum growth potential, minimum essential interregional balances.

Figure 6.9 A sequence of stages in spatial organization. *Source:* J. Friedmann (1966), *Regional Development Policy: A Case Study of Venezuela* (Cambridge, Mass.: MIT Press), fig. 2.1. Copyright © 1966 by the Massachusetts Institute of Technology. Reprinted by permission.

the areas of investment decisions and to the broader structure in which such decisions are set. Some alternative approaches to understanding location in space make up the chapters in Part Two of this book.

FURTHER READING

Hewings, G. J. D. (1985). *Regional Input-Output Analysis.* Newberry Park, Calif.: Sage.

Higgins, B., and D. J. Savoie (eds.) (1988). *Regional Economic Development: Essays in Honor of François Perroux.* Boston: Unwin Hyman, Part One.

Miernyk, W. H. (1965). *The Elements of Input-Output Analysis.* New York: Random House.

Myrdal, G. (1957). *Rich Lands and Poor.* New York: Harper & Row.

Pred, A. R. (1973). "The Growth and Development of Systems of Cities in Advanced Economies." In A. R. Pred and G. Tornquist (eds.), "Systems of Cities and Information Flows," *Lund Studies in Geography, Series B,* 38: 9–82.

TWO

Location in Space: Alternative Perspectives

At this point we leave behind those neoclassical insights that sustained economic geography as a discipline from the middle years of the 1960s to the later 1970s. Even now, almost two decades after the first edition of *Location in Space,* there is an attractiveness in the clarity and rigor that the neoclassical era brought to the discipline. Increasingly, however, as the 1970s unfolded, economic geographers began to find themselves confronted with issues for which location theory in its traditional form proved inadequate. For example, there was the growth of giant global enterprises. The realization that many major corporations had greater sales turnover than most nation-states and that few aspects of our lives were untouched by their activities raised questions about a theory that largely ignored them. The theoretical perspectives we have just reviewed had no place for such enterprises. There was, by assumption, no need for a theory of the firm. Similarly, the evidence of oligopoly—of small numbers of competitors responding knowingly to one another's moves—found no echo in free entry and perfect competition. These same issues had been addressed by a long and honorable tradition in behavioral economics, and economic geographers began to explore the growing literature on theories of the firm and the strategy and structure of corporations.

Taking this approach, of course, meant leaving behind many of the safe havens of neoclassical theory for relatively uncharted waters. But the evidence of daily experience, so far removed from the classical models, demanded that the voyage be attempted. In Chapters 7 and 8 we attempt to summarize some of the insights of this voyage of discovery. Inevitably this entails a retreat from the formal properties of space economy—from patterns and regularities modeled and verified. The new insights are directed more toward exploring the causality behind observed spatial outcomes, not from economic principles per se but from the intentionally rational behavior of corporate managers. Through this literature geographers have begun to build intellectual bridges to the management sciences.

Another source of growing dissonance for economic geographers during the 1970s was the discovery that neoclassical theory could not handle the kinds of spatial changes most commonly observed in the advanced capitalist nations from about 1970 onward. An era of recession had set in for advanced economies, replacing the buoyant sixties. Questions now being asked of economic geographers were about plant closings rather than openings, of employment losses rather than the multiplier effects of new ventures. In the 1960s there had been a rereading of Karl Marx's social and economic theories by a small but influential number of geographers. The coming together of new questions about decline, equity, and social justice with this rereading of Marx launched another voyage of discovery for economic geography. With orthodox economics unable to address questions of social as well as economic relations in the spatial domain, new insights were sought from the broader and longer-established field of political economy. Some of the earliest work in the industrial sphere began to have an impact in the United Kingdom, and meetings of British and American economic geographers started forming a community of interest in a Marxist-informed political economy of space and place.

Chapters 9 and 10 attempt to give a sense of this approach. They begin with a brief and highly simplified overview of Marx's social and economic theories as they relate to our subject. This is necessary because although the terms have crept into the literature during recent years, the essential context has often been stripped away. Two chapters are quite inadequate to explore what has probably been one of the most exciting and dynamic fields within the discipline during the past decade, but our objective is to make the approach more accessible to the wider audience rather than to produce a detailed professional review.

The aim of Part Two of our book, then, is to open a window on new insights and, with Part One, to provide a comprehensive overview of the changing perspectives in economic geography that have shaped our understanding of what the "space economy" is and what influences it.

Chapter
7

The Contemporary Business World: Economic Structure, Business Organizations, and the Competitive Process

THE COMPETITIVE STRUCTURE OF MODERN ECONOMIES

Whether we like it or not, we live in a competitive world. Indeed, the modern economy is competitive not only in a general sense but also at an increasing geographic scale. Whereas at one time a firm's main competitors were likely to be found in the same town or region, today they are just as likely to be found at the other end of the country or even on the other side of the world. Modern capitalist market economies, then, are intensively competitive in nature. But what form does that competition take, and what are its geographic implications?

Alternative Models of Competition

In the chapters of Part One, we adopted a particular model of competition known as *perfect competition*. A perfectly competitive industry has a number of interrelated characteristics:

1. There are a large number of producers or sellers of the product and a large number of buyers.

2. Each producer supplies such a very small quantity of the product relative to the industry's total output that the individual producer has no influence on the price received because this is set by the market. In Figure 7.1*a*, both price and quantity are set by the balance between supply and demand for the industry as a whole. This price ($32 in Figure 7.1) is the price at which each *individual* producer will see his or her product. In effect, this means that the demand curve for the *individual* producer is horizontal (Figure 7.1*b*). In other words, firms in perfectly competitive industries are price takers, not price makers. If the firm tries to set a price above the market price, the buyer will go elsewhere; if the firm cuts its price, it is likely to take a loss. This is because of point 3.

3. Costs of production are assumed to increase at a level of output that is small relative to the total market. In other words, each producer faces identical cost conditions.

4. Each firm in the industry produces an identical or homogeneous product to that of its competitors.

5. Entry into the industry for new firms is easy.

6. Each firm and each buyer is assumed to have perfect information about products and prices.

In the "pure" model of perfect competition, it is also assumed that no costs are involved in transporting goods or materials. But the essence of our spatial model in Chapter 1 and subsequent chapters was the very existence of such costs, which represent the frictional effect of geographic distance. Thus as soon as one admits geography or space into economics, one creates conditions of competition that are less than perfect. In fact, as we showed in Part One, the existence of geographic distance creates a particular kind of monopoly in which all customers within a particular geographic range of a point of supply purchase from that producer and from no other. This is the central feature of the market area models, including central place theory, discussed in Part One. So even if all the assumptions of perfect competition were to apply, perfect competition itself is impossible once the facts of geographic space and distance are introduced.

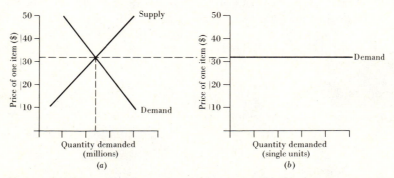

Figure 7.1 Demand and supply curves in a perfectly competitive industry.

In the economic literature, such "modified" perfect competition created by the geographic fragmentation of markets or by firms making very small degrees of differentiation in their product is known as *monopolistic competition*. There are still large numbers of producers, but because their products are not exactly homogeneous, the firm has some influence on price. In these circumstances, an individual firm faces a downward-sloping demand curve of the kind discussed in Chapter 1 (see Figure 1 in Box 1.1).

In its purest form, however, *monopoly* describes a situation characterized by three specific factors:

1. The output of an entire industry is controlled by one seller.
2. The single seller sets the price at which the good is sold to maximize profits; that is, the monopolist is a price maker.
3. Very strong barriers to entry prevent the monopolist from being threatened by potential competitors. These barriers may include economies of scale of production and product differentiation, among many others.

Just as it is difficult, if not impossible, to find real-world examples of perfect competition, so too pure monopolies are pretty rare beasts. More common is the situation in which an industry is dominated by a few very large firms. Such firms are known as *oligopolists*. Oligopolists, in fact, have some but not total monopoly power. An industry in which oligopolistic competition exists has the following characteristics:

1. A few producers dominate the industry, each one producing a significant share of the industry's total output.
2. A high degree of interdependence among such firms means that an action taken by one of them (say, reducing or increasing the selling price or moving into a new market) is likely to be met by a countermove by the others. Each firm seeks to second-guess what its competitors will do. It makes its decisions on its perceived evaluation of such likely actions.
3. Given such uncertainty, oligopolists tend to seek stability in their industries. Competition takes on forms that emphasize product differentiation (making a product seem to be different—and therefore superior—to competition through such means as advertising and packaging).
4. Hence there may be both competition and collusion in oligopolistic industries.
5. Oligopolists seek to maintain high barriers to entry to prevent the incursion of new firms that might upset stability.

Given these characteristics of oligopolistic competition, it is not surprising that considerable use has been made of game theory to aid understanding. The example used to illustrate locational interdependence in Chapter 5 is relevant in this respect and should be reviewed.

Figure 7.2 summarizes the four major alternative models of competition

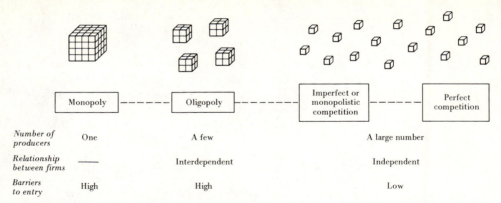

	Monopoly	Oligopoly	Imperfect or monopolistic competition	Perfect competition

Number of producers	One	A few		A large number
Relationship between firms	——	Interdependent		Independent
Barriers to entry	High	High		Low

Figure 7.2 The spectrum of different types of competition in a market economy.

along a spectrum of types. The question to be posed now is, which of them best fits the current situation in modern capitalist economies and therefore is most likely to affect the changing geography of economic activity? To answer this question, we need to examine some empirical evidence.

Trends in Business Concentration

One way of assessing the type of competitive structure that exists in an economy or in specific industries is to measure the share of sales, assets, or some other measure accounted for by a specified number of firms. This is known as the *business concentration ratio.* At the level of an entire national economy, like that of the United States, we would normally be interested in the share held by the top 50, 100, or 200 firms. At the level of a particular industry, such as automobiles or electronics, our focus would be narrower—say, the share of each industry accounted for by the leading four to eight firms.

In fact, it is not easy to obtain consistent data that allow us to establish whether or not business concentration has increased or decreased over time. The general trend throughout most of the twentieth century has been for a larger and larger share of economic activity to be performed by a relatively smaller number of large business organizations.

Figure 7.3, for example, shows that in 1910, the 100 largest manufacturing enterprises in the United States produced 22 percent of total net output; by 1970, the 100 largest firms produced 33 percent of the total. In the United Kingdom, the increase in manufacturing concentration was even steeper. The time series data do not tell us what has happened since the early 1970s. Anecdotal evidence suggests that concentration may have continued to increase, though possibly at a slower rate, and may even have peaked. But even if today's level of concentration remains at that of 1970 it would still be very high indeed. It would mean that at least 33 percent of national manufacturing output was being produced by a minute fraction—around 0.03 percent—of the total number of U.S. manufacturing firms. Without doubt, the U.S. economy (and

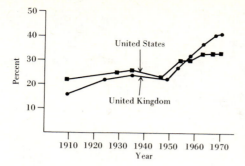

Figure 7.3 Shares of the 100 largest enterprises in manufacturing net output, United Kingdom and United States. *Source:* S. J. Prais (1976), *The Evolution of Giant Firms in Britain* (Cambridge: National Institute of Economic and Social Research, Cambridge University Press), chart 6.1.

that of the other industrialized nations as well) is highly concentrated. It is an oligopolistic economy dominated by a relatively small number of firms. As Richard Barber (1970) observed:

> The presidents of a hundred companies—a group suffficiently small to be seated comfortably in the reading room of the Union League Club in Philadelphia—represent almost as much wealth and control as large a share of the nation's economic activity as the next largest 300,000 manufacturers—a group that would completely fill four Yankee Stadiums. (p. 20)

Hunt and Sherman (1986) use a different analogy to express the same idea:

> If we were to describe America's industrial landscape, we would begin with a vast plain of millions of tiny pebbles, representing all the economically powerless, monopolistically competitive business firms. At the center of this enormous plain would rise a few hundred colossal towers, representing the important oligopolistic corporations. These few hundred towers would be so large that they would make insignificant the entire plain below them. (p. 330)

These observations refer to the entire national economy. What about individual industries? Do they all display equally high levels of business concentration? Again, there are data problems in trying to answer such questions, particularly problems of how industries are classified. For example, concentration ratios may be relatively low at the broad two-digit Standard Industrial Classification (SIC) level but very high at the five-digit level.*

Figure 7.4 shows the extent to which the four-firm concentration ratio (the percentage of industry output produced by the four largest firms in each industry) varied among U.S. industries in 1982. The variation is considerable. At one end of the scale, 92 percent of motor vehicle production was in the hands of only four firms. Within the food sector, the production of cereals was

---

* In the United States, for statistical purposes, industries are divided, first, into very broad groups such as "food and kindred products" or "primary metals industries," which are given a two-digit identification. These broad categories are then further subdivided. A five-digit industry within the "food and kindred products" group might be "ground coffee" (SIC 20951); within the "primary metals" group, an example would be "steel tubing" (SIC 33176).

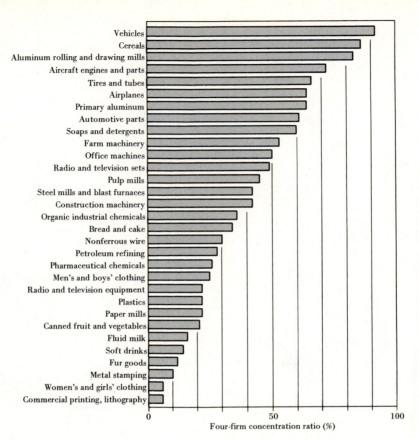

Figure 7.4 Concentration ratios in the U.S. manufacturing industry. *Source: U.S. Census of Manufactures, 1982.*

almost as highly concentrated (86 percent). At the other end of the spectrum were industries such as women's dresses and commercial lithographic printing where the leading four firms accounted for only 6 percent of each industry's output.

A high level of business concentration is not confined to the manufacturing sector. Similar patterns of increasing concentration have occurred in U.S. agriculture. There has been a steep decline in the number and relative importance of small farms and a commensurate increase in the relative importance of large farms. Between 1940 and 1982, the number of farms in the United States declined by two-thirds. The decline was greatest among small farms; in contrast, the number of farms of more than 1,000 acres increased by 62 percent. By 1982, 2.9 percent of all U.S. farms were in holdings of more than 2,000 acres, but they occupied almost 50 percent of all U.S. farmland. The fragmented, small-scale agricultural sector has been transformed into an *agribusiness* that is not only controlled by very large business corporations but is also tied closely into the manufacturing and service sectors of the economy.* Similar trends are evident in other parts of the economy, for example, in retailing. Retailing has traditionally been a small-enterprise sector, and this is still broadly true. Even so, it has become dominated increasingly by the big retail chains. More than one-fourth of total U.S. retail sales is now accounted for by the 50 leading retailers, who employ, in total, some 4 million workers.

The empirical evidence therefore suggests that probably no industries have a competitive structure at the two extreme ends of the spectrum shown in Figure 7.2. Industries in which a few leading firms are not significant (the ones with very low concentration ratios) are less numerous than those that can best be described as oligopolistic. This, then, is the dominant form of competitive structure in the economies of the United States and the other industrialized countries. However, the position is more complex than this because an important feature of the contemporary economy is the diversity of operations of many large firms. During the 1960s, in particular, many firms in the United States grew by entering many different product markets to emerge as *conglomerate enterprises*. Some of the excesses of this movement have since been reversed as firms have tended to pull back into more related areas of business activity. The fact remains, however, that most large firms in the United States are multiproduct rather than single-product operations.

This is one of the major differences between the few giant firms of the late nineteenth and early twentieth centuries and the more numerous giant firms of today. In those earlier days, if a firm had a name like U.S. Steel, U.S. Rubber, or American Tobacco, you could be pretty sure that its efforts would be entirely devoted to producing steel, rubber, or tobacco, respectively, or at least to processes very closely connected to them. Not anymore. A good deal of their energies is now devoted to a diverse range of activities such as domestic appliances, real estate, fertilizers, plastics, and many others far removed from their original activities. For such reasons, many firms have changed their names to rid themselves of too close an association with a particular product. Hence U.S. Rubber changed its name to Uniroyal, and American Tobacco has become American Brands.

The facts that the U.S. economy is predominantly oligopolistic and that most leading firms spread their energies across a wide range of product areas are significant. But the geographic significance of these developments arises from a further, closely related fact: all of the important firms and most of the medium-sized firms in virtually every sector of the economy are now multilocational or multiplant firms. Such firms consist of vast geographic networks of offices, factories, research laboratories, and the like, controlled ultimately from a single corporate headquarters. All 500 of the leading U.S. corporations listed each year in the *Fortune 500* directory operate multiple plants across the entire North American continent. What is more, they are all strongly multinational in their operations.

---

* Wallace (1985) presents a detailed discussion of agribusiness from a geographic perspective.

The rapid growth of multinational—or more appropriately, *transnational corporations* (TNCs)—is one of the most significant developments of the past few decades.* Just as individual national economies have become dominated by a small number of giant firms, so too the world economy is increasingly dominated by a few hundred transnational corporations. Roughly one-fourth of total world manufacturing production (outside the centrally planned economies) is performed by TNCs, even though, numerically, they are but a minute fraction of the total population of business enterprises in the world economy. A mere handful of giant global corporations is especially dominant. Almost three-fourths of U.S. foreign direct investment is performed by fewer than 300 firms. The scale of such firms is often compared to that of entire nation-states: "Of the 100 largest economic units in the world today, half are nation-states and the other half TNCs" (Benson and Lloyd, 1983, p. 77). Table 7.1 shows the tip of this iceberg: the leading industrial corporations in the world in 1987. Each of these companies has operations in scores of countries across the globe.

The story is similar in other sectors, such as finance, advertising, hotels, as well as in primary commodities like food, agricultural raw materials, and ores and minerals. For example, world commodity trade is especially strongly dominated by large transnational corporations, as Table 7.2 reveals. Similarly, we can see the emergence of transnational service conglomerates and, indeed, of firms whose national and international operations encompass activities across all economic sectors, from agriculture and mining through manufacturing to services (Claimonte and Cavanagh, 1984).

We live, then, in a world of oligopolies, a world in which hundreds of thousands of small firms still exist but a relatively small number of large enterprises predominate. In general, business concentration has increased over the course of the twentieth century across virtually all economic sectors. Most sectors are more concentrated in a structural sense than they were half a century ago. But this does not necessarily mean that the level of business concentration will continue to increase at the same rate or even at all. The logical end point would be total dominance by one firm in each industry: complete monopoly. Extrapolating historical trends can be dangerous. For example, in a classic study of the 1930s, Berle and Means (1932) argued that if certain observed trends of that time were to continue, the 200 largest industrial corporations would control 100 percent of American industrial assets by 1972. As we have seen, this has not happened. Indeed, in some cases, business concentration may actually decrease over time. But this does not so far seem likely to alter the prevailing picture of an oligopolistic economy.

The Concept of a Segmented Economy

We have stressed the extent to which most sectors of the economy are dominated by a relatively small number of large firms. But this does not mean that

---

* For a detailed discussion of transnational corporations in the context of global industrial change, see Dicken (1986).

Table 7.1   THE WORLD'S LEADING INDUSTRIAL CORPORATIONS, 1987

Corporation	Major activity	Headquarters country	No. of employees
General Motors	Motor vehicles	United States	813,400
Royal Dutch/Shell	Petroleum	Netherlands and United Kingdom	136,000
Exxon	Petroleum	United States	100,000
Ford	Motor vehicles	United States	350,320
IBM	Computers	United States	389,348
Mobil	Petroleum	United States	120,600
British Petroleum	Petroleum	United Kingdom	126,020
Toyota	Motor vehicles	Japan	84,207
IRI	Metals	Italy	422,000
General Electric	Electronics	United States	302,000
Daimler-Benz	Motor vehicles	West Germany	326,288
Texaco	Petroleum	United States	50,164
AT&T	Electronics	United States	303,000
Du Pont	Chemicals	United States	140,145
Volkswagen	Motor vehicles	West Germany	260,458
Hitachi	Electronics	Japan	161,325
Fiat	Motor vehicles	Italy	270,578
Siemens	Electronics	West Germany	359,000
Matsushita	Electronics	Japan	134,764
Unilever	Food	United Kingdom and Netherlands	294,000
Chrysler	Motor vehicles	United States	122,745
Philips	Electronics	Netherlands	336,672
Chevron	Petroleum	United States	51,697
Nissan	Motor vehicles	Japan	105,443
Renault	Motor vehicles	France	188,936

Source: Fortune, 25 April 1988; 1 August 1988

Table 7.2   CONTROL OF WORLD COMMODITIES TRADE BY TRANSNATIONAL CORPORATIONS

Commodity	Share of world trade of 15 largest TNCs (%)
Wheat	85–90
Coffee	85–90
Bananas	70–75
Pineapples	90
Forest products	90
Tobacco	85–90
Crude petroleum	75
Copper	80–85
Tin	75–80

Source: After F. F. Clairmonte and J. H. Cavanagh (1983), "Transnational Corporations and the Struggle for the Global Market," Journal of Contemporary Asia 13: 456.

smaller firms no longer exist. On the contrary, there are many hundreds of thousands of small firms throughout the economy. Thousands are born every year; thousands of small firms also die every year. Both birth rates and death rates in the small-firm sector tend to be very high. But just as not all large firms have identical characteristics, there are important differences within the small-firm population. This leads us to the notion of the segmented structure of the contemporary economy. In 1968, Robert Averitt described the U.S. economy as a *dual economy,* an economy consisting of two major types of business enterprise:

1. *Center firms:* the large, dominant firms of the economy, complex in organization and strong in resources, power, and influence.
2. *Periphery firms:* the relatively small firms of the economy, generally simple in organization and relatively weak in resources, power, and influence*

The important point about this distinction between center and periphery firms is that it does not relate only to size but to important functional characteristics of the two types as well. Small firms, for example, cannot simply be equated with a particular volume of output or number of employees. A small firm, more realistically, has the following features:

1. An identity between management and ownership
2. An absence of specialized staff for separate functions and of facilities designed specifically for research and analysis
3. An inability to finance itself by floating securities or to secure its funds through sources such as investment bankers
4. A personal relationship between owners and employers and customers
5. Affiliation with a local community
6. Chief dependence for its market on the local area

"These factors, when present in combination, make a small business recognizable even when its volume of business is substantial" (Kaplan, 1948, p. 37).

The small firms making up the periphery economy are not only small but also, in many cases, functionally subordinate to and dependent on center firms. Averitt (1968) divides periphery firms into three main types based on this relationship:

1. *Satellite firms* are functionally linked to center firms on either the input or the output side. Often they are specialist subcontractors, performing functions that the center firm does not choose to perform

---

* The idea of a dual economic structure also forms the basis of John Kenneth Galbraith's analysis of the U.S. economy, although his terminology differs from Averitt's. Galbraith used the terms *planning system* and *market system* in a way that suggests that they are the equivalent of Averitt's *center* and *periphery* firms, respectively.

itself. We shall discuss this kind of interfirm relationship in more detail later.

2. *Loyal opposition or competitive fringe firms* compete with center firms. They tend to have a relatively short life, often being absorbed by center firms.

3. Averitt regarded *free agent firms* as occupying a residual category operating "on the fringes of the raw material processing–finished manufacture–retailing continuum, filling in production cracks and crannies. . . . Manufacturing free agent firms often specialize in producing unique articles or unique batches of articles" (p. 66). Such firms may be able to operate, for example, if markets for particular products are geographically dispersed. Growing economies invariably contain numerous interstices—productive niches—that such small firms may be able to exploit.

The geographers Michael Taylor and Nigel Thrift (1982, 1983) have developed the concept of segmentation in greater detail. Their classification of firms is shown in Figure 7.5. Their scheme shares some common features with Averitt's, particularly the satellite and loyal opposition categories of small firms. But their scheme is more detailed and allows for greater variety of firm type within both major segments. Taylor and Thrift divide their smaller-firm segment (roughly equivalent to Averitt's periphery sector) into three major segments, each of which is further subdivided:

1. The *laggard* segment represents the "tail" of the small-firm population. Turnover of firms is high; a good many laggard firms continue to survive, though few seem to grow. Two types of laggard small firm are those owned by craftsmen (proprietors with a working-class

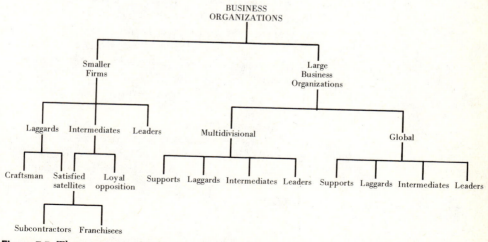

Figure 7.5 The segmented economy. *Source:* M. J. Taylor and N. J. Thrift (1983), "Business Organization, Segmentation and Location," *Regional Studies* 17: 445–465, fig. 4.

background) and satisfied smaller firms (whose owners consciously refrain from trying to get any bigger). Very often the life span of the laggard smaller firm is equivalent to that of the owner; the firm dies when the owner dies or retires.

2. The *intermediate* segment consists of Averitt's categories of satellite and loyal opposition firms.

3. The *leader firm* segment represents the dynamic element of the smaller-firm population, the innovators of new products and processes, the finders of profitable, if sometimes ephemeral, niches in the marketplace. This segment has the greatest potential for growth into larger firms but also a high rate of failure. Often such leader firms are taken over by bigger firms.

The other major business segment, which Taylor and Thrift call the large-business segment, can be divided into two subsegments: the *multidivisional corporation* segment and the *global corporation* segment. As Figure 7.5 shows, both of these can be further subdivided into leaders, intermediates, laggards, and supports. In effect, Taylor and Thrift are arguing that each large firm forms its own center and periphery.

One of the major forces generating both geographic and nongeographic change in the contemporary economy is, of course, technological change. Differing attitudes toward technology explain some of the differences between the various business segments described by Averitt and by Taylor and Thrift. We shall have more to say about technology in this and subsequent chapters, but at this point, it is useful to introduce another way of classifying business organizations: in terms of their attitudes toward technology. The scheme, devised by Christopher Freeman (1974), consists of six types (or segments) of firm:

1. *Offensive* firms aim for technical and market leadership through heavy research-and-development spending on the introduction of new products. They are extremely research-intensive.

2. *Defensive* firms also invest heavily in R&D but mainly to enable the firm to react and adapt to technological changes introduced by competitors. They exist in most oligopolistic markets.

3. *Imitative* firms pursue a follower strategy in established technologies. To succeed, they must have certain advantages, such as a captive market or low production costs.

4. *Dependent* firms are essentially satellites, tied to a dominant large-firm customer. Often they are subcontractors, frequently small or medium-sized. Technical initiative is often derived from their customers.

5. *Traditional* firms have unchanging technologies and traditional markets that are often localized. They are predominantly small firms—the "peasants" of industry.

5. *Opportunist* firms are headed by entrepreneurs who identify a new market opportunity (niche) and exploit it quickly.

We have spent some time elaborating on the various ways in which modern economies are structured in order to demonstrate the considerable diversity which exists. Such structural diversity, as we shall see, is also reflected in varying degrees of geographic diversity.

Businesses as the Primary Agents of Spatial Change

You may feel that little in this chapter so far has been obviously geographic. But to understand the geography of the contemporary economy, we need some understanding of the way in which the economy is structured in a competitive sense. Our position in this chapter is that we particularly need to understand how business organizations operate. As Taylor (1984) observes, "The business organization is the basic unit of the economy, the point of production, the crucible within which both macro- and micro-forces meet and are played out" (p. 8). Hence the business organization must occupy a primary position in our attempts to understand the economic geography of the world in which we live. We have established so far in this chapter that despite the important roles played by millions of small firms, the prime movers in the economy are the large corporate enterprises. Their behavior is effectively monopolistic; their competitive environment is characterized by oligopolistic structures. These large firms operate in both factor markets and product markets, sitting astride supply and demand, drawing in production factors and part-finished goods from other firms, supplying finished and part-finished goods to other firms and final consumers. In effect, then, the economy, both geographically and non-geographically, is articulated primarily through the decisions and actions of such enterprises. Clearly, we need a conceptualization of the firm to understand just how such articulation occurs. Again to use Taylor's (1980) words, "The geography of enterprise requires a theory of enterprise" (p. 151).

In this spirit, we can set out a series of basic propositions on which we can base our specific spatial questions:

1. The oligopoly is the dominant form of competitive structure in modern capitalist economies.
2. The major objects of interest are the "prime mover" corporate enterprises—small in number but immense in economic power.
3. The tools of analysis most appropriate for the study of the large oligopolistic enterprise can be derived from organizational, behavioral, and management theories.
4. The appropriate form of analysis is based on dynamic change in a complex and uncertain competitive environment.
5. The internal hierarchical structure of the corporate business is important for our understanding of its behavior and its social, economic, and spatial impact.
6. One of the keys to understanding the evolution and operation of the modern space economy is the recognition of the kinds of interrelationships which exist among firms of different sizes, types, and degrees of power within a segmented economic structure.

BUSINESSES IN A VOLATILE ENVIRONMENT: A TWO-WAY RELATIONSHIP

The Business Organization is Not a "Black Box"

In the jargon of the systems theorist, a *black box* is a component of a system whose structure we know nothing about. Its behavior is inferred from its input and output characteristics alone. This is the approach we adopted in Part One of this book toward the firm seeking optimal locations in space. In Chapter 4, for example, we introduced the concept of the production function, which simply stated that a firm's output would be some function of the particular combination of inputs used (capital, labor, land, and technology). But we said nothing about the function itself: how the inputs would actually be transformed into outputs. In other words, we adopted a black box approach to the firm (Figure 7.6). In so doing, we were simplifying a complex reality. But we were also implicitly following a long-established convention in economic geography: that the geographer's interest stops at the factory gate. Edward Ullman (1953) expressed this viewpoint forcibly in the early 1950s when he argued that geographers

> have no more interest in the production process inside a factory than has an economic theorist. What we are mainly interested in . . . is what comes in at the back door and what goes out the front door of a plant, so that we may know why it is located where it is and what its effect on the area under study is. The effects of changes in internal processes, however, are relevant only insofar as they affect these external relations. (p. 55)

However, simply looking at a plant's inputs and outputs as Ullman suggested may tell us very little about "why it is located where it is and what its effect on the area . . . is." How business organizations work, how they are structured, and how decisions are made are all critical areas of legitimate interest to the geographer and vital to our attempts to understand the economic geography of the world around us. The pioneer of such a perspective was Robert McNee, who in 1960 argued for a "more humanistic economic geography: the geography of enterprise." Such an approach explicitly recognizes that the way in which decisions are made will vary substantially among business types. So we do need to get inside the business organization; we cannot simply treat it as a black box.

Business organizations, from the simplest one-person operation to the giant global corporation, consist of a package of different functions that in combination enable the organization to achieve its specific objectives. The objectives themselves may be quite complex, although in economic models,

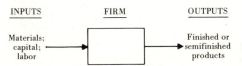

Figure 7.6 The concept of the business organization as a "black box."

INPUTS   FIRM   OUTPUTS

Materials;   Finished or
capital;   semifinished
labor   products

business goals are often simplified to a single goal of profit maximization (this is not as simple a concept as it sounds). Clearly, all firms in a capitalist market system are driven by the profit motive. A firm's annual rate of profit is the "bottom line," the key barometer of its business health. Any firm that fails to make a profit over a period of time is likely to go out of business altogether. But within the constraint of basic survival, firms may pursue a whole variety of specific goals. For example, pursuit of a bigger market share is a commonly sought goal on the theory that the greater a firm's market share, the greater its competitive power.

Not all firms are profit maximizers or seekers after market leadership or avid growers. Our discussion of the types of firms in various business segments of the economy suggested that some firms are *satisficers* with a fairly low level of aspiration above that of sheer survival. Indeed, the organizational theorist Herbert Simon (1960) argues that decision makers in general tend to be satisficers simply because they lack the information (and the ability) to optimize in a highly uncertain world. Simon's argument is that decision makers seek to do the best they can on such information as they can acquire; that is, they are boundedly rational in their behavior.

The more complex the business organization, the more complex is likely to be the setting of specific goals because different interest groups within the organization will attempt to impose their particular viewpoint and aspirations.* Figure 7.7 shows that within the larger business organization, the goals and objectives of the organization itself will be the outcome of a power struggle among the three major elements: individuals, stakeholders, and coalitions. These will in turn be influenced by forces outside the firm, the broader economic, social, and cultural environment.

We observed earlier that the business organization is, essentially, a package of functions. What are these functions, and how do they relate to each other? A particularly useful framework for examining this question is Michael Porter's *value chain* (Figure 7.8). As Porter (1985) states:

> Every firm is a collection of activities that are performed to design, produce, market, deliver and support its product. . . . Value activities are the physically and technically distinct activities a firm performs. They are the building blocks by which a firm creates a product valuable to its buyers. (pp. 36, 38)

---

* Galbraith (1974) argued that in the modern business corporation, the dominant interest group has become the management—what Galbraith calls the *technostructure*. This situation has arisen because over the years, the ownership of the corporation and the control of the corporation have become separated. The legal owners of a publicly quoted company are the stockholders, but with the spread of stock ownership among very large numbers of individuals and institutions, they have no direct control over the running of the firm. This is in the hands of the executive group of management, who effectively control how the firm operates, subject to the annual stockholders' meeting. The interests of the management group in protecting their own position may influence the specific goals that the organization pursues. Galbraith drew heavily on the classic 1932 work on the American corporation by Berle and Means. For a useful general discussion of business goals and objectives, see Ansoff (1987) and Johnson and Scholes (1988).

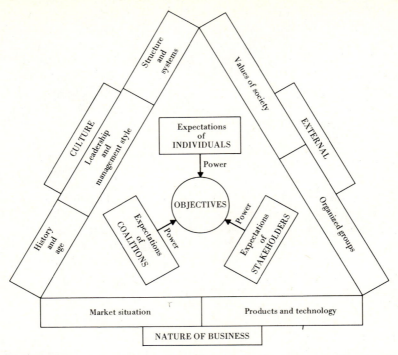

Figure 7.7 Forces influencing the decision-making process. *Source:* G. Johnson and K. Scholes (1988), *Exploring Corporate Strategy,* 2d ed. (Hemel Hempstead, England: Prentice-Hall), fig. 5.1.

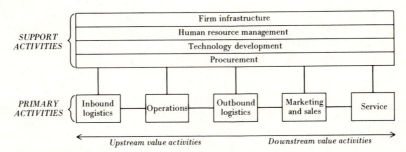

Figure 7.8 The major functions performed in the production of goods or services: the value chain. *Source:* From M. E. Porter (1986), "Competition in Global Industries: A Conceptual Framework," in M. E. Porter (ed.), *Competition in Global Industries* (Cambridge, Mass.: Harvard Business School Press), p. 24, fig. 1.3. © 1986 President and Fellows of Harvard College. Reprinted by permission.

The firm's value chain consists of two major sets of activities: *primary* activities and *support* activities. Figure 7.8 shows that there are five types of primary activity:

1. *Inbound logistics* are activities that receive, store, and distribute the inputs needed in the firm's production process. They involve such activities as handling materials, warehousing, controlling inventory, and transportation.
2. *Operations* are activities that transform these inputs into the firm's product (good or service). Such activities obviously vary according to the kind of production being performed. In a manufacturing firm, they would include machine manufacturing processes, assembly, testing, and packaging.
3. *Outbound logistics* are activities that collect, store, and distribute the firm's products to its customers.
4. *Marketing and sales* are activities that inform the potential customers of the product's existence and make their purchase possible (by advertising, sales representatives, etc.).
5. *Service* consists of activities that help to maintain or improve the value of the firm's product in its use (after-sales service and maintenance).

Each of the five primary activities is linked to a set of support activities (Figure 7.8):

1. *Firm infrastructure* activities are the management activities of planning, financial control, accounting, legal affairs, and the like.
2. *Human resource management* activities involve recruiting, training, motivating, and controlling the firm's labor force.
3. *Technology development* is concerned with such activities as research and development, process, and product design. However, all the firm's value activities, no matter how mundane, have a technological basis in the form of know-how.
4. *Procurement* activities are concerned with purchasing or acquiring the firm's inputs.

The value chain describes all the activities and functions that have to be performed to produce a particular product or service. But where will these activities be performed in an organizational sense? (The same question in a geographic sense will be discussed in Chapter 8). All five of the primary activities may be performed within a single firm. Such a firm is said to be *vertically integrated*. At the other extreme, a firm might specialize in just one of the primary activities in the value chain. Between these two extremes lies a whole spectrum of alternative arrangements. Since each firm has its own range or *economic scope* of value activities, a major question is, what determines the boundary of the firm? In other words, how is it decided which of the value activities a firm will perform for itself and which it will leave to other firms to perform and purchase the product or service as a market transaction? The key to answering this question lies in the concept of *transaction costs*.

The performance of all economic activities involves a whole maze of transactions or exchanges: between a firm and its workers, its suppliers, its customers, and so on. More than half a century ago, Ronald Coase (1937) suggested that a firm would decide which transaction to perform for itself on the basis of its *marginal cost*. That is, a firm would carry out a particular transaction up to the point at which "the costs of organizing an extra transaction within the firm are equal to the costs involved in carrying out the transaction in the open market, or to the costs of organizing by another entrepreneur" (p. 394). Beyond this point, it makes more sense for the firm to have the particular transaction performed outside.

The boundary of the firm, therefore, is the point at which the internal transactions of the firm are replaced by the external transactions of the market. Oliver Williamson (1975, 1986) suggested that economic transactions are organized and coordinated differently in these two contexts. Transactions that occur within the firm tend to be organized in a *hierarchical* manner, whereas transactions in the market tend to be organized in a *horizontal* manner. Generally speaking, the greater the degree of uncertainty—whether over the availability, price, or quality of supplies of inputs or the price obtainable for the firm's outputs—the greater the incentive for the firm to control these transactions itself. However, the interface between internalization and externalization is far from static. A firm may decide to increase the degree of integration of its operation by internalizing either "upstream" or "downstream" (Figure 7.8). Alternatively, it may decide to reduce its degree of integration—to deintegrate some of its functions by hiving them off to other firms. As we shall see in subsequent chapters, this shifting process of integration and disintegration is of great significance in the changing geography of economic activities.

The business organization, then, is not a black box. It is made up of a complex structure of interrelated activities. Far from being irrelevant to the economic geographer, the ways in which firms organize their activities internally are of fundamental importance, as we shall see in subsequent chapters. At this stage, however, we need to place the business organization in a broader context, that of its external environment.

An Uncertain and Volatile Environment for Business

We borrowed the term *black box* from systems theory. Continuing this strand of thought, we can regard each business organization as an *open system*, one that is connected to and interacts with an *external environment*. All business organizations exist in an environment that is both complex (in that it consists of several layers and many elements) and dynamic (in that it is in a constant state of flux). A major task of all business organizations, therefore, is to try to achieve a good fit between itself and its environment.

What do we mean, then, by the term *environment*? To the traditional geographer, *environment* is usually taken to mean the natural or physical environment. But the term as used here means a great deal more than this. It encompasses everything outside the boundary of the firm. But this isn't a very helpful definition. What matters most are the parts of the external environment that either directly or indirectly impinge on the organization.* Figure 7.9 depicts the environment as consisting of a series of layers, rather like the skins of an onion. At the outer, most general level, business organizations exist within a *macroenvironment,* subdivided into a *global* and a *societal* environment. This is the part of a firm's environment that is the source of general sociocultural influences such as social values and norms as well as the general technological and educational dimensions of society. More directly relevant for our purposes are the other two categories of environment shown in Figure 7.9, the *domain* and the *task environment.* The domain is "the group of organizations with which a focal organization *could* interact, the points within the societal environment with which that organization might *potentially* have contact, given the activity and range of activities it undertakes" (McDermott and Taylor, 1982, p. 18). The domain, therefore, consists of *possible* sources of interaction and influence. The task environment, however, consists of *actual* interactions between the firm and its environment. It is defined as "the set of organizations with which a focal organization *actually* establishes relationships. Such exchanges can be of materials, goods or information and can be considered as fundamental to the survival of the organization and its decision making" (p. 18). The task environment, therefore, incorporates the kinds of linkages discussed in Chapter 5 but also involves the notion of uneven power relationships. Figure 7.10 displays the major components of a firm's task environment. The environmental pressures exerted by organizations with which the firm interacts are divided into three broad types:

1. *Competitors* are existing competitors in the same industry, potential new entrants, and the threat of substitute products.
2. Groups with *bargaining power* are either suppliers of the firm's inputs or customers of its outputs with the potential capacity to exert leverage over the firm in the terms (including the price) of the transactions.
3. *Regulators* include the regulatory and legislative actions of governments (taxation, safety requirements, supervision of entry into the industry, etc.) and the actions of labor unions in influencing such factors as wage levels, conditions of work, and terms of employment.

The precise nature of these external forces will tend to vary according to the particular industry in which the firm is involved.† In some economic activities, such as agriculture and tourism, the state of the natural (physical) environment, notably weather conditions, will be especially significant. Nobody in agricultural America will soon forget the immense problems created

---

* For a detailed discussion of the concept of environment within the structural contingency framework of organization theory, see McDermott and Taylor (1982).

† Porter (1985) analyzes the competitive forces in some detail.

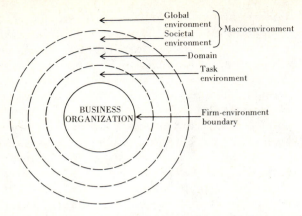

Figure 7.9 The external environment of a business organization. *Source:* After P. J. McDermott and M. J. Taylor (1982), *Industrial Organization and Location* (Cambridge: Cambridge University Press), fig. 3.1a.

by the massive drought and record heat waves of 1988. But it should not be forgotten that farmers and tourist facility operators also have to cope with the socioeconomic environment we have been discussing.

Use of Game Theory to Illustrate Uncertainty in Decision Making

The nature of the firm's external environment is a source of substantial *uncertainty.* To cope with such uncertainty, firms have to calculate or guess what the

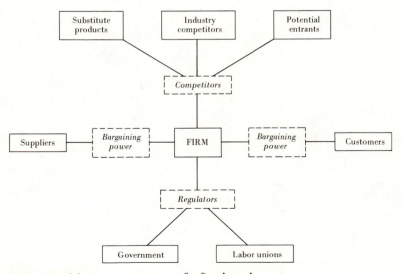

Figure 7.10 Major components of a firm's task environment.

future state of the environment might be and choose its actions accordingly. We can illustrate this by drawing again on simple game theory (we first met this technique in Chapter 5 when discussing agglomeration). Here we use game theory to explore two hypothetical cases: that of a small manufacturing firm needing to expand its operations and that of a farmer trying to decide which combination of crops to grow. In both cases, the environment is regarded as the other player in the game; its strategies take the form of alternative future states of the environment. In our first example, the uncertain environment takes the form of possible government actions that will affect the locational behavior of a manufacturer. In the second example, the environment takes the form of possible weather conditions that will affect crop yields.

1. Manufacturer Location Decision as a Game Theory Problem

Imagine the case of a small-scale manufacturer of electrical machinery who is located in rather cramped, though low-rent, premises close to the center of a large city. Demand for the firm's products is increasing, and the firm needs to install additional machinery to produce more output. This could be done in one of three ways; in game theory terms, the manufacturer identifies three alternative strategies:

>*Strategy 1.* Remain at the existing location and install new equipment to increase output.

>*Strategy 2.* Close down the central city plant and move to a suburban site in the same city.

>*Strategy 3.* Close down the central city plant and relocate in a different part of the country where labor is cheaper.

Suppose, again for the sake of simplicity, that the future state of the environment can take one, and only one, of three alternative forms. The manufacturer is uncertain as to which one will actually occur.

>*State of the environment I.* The government introduces a program of equipment grants for small firms.

>*State of the environment II.* A new suburban expressway is built, linking suburban industrial areas with the national interstate highway system.

>*State of the environment III.* The government introduces a regional development policy with major financial assistance for manufacturing firms that locate in designated areas of high unemployment.

We can now put these two sets of strategies together in the form of a matrix (Table 7.3) in which the numbers in each cell represent the payoffs or values of each strategy. The values in the cells, of course, are hypothetical; they are designed to represent differences in the value to the manufacturer of the various location choices. Each of these alternative strategies has its advantages and disadvantages. Staying put is the "least effort" solution. But the premises may be rather obsolescent, and city congestion may be increasingly costly.

Table 7.3   PAYOFF MATRIX FOR THE ELECTRICAL MACHINERY MANUFACTURER

		States of the environment		
Manufacturer's alternative strategies		Government equipment grants I	New suburban motorway II	Government regional policy III
Reequip at existing location	1	200	155	145
Move to suburbs	2	130	220	130
Move to surplus labor area	3	118	118	225

Moving the plant either to the suburbs or to a more distant location may be an expensive and inconvenient process but may offer the prospect of better factory premises, perhaps a more reasonable living area, and, in the more distant case, the prospect of lower labor costs. However, which strategy is actually adopted will depend on more than these kinds of considerations. It will depend on the manufacturer's perception of future events and attitude toward risk and uncertainty. Let us consider some possible cases.

(*a*) The *cautious* manufacturer does not like to take chances and opts for the strategy that will minimize the possible disadvantages. In game theory terms, a *maximin* strategy is adopted; that is, each row of the payoff matrix is examined and the worst possible outcome for each strategy is identified. In Table 7.3 these are

145 for strategy 1

130 for strategy 2

118 for strategy 3

The manufacturer thus assumes that the worst will happen and chooses the "best of the worst," strategy 1, which is to stay put and install new equipment. Whichever state of the environment actually occurs, therefore, the cautious manufacturer will achieve a payoff of at least 145, which is the lowest payoff for strategy 1, and could achieve a good deal more. If environment state II occurs, he would get a payoff of 155, or 200 if state I occurs.

(*b*) The *reckless* manufacturer, at the opposite end of the attitude-to-uncertainty scale, is prepared to stake all on the chance of gaining the largest possible return. In game theory terms, we can imagine the manufacturer scanning the payoff matrix and choosing the largest payoff for each possible state of the environment:

200 for strategy 1

220 for strategy 2

225 for strategy 3

Being an optimist, the manufacturer will choose the strategy associated with the largest possible payoff (strategy 3) and relocate the plant in a surplus labor area. We would call this a *max-max* strategy.

(*c*) The *calculating* manufacturer tries to assess the relative probabilities of each possible environmental state. Some environmental states may be regarded as more likely to occur than others; that is, the manufacturer may be able to weight the value of each strategy by guessing at the relative likelihood or probability of each state of the environment occurring. Let us make the following assumptions:

> The probability of state I of the environment occurring is .2.
>
> The probability of state II of the environment occurring is .5.
>
> The probability of state III of the environment occurring is .3.

(Note that the values add up to 1 because it is assumed that no other alternatives exist.) Using these probabilities, the manufacturer can calculate the total expected value for each of the three strategies by multiplying each payoff value by the appropriate probability and adding the results:

> Strategy 1: .2(200) + .5(155) + .3(145) = 40 + 77.5 + 43.5 = 161
>
> Strategy 2: .2(130 + .5(220) + .3(130) = 26 + 110 + 39 = 175
>
> Strategy 3: .2(118) + .5(118) + .3(225) = 23.6 + 59 + 67.5 = 150.1

Clearly, strategy 2, a move to the suburbs, is most appropriate given the manufacturer's assessment of the relative likelihood of future events.

(*d*) The *naive* manufacturer takes a less sophisticated view of the future. Calculation of different probabilities for each environmental state may be difficult. A less sophisticated method would be to assume that there is an equally good chance of each state of the environment occurring, that is, to give an equal probability weighting to each payoff value. The overall values for each strategy would then be as follows:

> Strategy 1: .33(200) + .33(155) + .33(145) = 66 + 51.15 + 47.85 = 165
>
> Strategy 2: .33(130) + .33(220) + .33(130) = 42.9 + 72.6 + 42.9 = 158.4
>
> Strategy 3: .33(118) + .33(118) + .33(225) = 38.9 + 38.9 + 74.25 = 152.05

Thus strategy 1 is the best one to adopt with this attitude toward environmental uncertainty.

(*e*) The *provident* manufacturer wishes to "minimize regret." Most of us at some time or other, having made a particular decision and observed its outcome, wish we had made a different decision. In other words, we suffer from a degree of regret whose intensity is reflected in the difference in the payoff between the decision we made and the decision we *would have made* had we only known better at the time. We can apply this *regret criterion* to our manufacturer's decision-making problem.* Suppose our manufacturer chose

---

* Such a measure of regret is analogous to the economist's concept of *opportunity cost,* which is a measure of the cost of one alternative in terms of the alternatives not taken up (see the note on p. 57).

strategy 3 and state I of the environment occurred. The payoff amounts to 118. But if it was known that state I was going to occur, the manufacturer would have chosen strategy 1, which has the highest payoff value for state I of the environment. The difference between what was actually received—118—and what could have been received—200—is a measure of regret (in this case, the regret value is 82). If we apply this line of reasoning to all the other payoff values—that is, if we compare each payoff value with the maximum value in its respective column of the payoff matrix (Table 7.3)—we can construct a regret matrix, as shown in Table 7.4. A manufacturer wishing to minimize regret, therefore, would adopt a *minimax* strategy by identifying the largest regret value for each strategy:

> 80 for strategy 1
> 95 for strategy 2
> 82 for strategy 3

and choosing the lowest of these values, 80. Thus strategy 1 would be the chosen strategy.

   2. *Farmer's Crop Decision as a Game Theory Problem*   A major question facing the farmer is what crops to grow. Apart from having to take consumer preferences into account, a key issue is the likely future state of the weather. Should a single crop be grown or a combination of crops? Which crops are most suitable? Again, we can look at this problem in game theory terms to illustrate the way in which environmental uncertainty can influence the decision made. Imagine the case of a farmer choosing among three possible crop types and facing an environment that might take one of the three forms: wet, dry, or average rainfall. Table 7.5 sets out the payoff matrix, and Table 7.6 summarizes the outcome of the different strategies based on the same five attitudes toward uncertainty discussed for the small manufacturer. Table 7.7 is the farmer's regret matrix.

   Table 7.6 shows which of the three crops the farmer should grow, depending on his attitude toward uncertainty. In each case, the solution is for one

Table 7.4   REGRET MATRIX FOR THE
            ELECTRICAL MACHINERY
            MANUFACTURER

Manufacturer's strategies	States of the environment		
	I	II	III
1	0	65	80
2	70	0	95
3	82	102	0

Table 7.5   PAYOFF MATRIX FOR THE FARMER

Farmer's crop strategies		States of the environment		
		Dry I	Average II	Wet III
Wheat	1	23	18	10
Barley	2	13	16	20
Rice	3	11	20	21

crop out of the three possibilities. But it may be better for the farmer to mix strategies, either over a period of years (sometimes growing one crop, sometimes another) or by dividing up the farm and growing a combination of crops each year. One very simple way of deciding on the correct mixture of strategies is to use the graphical technique of Figure 7.11. This involves reducing the alternatives of one of the two players to two from three. If we do this for the environment and assume that the weather will either be dry or wet, we can arrange the farmer's payoffs into two vertical columns and join the appropriate values by straight lines. For example, Table 7.5 shows that the payoff for wheat is 23 in a dry year and 10 in a wet year; in Figure 7.11, the value of 23 on the dry scale is connected to the value of 10 on the wet scale. The same procedure is carried out for the other two crops, barley and rice. If we then focus on the upper boundary of the graph and, in particular, on the lowest point of the upper boundary, we can identify the optimal crop combination, which happens to be wheat and barley. This is the minimax solution.

The graph tells us which is the best crop combination; it does not tell us in which proportions the crops should be grown either over a period of time or at a single point in time. Table 7.8 shows how this may be resolved (see also Gould, 1963). For each of the two crops, the difference between the high and the low payoff value is taken (regardless of sign). This difference is allocated to the other strategy because it is an indication of the size of loss experienced if the

Table 7.6   SUMMARY OF SOLUTIONS TO THE FARMER'S PROBLEM FOR DIFFERENT ATTITUDES TOWARD UNCERTAINTY

	Attitude	Strategy	Payoff anticipated
(a)	Cautious	2. Barley	13
(b)	Reckless	1. Wheat	23
(c)	Calculating (probabilities: .3 dry, .4 average, .3 wet)	3. Rice	17.6
(d)	Naive (probabilities: .33 dry, .33 average, .33 wet)	3. Rice	17.2
(e)	Provident	2. Barley	(10 regret)

Table 7.7   REGRET MATRIX FOR THE
            FARMER

Farmer's crop strategies	States of the environment		
	I	II	III
1	0	2	11
2	10	4	1
3	12	0	0

wrong choice had been made. Table 7.8 reveals that the ratio between wheat and barley should be 35 : 65. This means either that the farmer should, over a period of years, grow barley roughly two years in every three and wheat the third year or, more realistically, that wheat and barley should be grown every year, with 35 percent of the land being put under wheat and 65 percent under barley.

Thus the problem facing all business organizations in an uncertain environment is not only the uncertainty of possible future outcomes but also the assessment of what those outcomes might be. An important element here is the way in which information is received about the external environment. Unlike the decision makers in Part One of this book, those in the more complex world have access only to limited and biased information.

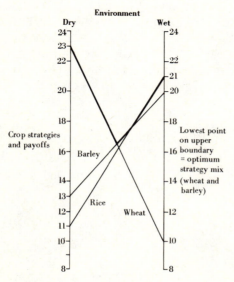

Figure 7.11 Graphical solution for the game theory crop combination problem.

Table 7.8   SOLUTION TO THE CROP COMBINATION PROBLEM

	Dry	Wet			
Wheat	23	10	$\|23 - 10\| = 13$	$\dfrac{7}{13 + 7} = \dfrac{7}{20} = 35\%$
Barley	13	20	$\|13 - 20\| = 7$	$\dfrac{13}{13 + 7} = \dfrac{13}{20} = 65\%$

Information and Knowledge of the External Environment

The business organization's behavior within its external environment is shaped not by the objective nature of that environment (what actually exists) but by the organization's *perception* of the environment (what is believed to exist). A key element in this complex process is the kind of information available to the organization. All decision makers, from the single individual to the complex large-scale organization, acquire information selectively. On the one hand, they neither acquire nor have access to "complete" or "perfect" information (if such a commodity exists). On the other hand, they interpret the information to which they do have access in terms of their own *coding mechanism*. Individuals and organizations "can react only to those information signals to which they are attuned. . . . They develop their own mechanisms for blocking out certain types of alien influence and for transforming what is received according to a series of code categories" (Katz and Kahn, 1966, pp. 22, 60).

Thus all individuals and organizations are aware of only a limited part of what we might call the total *objective environment*. William Kirk has called the part of the objective environment of which individuals are aware the *behavioral environment*. This is the segment of the objective environment about which information signals are received and interpreted. In the terminology used earlier, the behavioral environment consists of the task environment and parts of the domain of the organization. Only a limited proportion of the information transmitted from the objective environment is effectively received. This information is relevant for the organization's conscious and purposive behavior. Of course, events and phenomena that are unknown to a particular individual or organization can ultimately have an effect on them. We shall discuss an example of this shortly.

The precise form of the behavioral environment will, by definition, vary from individual to individual and from organization to organization. However, "similar" individuals will have knowledge and viewpoints in common because

> the vast bulk of our knowledge of fact is not gained through direct perception but through the second-hand, third-hand, and nth-hand reports of the perceptions of others, transmitted through the channels of social communication. Since these perceptions have already been filtered by one or more communicators, most of whom have frames of reference similar to our own, the reports are generally

consonant with the filtered reports of our own perceptions, and serve to reinforce the latter. (March and Simon, 1958, p. 153).

Figure 7.12 sets out some of the major influences on this process of environmental perception and awareness. Individuals having several of the variables in common will tend to have a broadly similar image of the world.

In this respect, geographic location is an important variable. Information is not available everywhere in the same quantity and quality, despite the spread of telecommunications. Some places are more "information-rich" than others in that they are more centrally connected into physical communications networks (see Chapter 3). Similarly, position in group communication networks is an important influence because information flow between members of the same social or business group is likely to be substantially greater than information flow between different groups. The five hypothetical situations in Figure 7.13 illustrate the relationship between group membership and information flow. Difficulty of communication increases from A to E. Information flows most readily when both the possessor of an item of information and the seekers of such information belong to the same group (situation A) and with

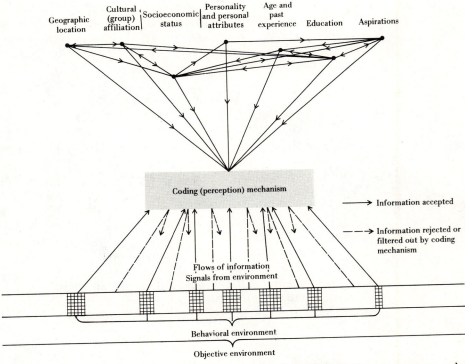

Figure 7.12 Elements in the perception process: The behavioral environment and the objective environment. *Source:* Portions after D. L. Huff (1960), "A Topographical Model of Consumer Space Preferences," *Papers and Proceedings of the Regional Science Association* 6: 159–173, fig. 3.

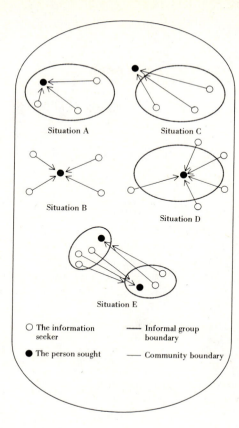

Situation A

Situation C

Situation B

Situation D

Situation E

○ The information seeker

● The person sought

—— Informal group boundary

—— Community boundary

Figure 7.13 The effect of social groups on the flow of interpersonal communication. *Source:* H. F. Lionberger (1960), *Adoption of New Ideas and Practices* (Ames: Iowa State University Press), fig. 5. Copyright © 1960. Reprinted by permission.

greatest difficulty when they are members of different and mutually exclusive groups. Such considerations are certainly important in the decision-making processes in large organizations since different interest groups within the management structure, as well as other interest groups, are involved (see Figure 7.7).

An ability to overcome the frictional effects of geographic space and to participate in a diversity of groups and associations will help to reduce many of the restrictive influences on environmental awareness. At the personal level, socioeconomic status is an especially important influence: the affluent, the well-educated, and the highly skilled tend to have more extensive environmental knowledge than the poor, the illiterate, and the unskilled. As Hagerstrand (1967) has observed:

> From daily experience, we know that the links in the network of private communications must differ in spatial range between socio-economic groups. . . . Some individuals are wholly bound to the local plane, others operate on the regional and local planes, and still others operate more or less on all three. Those belonging to a wider range and at the same time having links in common with lower

ones form the channels through which information disseminates among the planes. (p. 8)

Although Hagerstrand refers to individuals, exactly the same points apply to business organizations. The small firms in the periphery economy discussed earlier in this chapter are confined in their spatial range mostly to the local or regional spatial scales. The extent of their environmental awareness is consequently minuscule compared with that of the corporation operating across the globe and having the information-gathering capacity of *global scanners* (Vernon, 1979). Figure 7.14 shows this structure of different spatial levels of interaction.

A Turbulent Environment: Changing Markets and Technologies

All business organizations, therefore, face conditions of uncertainty in their external environments. Such uncertainty is intensified as environments become both more complex and more dynamic—in Emery and Trist's terms, more turbulent (Figure 7.15). Increased turbulence may arise both in the business organization's direct transactions with other organizations in its environment and in the overall "causal texture" of the environment. Emery and Trist (1965) provide an appropriate example to illustrate this latter point. A major food-processing company in the United Kingdom in the 1950s had a dominant position (65 percent of the market) in its major product, a canned vegetable. At the end of World War II, the company made a massive investment in a new, highly automated factory dedicated to very long production runs and high-volume output for its existing large market. Unfortunately, "the character of the environment . . . began to change while the factory was being built."

A number of separate developments in different parts of the firm's environment began to come together to exert a devastating effect on the company. A

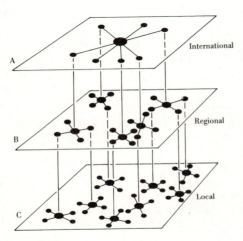

Figure 7.14 Diffusion of information between individuals at different geographic levels of interaction. *Source:* T. Hagerstrand (1967), "A Monte Carlo Simulation of Diffusion," in W. L. Garrison and D. F. Marble (eds.), "Quantitative Geography: Part 1. Economic and Cultural Topics," *Northwestern University Studies in Geography* 13: 1–32, fig. 3. Reprinted with permission.

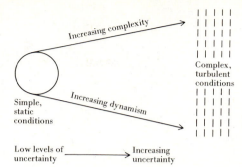

Figure 7.15 Increasing environmental uncertainty. *Source:* After G. Johnson and K. Scholes (1988), *Exploring Corporate Strategy*, 2d ed. (Hemel Hempstead, England: Prentice-Hall), fig. 3.3.

number of new, small canning firms began to emerge on the basis of the availability of cheaper cans after the removal of postwar controls, the availability of cheaper imported fruit and vegetable crops, and major changes in retailing, particularly the rapid growth of supermarkets, which wanted to sell own-brand products. These retailers placed large bulk orders with the small canners and captured a large share of the traditional market. At the same time, increasing consumer affluence led to demands for a much wider range of processed foods while food-processing technology, particularly the development of quick-freeze methods, transformed the market even further. The problem for the existing market leader was that its new automated factory could not be adapted to the new market conditions. The firm had "failed entirely to appreciate that a number of outside events were being connected with each other in a way that was leading up to irreversible general change" (Emery and Trist, 1965, p. 24).

For business organizations, therefore, two of the major sources of environmental turbulence are shifts in market conditions and shifts in technology. Although related in some respects, the two are not synonymous. A key element in changing market conditions is the way in which demand for a product tends to change over time. A common way of describing such change in the marketing literature is that of the *product life cycle.** Figure 7.16 shows how this idealized sequence of product change is believed to proceed.

The essence of the product cycle model is that the growth in sales for a particular product follows a systematic path, from initial introduction onto the market through a series of stages of development, growth, maturity, decline, and obsolescence. When a new product is first introduced, the total volume of sales tends to be low because customer knowledge is limited and customers are generally uncertain about the product's quality and reliability (of course,

---

* This concept has also been widely adopted in the geographic literature to explain spatial changes in economic activity. As such, we will refer to it again later. However, there are major problems in translating the product cycle concept from its original nonspatial form to a spatial context (see Storper, 1985; Storper and Walker, 1983; Taylor, 1986). At this stage, we are interested in the product cycle purely as a means of describing changes in demand for a product over time.

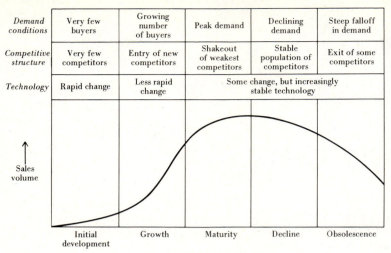

Demand conditions	Very few buyers	Growing number of buyers	Peak demand	Declining demand	Steep falloff in demand
Competitive structure	Very few competitors	Entry of new competitors	Shakeout of weakest competitors	Stable population of competitors	Exit of some competitors
Technology	Rapid change	Less rapid change	Some change, but increasingly stable technology		

Sales volume →

| Initial development | Growth | Maturity | Decline | Obsolescence |

Stages of product development

Figure 7.16 The product life cycle.

some people will always be willing to try the latest product immediately). If the new product gains a foothold in the marketplace (most new products do not), it enters a phase of rapid growth as overall demand increases. At this stage, the number of competing firms in the industry tends to increase. Eventually, however, demand for the product is likely to flatten out as the market becomes saturated and a high proportion of demand takes the form of replacement rather than new demand. The weaker competitors tend to be shaken out as competitive conditions tighten, leaving a core of dominant producers. Eventually, it is assumed, demand will fall and the product will become obsolescent.

Like all stage or cyclic models, the product life cycle is highly generalized. Not every product will follow the cycle precisely. The model tells us nothing about the relative length of each stage; it is merely programmatic. But it has some real value as a general approximation of changing market conditions for a product. Most business organizations believe that a product cycle exists and behave accordingly. The development path suggested by the product cycle model has very important implications for the growth and survival of firms. It implies that all products have a limited life, that obsolescence is inevitable. Of course, the rate at which the cycle proceeds will vary greatly from one product to another. For some ephemeral products, the cycle may run its course in a single year or even less. In others, the cycle may be very long indeed. However, the average length of product cycles is generally believed to be getting shorter and shorter, thus increasing pressure on firms to find new products.

Quite apart from changes in market demand related to the product cycle, the turbulence of a firm's environment is affected by the more general pace and direction of technological change. As Figure 7.16 shows, the nature of production technology tends to change during the product cycle, though not in the deterministic and linear way sometimes suggested. In addition, production technologies—like the transport and communications technologies discussed in Chapter 3—are changing very rapidly. Both kinds of change are contributing to the increasing turbulence of organizational environments. Most technological change is gradual and incremental. But from time to time, truly radical changes occur in the technology system. Christopher Freeman (1987) calls such radical shifts *changes in the technoeconomic paradigm*. In Freeman's words, these are the

> "creative gales of destruction." . . . They represent those new technological systems which have such pervasive effects on the economy as a whole that they change the "style" of production and management throughout the system. The introduction of electric power or steam power or the electronic computer [typifies] such deep-going transformations. A change of this kind carries with it many clusters of radical and incremental innovations, and may eventually embody several new technology systems. Not only does this . . . type of technological change lead to the emergence of a new range of products, services, systems and industries in its own right, [but] it also affects directly or indirectly almost every other branch of the economy. . . . The changes involved go beyond specific product or process technologies and affect the input cost structure and conditions of production and distribution throughout the system. (p. 130)

The most far-reaching changes currently affecting the general technological environment are those based on the *information technologies* (IT). IT is the result of the relatively recent convergence of two formerly distinct technological streams: telecommunications technology, which is concerned with transmitting information, and computer technology, which is concerned with processing information. Each of these technologies has created rapid and far-reaching changes in economy and society. But the really fundamental implications arise from their convergence. Common to both, of course, is the microelectronics industry, particularly "the continuing and dramatic improvement in large-scale integration of electronic circuits, and the continuing fall in costs which this permits" (Freeman, 1987, p. 132). This convergence facilitates their incorporation in a remarkable range of products both for consumers and for producers and has a direct effect on production technologies.

The new technoeconomic paradigm based on the information technologies has three key features:

1. A trend toward information intensity, rather than energy or materials intensity, in production
2. Enhanced flexibility of production, which transforms the old relationship between scale or output and cost of production
3. A major shift in labor requirements as a result of the combination of increased automation and increased flexiblity

Obviously, these revolutionary and pervasive changes in technology will alter the form and structure of economies at all geographic scales. To organizations with access to the new technologies they offer an enlarged range of possibilities in terms of what to produce, how and how much to produce, and where to produce it. For business organizations in general, such changes help to make the external environment more volatile and increase competitive pressures.

ADVANTAGES OF THE LARGE BUSINESS ORGANIZATION

In general, larger firms have greater leverage over their uncertain and volatile environments than smaller firms. Larger firms possess a higher degree of adaptability to environmental influences through the amount of power they can exert. By internalizing transactions with suppliers and customers or by exerting their greater strength over externalized transactions, large firms place themselves in a dominant position within the segmented economy. Through its greater possession of power, the large firm can gain easier access to government. It may well be able to secure price reductions from suppliers because few suppliers are willing to lose a large customer. The larger a firm's share of a market, the greater its ability to set prices at the level the company wants rather than at the level set by the market.

Undoubtedly, one of the major reasons why business enterprises have become increasingly large is the existence of economies of scale. When we discussed these in Chapter 5, it was in the restricted framework of economies at the level of the individual plant. However, similar, though even greater, economies may be present in firms operating more than one plant or performing the whole sequence of value activities. In other words, the very existence of a multiplant or multifunction organization may further increase the economies of size in a way shown by the shaded area in Figure 7.17 (compare this with Figure 5.5). But why should firms operating several plants be supposedly more efficient (in cost terms) than single-plant firms? Part of the answer lies in the broader relevance of the components identified in Chapter 5: economies of specialization, mass reserves, and large-scale purchasing.

Managerial economies—the better use of managerial skills and the benefits of a more sophisticated organizational structure—are claimed to be especially significant. The creation of specialist departments to carry out research and development, organize marketing, and arrange finance and expenditure is the hallmark of the multiplant enterprise. Marketing is especially important in consumer goods industries, where very large enterprises can not only operate a larger and more comprehensive sales organization but also engage in sales promotion and advertising on a scale beyond the scope of the small firm. For example, it has been estimated that roughly half of all the costs that make up the final price to the consumer are marketing costs. In this respect, we are talking as much of economies of scope as of economies of scale, an issue related to our discussion earlier in this chapter of the firm's value chain.

The concept of mass reserves also applies to the multiplant enterprise in the sense that such an organization can transfer financial and other resources between its component units to bolster flagging performance in one or to

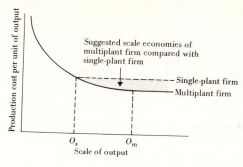

Figure 7.17 Hypothetical scale-cost curves for a single-plant firm and a multiplant firm.

increase the enterprise's overall performance in ways that the small firm simply cannot do. Likewise, in terms of large-scale or bulk purchasing, very large orders for materials and services generally bring lower prices. In the case of finance, the very large enterprise can generally acquire funds for expansion on cheaper terms than small firms.

However, not everybody would accept these advantages of multiplant firms without reservation. As with plant economies, it is argued that beyond a certain size of firm, diseconomies of scale will set in. One reason is the *diminishing control* created by the overload of responsibility and information and the remoteness of top decision makers from the day-to-day operations of the enterprise. But this does not necessarily imply that there is an absolute limit on firm size imposed by such diseconomies of coordination. Indeed, as we shall see in the next chapter, one of the major activities of large, growing enterprises has been to devise new organizational structures that can cope with problems of coordination and control in very large organizations.

FURTHER READING

Dicken, P. (1986). *Global Shift: Industrial Change in a Turbulent World*. New York: Harper & Row, chaps. 3, 4.

Gould, P. R. (1963). "Man Against His Environment: A Game Theoretic Framework," *Annals of the Association of American Geographers* 53: 290–297.

Johnson, G., and K. Scholes (1988). *Exploring Corporate Strategy* (2d ed.). Hemel Hempstead, England: Prentice-Hall, chaps. 3–5.

McDermott, P. J., and M. J. Taylor (1982). *Industrial Organization and Location*. Cambridge: Cambridge University Press.

Porter, M. E. (1985). *Competitive Advantage*. New York: Free Press, chap. 1.

Taylor, M. J., and N. J. Thrift (1983). "Business Organization Segmentation and Location," *Regional Studies* 17: 445–465.

Chapter 8

Strategic Behavior of Large Business Organizations: A Geographic Perspective

THE IMPORTANCE OF STRATEGIC DECISIONS

Traditionally, the primary focus of the economic geographer has been on one type of business decision: the location decision. The major research question has dealt with why a firm chooses to locate its activities in a particular place. In effect, the location decision has been extracted from the whole morass of decisions that business organizations have to make and looked at in virtual isolation. Quite apart from the immense empirical difficulty of obtaining a satisfactory answer to such a question (due to historical influences or the inability to interrogate the decision makers), it is an inadequate question for several reasons. First, the location decision in itself is not the only type of business decision that affects the geography of economic activity. The majority of business decisions have some spatial impact, including ones that are not explicitly and deliberately intended to do so. Second, the location decision is not made in isolation. It is part of the general body of investment decisions concerned with the allocation of a business's resources to various product and geographic markets. Thus to concentrate only on location decision making is as narrow and restrictive as Ullman's refusal to look inside the business organization (see Chapter 7).

We can identify three major types of business decisions (Ansoff, 1987): strategic, operating, and administrative.

Strategic decisions are the broad policymaking decisions aimed at enabling the business to achieve a good fit with its external environment. The term *strategy* comes from the military literature. Originally, it meant "the art of projecting and directing the larger military movements and operations of a campaign." It was distinguished from *tactics,* which were regarded as "belonging only to the mechanical movement of bodies set in motion" by the higher-level strategic decisions. The business historian Alfred Chandler (1962) has provided one of the most comprehensive definitions of *business strategy:*

> the determination of the basic long-term goals and objectives of an enterprise, and the adoption of courses of action and the allocation of resources necessary for carrying out these goals. Decisions to expand the volume of activities, to set up distant plants and offices, to move into new economic functions, or become diversified along many lines of business involve the defining of new basic goals. New courses of action must be devised and resources allocated and reallocated in order to achieve these goals and to maintain and expand the firm's activities in the new areas in response to shifting demands, changing sources of supply, fluctuating economic conditions, new technological developments, and the actions of competitors. (p. 13)

Operating decisions constitute the bulk of a firm's decision-making activities. They are set within the broad framework established by strategic decisions and are concerned with the "nuts and bolts" of making the strategy work effectively. The objective of operating decisions is to

> maximize the profitability of current operations. The major decision areas are resource allocation (budgeting) among functional areas and product lines, scheduling of operations, supervision of performance and applying control actions. The key decisions involve pricing, establishing marketing strategy, setting production schedules and inventory levels, and deciding on relative expenditures in support of R&D (research and development), marketing and operations. (Ansoff, 1987, p. 24)

Administrative decisions are concerned with the coordination of the organization's multifarious activities, the way in which the organization's activities are linked internally, and lines of authority and communication.

The organizational theorist Herbert Simon (1960) suggested that the modern business organization could be regarded as being like a "three-layered cake":

> In the bottom layer we have the basic work processes—in the case of a manufacturing organization, the processes that procure raw materials, manufacture the physical product, warehouse it, and ship it. In the middle layer, we have the programed decision-making processes, the processes that govern the day-to-day operation of the manufacturing and distribution system. In the top layer, we have the nonprogramed decision-making processes, the processes that are required to

design and redesign the entire system, to provide it with its basic goals and objectives, and to monitor its performance. (p. 40)

Each of these three broad decision categories is of great geographic significance. This is especially true of strategic decisions because these determine the overall functional and geographic shape of the business. Operating decisions are also extremely important geographically, but since they are set within the firm's strategic framework, they can be regarded as being subsidiary to and dependent on strategic behavior. In this chapter, therefore, our primary attention is on strategic decision making.

Strategy itself may be further divided into corporate strategy and competitive strategy. *Corporate strategy* is the broad-brush level of decision making, the kind described in Chandler's quotation. It is concerned with the overall scope of the organization: the kinds of business that the firm decides to be involved in (including which elements of the value chain—see Figure 7.8—are to be performed). *Competitive strategy,* by contrast, "is about how to compete in a particular market. . . . Whereas corporate strategy involves decisions about the organization as a whole, competitive strategy is more likely to be related to a unit within the whole" (Johnson and Scholes, 1988, p. 9).

Insofar as strategy is concerned with matching the business with its external environment, the nature of that environment—especially its degree of volatility and uncertainty—plays a key role. Stable or slowly changing environments are likely to be matched by relatively stable strategies or strategies that evolve gradually. Turbulent and uncertain environments need to be met by dynamic strategies. If the contemporary business environment is increasingly turbulent, as we suggested in Chapter 7, we would expect strategic decision making to become increasingly dynamic and for major strategic decisions to be made that alter not just the internal shape of the business itself but also its geography and its geographic impact. But it would be wrong to regard the influence of the external environment on a firm's behavior as totally deterministic. Indeed, facing the same external environment, firms in the same industry may respond differently. Also, we should not suppose that strategic behavior is confined solely to the very large business, although it is more likely to be more highly formalized in such an organization. All business organizations need to have a strategy of some kind, although in many firms, especially small ones, this may be implicit rather than explicit. For many less powerful firms, particularly those in the peripheral sector of the economy (see Chapter 7), formal strategic behavior is a luxury pushed out of the way by the day-to-day needs of survival. To many such firms, not only is the long run a long way off, but they would probably also go along with the aphorism "In the long run, we are all dead" (all too true for the majority of small firms). But for the large, oligopolistic firms that dominate the modern economy, strategic behavior is a continuous, high-profile activity. If we are to understand the geography of such organizations and their broader geographic impact, we need to explore how strategies are devised and implemented.

COMPETITIVE STRATEGIES IN TODAY'S WORLD

"Competitive strategy," says Michael Porter (1985), "is the search for a favorable competitive position in an industry, the fundamental arena in which competition occurs. Competitive strategy aims to establish a profitable and sustainable position against the forces that determine industry competition" (p. 1). The need for strategic decision making arises from the business's recognition or perception of either a threat to its existing operations or an opportunity to develop in new directions. Ideally, businesses constantly monitor both their external environments and their internal affairs. They engage in what the business analyst calls SWOT analysis (analysis of strengths, weaknesses, opportunities, and threats). But such monitoring tends to happen only in the better-managed firms. For the majority, the need for a strategic decision is usually triggered by a specific event, for example, the sudden emergence of a competitor, a major fall in sales or profits, a major technological development, or a crisis in the firm's internal operations.

Types of Competitive Strategies

Porter (1985) identifies three basic types of competitive strategies that a business might choose to pursue (Figure 8.1). Each of these is intimately related to the value chain shown in Figure 7.8.

Cost Leadership   One way for the firm to strive for a competitive advantage is to aim to be the low-cost producer of a good or service. Generally speaking, firms that pursue a cost leadership strategy produce a standardized product.

Differentiation   A second basic competitive strategy is for the business to make some dimension of its activity different from that of its competitors. In the marketplace, differentiation may carry a higher price, which customers are prepared to pay because of the special qualities being offered. At the same time, however, a firm pursuing a differentiation strategy cannot afford to ignore costs completely. But they have a lower priority than in a strategy of cost

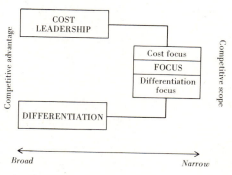

Figure 8.1 Major competitive strategy types.

leadership. A differentiation strategy, like one of cost leadership, can be derived from some or all parts of the value chain or from ways in which the parts are connected.

Focus   The fact that the competitive strategies of cost leadership and differentiation may be applied either very broadly or more selectively has led to a third type of competitive strategy. In the focus (or niche) strategy, a specific segment of a market is targeted. This might be a particular type of customer, a particular geographic market area, or a particular segment of a production process. A focus strategy, as Figure 8.1 shows, may be based on either of the two generic types of competitive strategy, cost or differentiation.

Strategic Directions

These basic or generic competitive strategies can be pursued along a whole variety of strategic directions, each of which has a clear geographic dimension.* Figure 8.2 summarizes the major categories, classified, initially, along two dimensions: product (present and new) and market (present and new). A strategy of *market penetration* (the upper left quadrant of Figure 8.2) involves attempting to increase the firm's share of its existing product in its existing market. In a strategy of *market development* (lower left quadrant in Figure 8.2), the firm stays with its existing products but attempts to develop them in new markets. These might consist of new geographic markets, new market segments, or new uses for the existing product. The third strategic direction shown in Figure 8.2 is that of *product development,* which involves the firm in developing new products for its existing markets. Given the existence of the produce life cycle (see Chapter 7) and the progressive shortening of such cycles, a strategy of product innovation and development is vital for the survival and growth of the large business. Christopher Freeman (1974) argued that for the large firm, not to innovate is to die, and he depicted large businesses as being trapped on an "innovative treadmill." But as we saw in our discussion of the segmented economy in Chapter 7, businesses differ markedly in their attitude toward technological change (review Freeman's six types of technological orientation in Chapter 7).

The fourth strategic direction shown in Figure 8.2 is more complex than the others. This is the strategy of *diversification.* Each of the other three strategies keeps the firm tied either to its existing markets or to its existing products. A strategy of diversification breaks one or both of these ties, as the enlarged lower right quadrant of the figure shows. A strategy of related diversification offers the kinds of strategic alternatives shown in Figure 8.3. Essentially, it involves the firm in integrating into activities, within, or close to, its

---

* Johnson and Scholes (1988), Chapter 6, contains an extremely clear and detailed discussion of strategic directions. Ansoff (1987) is particularly strong on discussion of diversification strategies.

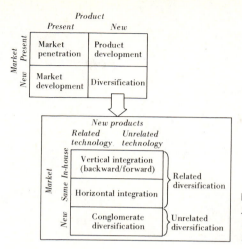

Figure 8.2 Strategic directions. *Source:* After H. I. Ansoff (1987), *Corporate Strategy* (Harmondsworth, England: Penguin), fig. 6.1.

value chain. *Vertical integration* consists of the extension of the firm's activities either "upstream" (*backward integration*) or "downstream" (*forward integration*). Thus the manufacturing firm shown in Figure 8.3 might decide to diversify into the production of its own components or of its own raw materials (backward integration). Alternatively, or in addition, it may decide to control its downstream activities of distribution or marketing by integrating forward in the value chain. In the terminology used in Chapter 8, this process of integration involves the internalization of some of the firm's transactions. The major reasons for a firm wishing to pursue a strategy of vertical integration include these:

1. An increased ability to control the quantity, quality, and price of material inputs, both raw materials and components
2. An increased ability to control its markets
3. The possibility of savings in production costs by linking successive stages in the production process rather than performing them separately

The other kind of related diversification shown in Figures 8.2 and 8.3 is *horizontal integration*. This involves the firm in moving into activities that are very closely related to its current activities, for example, into the production of complementary products, competitive products (in order to squeeze out competitors), or by-products. In general, strategies of related diversification (both vertical and horizontal) may offer greater stability or growth of corporate profits and the spreading of risks, on the grounds that not all activities in the diversified firm's portfolio will follow an identical cycle of demand. However, a strategy of related diversification keeps the firm "close to home," though not necessarily in a geographic sense.

An alternative strategy is that of *unrelated diversification*, which results in the emergence of conglomerate business organizations. The term *con-

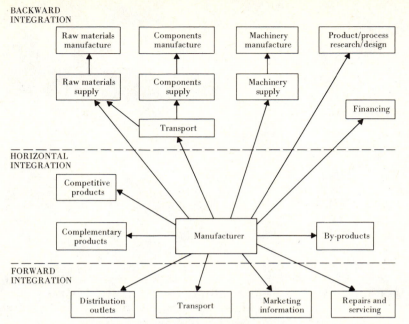

Figure 8.3 Strategies of related diversification. *Source:* G. Johnson and K. Scholes (1988), *Exploring Corporate Strategy* (Hemel Hempstead, England: Prentice-Hall), fig. 6.4.

glomerate is based on the geologic analogy of a rock made up of fragments of material of unrelated and diverse origins. The conglomerate form of business organization developed especially strongly in the United States during the 1960s when it became the most common type of strategy adopted by fast-growing firms. In 1949, more than one-third of the 500 largest U.S. corporations were single-business firms; by 1969, only 6 percent were of this type. Conversely, the proportion engaged in unrelated business activities—the true conglomerates—increased from 3 percent of the total in 1949 to almost 20 percent in 1969 (Rumelt, 1974). The usual explanation for diversification is *synergy*: the idea that the whole is greater than the sum of the parts or, expressed in shorthand terms, that 2 + 2 = 5. Some of the enthusiasm for conglomerate diversification seems to have waned in recent years as some of the firms that grew very rapidly through such a strategy in the 1960s and 1970s experienced disappointing results. The idea that good management can manage anything under the sun is no longer as popular as it was. A significant number of conglomerates, therefore, have disposed of some of their unrelated lines of business and begun to concentrate on activities more closely related to their core businesses, in terms of technology or product market. Nevertheless, the conglomerate form of business organization, based on a strategy of unrelated diversification, is still extremely common.

In discussing these various competitive strategies, we have adopted a strict two-dimensional view. In fact, business strategies are more complex than this. In Figure 8.4, *geography* is added specifically as a third dimension. Thus the directions in which corporate strategies may be developed are market type, product technology, and market geography.

Figure 8.4 shows three hypothetical cases. Firm A adopts a status quo strategy, remaining in its existing markets and technology. Firm B is at the opposite extreme; it has modified its strategies along all three dimensions. Firm C pursues an intermediate strategy whereby the firm continues to serve the same type of customer (market type) using the same product technology but expands into new geographic markets. In the earlier years of the twentieth century in the United States, such geographic expansion was primarily intra-national, to different parts of the country. Increasingly, however, geographic expansion by large firms has become international as formerly domestic enterprises have become transnational corporations. Such geographic expansion may take one or more of the strategic directions discussed here. In general, when a firm first ventures overseas, it tends to replicate its domestic operations. But with increasing international experience and as it develops a more complex and sophisticated international production network, its international strategies tend to become more adventurous.

Our discussion of competitive strategies has concentrated so far on strategies of expansion. But strategic decision making is rarely purely expansionary in every respect. An integral part of all strategic behavior is the possibility of *strategic withdrawal* from a particular business activity. A firm may decide to disinvest from a line of business because it no longer meets corporate objectives—for example, in terms of profitability or market growth. A product that has reached the late stage of its life cycle may well be dropped to make room for new products. In some cases, a firm may be forced out of a market niche by more powerful competitors. Indeed, one of the dangers of pursuing a narrowly focused niche strategy, especially for the smaller firm, is that a larger competitor will do exactly that. Withdrawal and disinvestment, therefore, are just as integral to corporate competitive strategy as the other alternatives discussed.

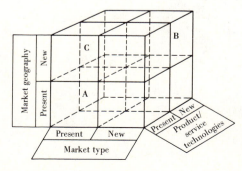

Figure 8.4 Business strategies in three dimensions. *Source:* H. I. Ansoff (1987), *Corporate Strategy* (Harmondsworth, England: Penguin), fig. 6.2.

Methods of Implementation

Each of these business strategies, whatever their geographic dimension, can be implemented in one or both of two main ways (Figure 8.5). One method is for the firm to grow by internal means, by building new production and distribution facilities or by expanding its existing facilities. Although this method has many advantages, it may not always be the most suitable. For example, a firm may wish to move into a new product market but find that there are major barriers to entry. These might be technological barriers (the firm does not have the necessary know-how); there might be marketing barriers (the firm does not have a presence or "brand image"); There might be distributional barriers (the firm does not have access to distribution channels and marketing outlets); and so on. Each of these could probably be built up internally over time. But time might be at a premium; the firm may have to move quickly. In these circumstances, the way into the new product market is to acquire or merge with a firm already in that line of business—that is, to grow by external rather than internal means. Acquisition and merger are especially likely to be the preferred method of expansion for a firm adopting diversification strategies.

There is no doubt that a very large proportion of the growth of the really large businesses in the United States and other industrialized economies can be attributed to the processes of acquisition and merger. Figure 8.6 shows that the process tends to occur in waves, with high peaks of intensive merger activity separated by troughs of relative quiescence. Between the 1890s and the early 1970s, for example, three major peaks of acquisition and merger activity occurred in the United States. The merger activity of the turn of the century formed the basis on which the modern industrialized structure of the United States was built. A structure that was strongly competitive (it consisted of large numbers of relatively small firms) was transformed into an oligopolistic structure (one dominated by a relatively small number of very large firms).

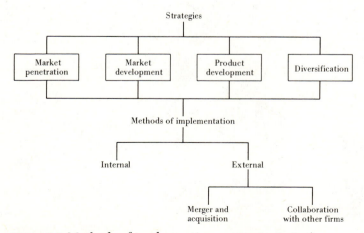

Figure 8.5 Methods of implementing corporate strategies.

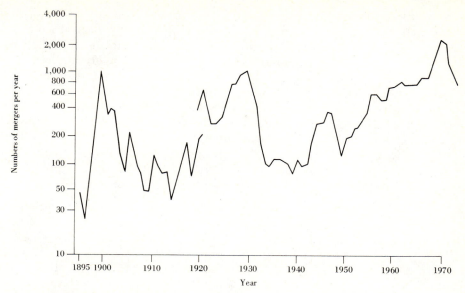

Figure 8.6 Merger activity in the United States, 1895–1972. *Source:* U.S. Federal Trade Commission.

Between 1897 and 1905, an average of 352 firms per year disappeared through merger, the peak year being 1899, when 1,208 firms were absorbed by other firms. During this period the first really gigantic business organizations came into being, firms such as U.S. Steel, American Tobacco, Du Pont, Corn Products, and Anaconda Copper.

The second major peak of merger activity in the United States occurred in the latter half of the 1920s. Between 1925 and 1929, almost 4,500 firms disappeared through merger, an average of 890 per year, with more than 1,200 disappearing in 1929 alone. This was the period in which firms like General Foods were formed by the merger of several food companies. The deep recession heralded by the 1929 Wall Street crash brought this second merger wave to a close. As Figure 8.6 shows, the high levels of these two waves were not surpassed until the 1960s, even though, of course, mergers continued to occur. The 1960s merger peak was even higher than its two predecessors. Between 1959 and 1968, a total of 11,142 manufacturing and mining firms were acquired with a total value of $54.55 billion. More and more firms began to spend their investment capital on buying existing firms than in creating new productive capacity. According to the U.S. Federal Trade Commission, in 1952, less than 4 percent of total corporate spending on new capital investment was devoted to the acquisition of other companies; by 1968, this figure had risen to 55 percent. In other words, 55 cents of every corporate investment dollar went into acquisition. For many firms, all their capital investment was acquisition investment. Many of the very big firms grew entirely this way.

Each of these merger waves was rather different in its emphasis. The first one was characterized by mergers that were predominantly horizontal, the aim being to reduce competition by swallowing up firms in the same lines of business. The 1920s wave continued this trend but expanded to encompass vertical mergers as firms acquired businesses upstream and downstream of their existing activities. The 1960s merger wave was distinguished most of all by diversified-conglomerate mergers. The pace of conglomerate mergers was frantic as firms moved into and out of different businesses at will. In some cases, companies or parts of companies were bought and sold in such a way that a *Business Week* writer was moved to observe that conglomerates were servicing industry in the way that Bonnie and Clyde serviced banks!

This temporal pattern of merger activity tends to support the notion of a kind of three-stage evolutionary sequence (Bannock, 1971):

1. Early growth emphasized the removal of competition by absorbing competing firms, leading to oligopolistic structures.
2. The next logical step for the large oligopolistic firm was to protect both its source of supply and its markets by vertical integration.
3. The third step, having ensured protection of supplies and outlets and a high market share, was to expand and to diversify into new product markets.

Of course, these are generalized stages, and not every large firm followed the precise sequence.

Merger activity since the early 1970s is less well documented, but there is no doubt that the same general wavelike pattern has continued. After the trough of the 1970s and early 1980s in the wake of the 1974 and 1979 oil crises, the mid-1980s witnessed a huge volume of merger activity and the emergence of megamergers involving staggering financial assets. A particular feature of recent years has been the huge increase in overseas acquisitions, a reflection of the increasing internalization of business activity. American firms have long experience in taking over firms in Europe, for example. But since the early 1970s, this tide has turned, and more recently, there has been a tidal wave of European firms buying up domestic firms in the United States. Figure 8.7 shows the increase in acquisitions of American companies by British firms between 1976 and 1986. In fact, acquisition has been the dominant mode of entry by British firms into the United States. Around 60 percent of the total number of investments made in the ten-year period were acquisitions.

The other external method of pursuing a particular strategy, shown in Figure 8.5, is that of *collaboration* with other firms, but in a form that maintains each firm's independent status. Collaborative ventures seem to be especially important in certain kinds of economic activity. They are very common, for example, in large-scale construction projects, such as the building of dams or airports, projects that are too large for individual firms to undertake alone. They are very common in research and development, especially in the very expensive areas of aerospace, telecommunications, and the like. In some industries, such as automobiles, firms enter into collaborative agreements to produce major components like transmissions or engines.

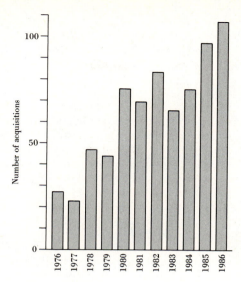

Figure 8.7 Number of acquisitions of U.S. firms by British firms, 1976–1986. *Source:* U.S. Department of Commerce.

A common type of interfirm collaboration to allow a firm to penetrate a new market is *licensing,* whereby a firm licenses its product for manufacture and distribution by another firm already in that market. The originating firm is paid a licensing fee and royalties on sales but keeps control of the product and its technology. The license applies only in the specified market and for a specified period of time. Another common form of interfirm collaboration is *franchising,* an arrangement whereby firms combine to exploit a particular branded product. The owner of the brand provides the basic product, its brand name (the most valuable property in franchising), and mass advertising. The franchisee provides capital, local production, packaging, and selling. Franchising is especially common in service industries (particularly fast-food sectors) and in such industries as soft drinks (Coca-Cola operates an immense international franchising operation). Geographers have tended to ignore collaborative ventures between firms, yet they are enormously important. Just think how different the landscape would look without the McDonald's, Burger King, and thousands of other franchise operations that exist throughout the land. But these are merely the most visible expressions of a much more pervasive phenomenon.

STRATEGY AND STRUCTURE: HOW ORGANIZATIONS ARE ORGANIZED AND REORGANIZED

"Structure Follows Strategy": Types of Organizational Structure

Corporate and competitive strategies are devised to enable business organizations to deal with the volatility and uncertainty of their external environments and to meet their goals and objectives. But the adoption of particular strategies has implications for the firm's organizational structure. Indeed, to be able to pursue certain strategies effectively, the firm has to design an appropriate organizational structure. This is the line of argument Alfred Chandler introduced in his classic (1962) studies of the evolution of the American Corporate economy:

> As the adoption of a new strategy may add new types of personnel and facilities, and alter the business horizons of the men responsible for the enterprise, it can have a profound effect on the form of its organization. Structure can be defined as the design of organization through which the enterprise is administered. This design, whether formally or informally defined, has two aspects. It includes, first, the lines of authority and communication between the different administrative offices and officers and, second, the information and data that flow through these lines of communication and authority. Such lines and such data are essential to assure the effective coordination, appraisal, and planning so necessary in carrying out the basic goals and policies and in knitting together the total resources of the enterprise. These resources include financial capital; physical equipment such as plants, machinery, offices, warehouses and other marketing and purchasing facilities, sources of raw materials, research and engineering laboratories; and, most important of all, the technical, marketing and administrative skills of its personnel. The theories deduced from these several propositions is, then, that structure follows strategy and that the most complex type of structure is the result of the concatenation of several basic strategies. (pp. 13–14)

Chandler was able to demonstrate, through his detailed historical study of the evolution of the large industrial enterprise in the United States, a developmental sequence of organizational structures. Significantly for us, geography plays a central role in this sequence along with the increasing complexity of business activity.

Each of the three major stages of development Chandler identified (Figure 8.8) represents an increase in organizational complexity. In stage I, the dominant form of business organization was the extremely simple single-function, single-product, single-plant firm. This firm is characteristic of perfectly competitive and monopolistic markets (see Figure 7.2). Most businesses in the peripheral segment of the modern economy are still of this type. During the mid-nineteenth century, it was the characteristic form of business organization throughout the United States and Europe. In this kind of organization, there is no clear differentiation among the three "layers" of decision making (strategic, operating, and administrative). Such an organizational structure is quite adequate for a modest scale of business operations. But as enterprises expanded, both in sheer size of output and, more important, in the geographic spread of their operations, the need arose for an organizational division of labor in which specialist departments were created to undertake specific functions. Such functional specialization, together with the operation of a number of geographically dispersed plants rather than just one, demanded a far greater degree of central control. Thus developed the separate headquarters unit, whose basic function was to plan the overall policy of the firm as a whole and to coordinate the activities of all the other organizational units. Chandler and Redlich (1961)

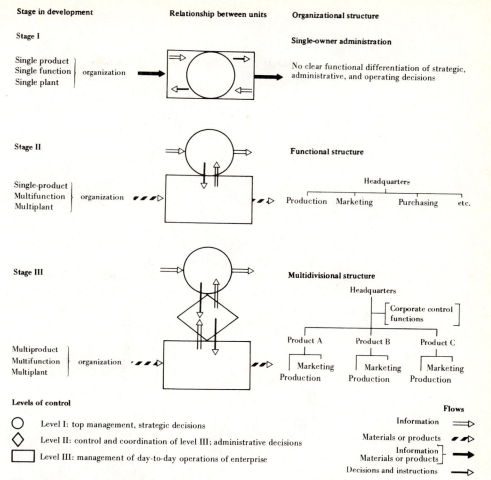

Figure 8.8 Stages in the development of organizational structures.

are quite categorical in attributing primary importance to the role of geography in this organizational development (stage II in Figure 8.8):

> Modern structure and administration of industrial enterprise began in the United States with the geographical dispersion of such firms. That is, it began when manufacturing enterprises came to possess a number of factories by building or buying new units or by combining with other firms. Geographical dispersion was the initial step in making modern industrial enterprise, because it made necessary the distinction between headquarters and field. (p. 6)

The multifunctional, multiplant, but still single-product firm required a functional type of organizational structure. But such a structure is too rigid and inflexible for a strategy that emphasizes increasing product diversification. Hence a third kind of organizational structure evolved: the *multidivisional* structure (stage III in Figure 8.8). Instead of having departments organized on functional lines, each major product is centered in a separate division. In the multidivisional firm, each division is responsible for a specific product within the firm's portfolio. The main advantage of this structure is that each division deals with a specific business environment—the industry in question—with its particular complex of market, technological, and regulatory conditions. In a multidivisional structure, therefore, organization is by product rather than by function. Each product division is responsible for its own basic functions of production and marketing, although some functions—notably finance—tend to be performed centrally by the headquarters for the entire corporation. The main advantage of the multidivisional structure was seen to be its ability to cope with a diversity of products. Not surprisingly, therefore, as American firms became increasingly diversified in the 1960s, they also tended to adopt this organizational structure. In 1949, only 20 percent of the 500 largest U.S. firms were using this form of organization; by 1969, more than 75 percent had adopted it (Rumelt, 1974).*

An important consideration in modern business organizations is the degree of interdependence between individual activities and functions, that is, the extent to which elements in the firm's value chain need to be tightly or loosely linked in an organizational sense. Again, the particular strategy being pursued is very important in determining the extent of such interdependence:

> The specific strategy that any multiunit firm decides to follow will . . . define the degree of diversity the firm must manage by virtue of the range of market, technological, and economic conditions faced by its various divisions in their respective industrial environments . . . [and] it will determine the extent of required and/or potential interdependence among the various divisions and the corporate headquarters. (Lorsch and Allen, 1973, p. 10)

The organizational theorist J. D. Thompson (1967) developed the concept of *internal interdependence* within organizations. He showed how organizational structure could be used to meet contingencies in the organization's task environment (see Chapter 7). Thompson identified three types of internal interdependence: pooled, sequential, and reciprocal. These reflect different organizational strategies and define the likely degree of autonomy or interdependence of individual units within the organization. They can be broadly equated with the strategies we have been discussing in this chapter. Figure 8.9 depicts the control and interdependence relationships among units in such organizations.

Organizations pursuing a conglomerate strategy, with a high degree of product diversification and a multidivisional structure, tend to display *pooled* interdependence. In such organizations, there is little or no direct connection

---

* The multidivisional structure remains the dominant form of modern business organization, although there are several variants on the theme and also other attempts at more complex types, such as matrix structures. For a detailed discussion, see Johnson and Scholes (1988).

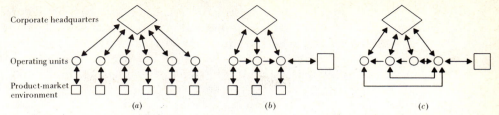

Figure 8.9 Interdependence relationships in different types of business organizations: (*a*) Pooled interdependence (a conglomerate enterprise); (*b*) Sequential interdependence (a vertically integrated enterprise); (*c*) Reciprocal interdependence (a large single-product enterprise.) *Source:* P. Dicken (1976), "The Multiplant Business Enterprise and Geographical Space: Some Issues in the Study of External Control and Regional Development," *Regional Studies* 10: 401–412, fig. 2.

between units operating within different product-market environments, although each unit makes a discrete contribution to the overall enterprise and each is sustained by the whole enterprise (Figure 8.9*a*). Firms that are vertically integrated (Figure 8.9*b*) have a much greater degree of interdependence among their component product units. Such interdependence is termed *sequential* because the output of one unit within the firm is the input for the next unit, and so on. The third type of internal interdependence, shown in Figure 8.9*c*, is *reciprocal* interdependence.

The essence of the argument in our discussion of competitive strategies and of evolving organizational structures is that business organizations are learners. But organizational learning is much more than just getting better at producing a specific product or performing a specific process. That is the kind of learning curve usually discussed in the industrial economics literature whereby the more of a given product a firm produces over time, the less each unit will cost to produce because of learning. Here our concern has been broader. We have been concerned with the ways in which business organizations learn to devise new strategies to deal with volatile and uncertain environments—how they learn to devise new forms of organizational structure. Both types of learning process develop either directly out of the firm's own experiences or indirectly through observation of the experiences of other firms. As in all populations, of course, there are good learners and bad learners; the latter tend not to survive in the harsh competitive environment of the contemporary economy.

The Geography of the Large Business Organization

As Robert Haig (1926) pointed out, "Every business is a package of functions, and within limits, these functions can be separated and located at different places" (p. 416). We have seen that geography has played an extremely important role in shaping the organizational structure of the modern business enterprise. Indeed, Chandler and Redlich (1961) claimed that geography was instrumental in the transformation from a simple, undifferentiated structure to a more complex functional structure. At the same time, however, the modern business organization, in implementing its specific strategies and adopting specific organizational structures, creates a new and complex geography of its own. In fact, the geographic structure of the large business organization is far more complex than the multidivisional structure implies. The kind of organization chart shown in the rightmost column of Figure 8.8 is merely the skeleton of the lines of control and communication within the organization. Each individual division may well consist of scores of individual offices, factories, laboratories, and marketing units. The geographic structure of such complex systems is a mixture of the historical and the very recent, of decisions made in the long-distant past as well as those made yesterday. It is a mixture, too, of locations consciously planned by the firm itself and of locations the firm has acquired from other firms it has taken over. Given such complexity, can we identify any general patterns in the geography of large business organizations? To a certain extent we can, but we should beware of assuming that such generalizations apply in each and every case. They are tendencies, not universals.

In effect, the firm has to decide on a location for every element in its value chain, both primary and support activities (see Figure 7.8). Of course, such decisions are not made on a blank map. For firms already operating a number of plants, the question will be a mixture of reallocation of activities between existing units and choosing totally new locations. Two related locational questions are involved:

1. Where in geographic space should a particular activity be located?
2. To what extent must the unit be located near other elements in the firm's value chain? That is, how far can the elements in the business package be separated and located in different places?

The first question relates to the kinds of variables we examined in Part One of this book: spatial variations in the availability and cost of production inputs or in access to markets. The second question relates to the extent to which a firm's separate activities need to be concentrated or dispersed organizationally. Do successive stages in the value chain need to be located next to one another, or can they be located separately in order to take advantage of geographic variations in input or output conditions?

Figure 8.10 depicts some of the possible arrangements. It shows that each of the three major layers in the "organizational cake" has different internal and external relationships (Figue 8.10*a* and 8.10*b*). It also shows three hypothetical geographic arrangements of the three organizational levels (Figure 8.10*c*). At the left in Figure 8.10*c*, the headquarters unit is located separately from all other units but each production unit (III) is located in conjunction with an administrative unit (II) that looks after its day-to-day administration. At the right in Figure 8.10*c*, a rather different spatial arrangement is shown. Here both the headquarters and the lower-level administrative units are located together while the firm's production units are dispersed. In the central example, each of the three levels is geographically separated from the others.

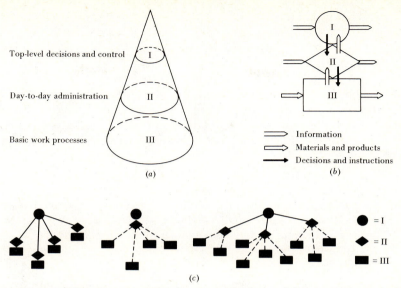

Figure 8.10 Alternative geographic arrangements of the operations of multiplant enterprises: (*a*) the internal hierarchy of business functions; (*b*) the major types of flows within a business organization; (*c*) some possible alternative geographic arrangements of a firm's business functions.

Since Robert Haig described the package of business functions more than six decades ago, technological developments in transportation and communications and in production and organizational technology have greatly increased the potential ability of business organizations to arrange their administrative and operating units in ways that fit their overall corporate requirements. Haig's limits have been greatly extended. In particular—largely because of innovations in the transmission and processing of information—efficient contact between administrative and operating units no longer necessarily depends on their being located at the same place. Each can be located according to its own specific needs. The point is that each activity in the firm's value chain does have its own specific locational requirements and that these can be satisfied in various types of geographic location. Each activity, therefore, tends to develop rather distinctive spatial patterns.

The Geography of Corporate Control and Administrative Functions
Geographers have devoted a good deal of attention to the spatial tendencies and changing geographic patterns of corporate control and administrative functions in the United States.* The corporate headquarters is the locus of

---

* See, for example, Borchert (1978); Goodwin (1965); Harper (1987); Pred (1974); Semple and Phipps (1982); Stephens and Holly (1980).

overall control of the entire business organization. Its functions are to make the strategic decisions that shape and direct the entire enterprise. Its concern is with the major allocative, investment, and disinvestment decisions. Its most important role is financial: it holds the purse strings for the entire organization. Much of its activity also involves high-level negotiations with other, similar organizations. In other words, corporate headquarters functions deal, for the most part, with nonroutine situations, including negotiating and bargaining with top executives in other organizations and meeting with government and other public officials. Tornquist (1968) argues that these important contacts

> cannot be maintained with adequate efficiency by letter and telecommunications but demand direct personal contacts between personnel and thus passenger transportation. That these contacts demand the personal attendance of often highly expert personnel is probably bound up with the fact that contacts in many cases involve considerable elements of what we might call problem-solving planning, keeping an eye on the course of events, pulse-feeling and reconnaissance. The contacts often take the form of talks and discussions in which personal effort is of great importance. (p. 101)

These face-to-face contacts are maintained and developed most easily if such contact-intensive activities agglomerate in information-rich locations. Hence the top control functions of modern business organizations show a very strong tendency to concentrate in a few major metropolitan complexes. These are locations that offer three particular advantages (Pred, 1974):

1. A greater potential for interorganizational face-to-face contacts
2. The availability of specialized business services, including financial, legal, and advertising
3. High levels of accessibility to other metropolitan areas

As a result of these characteristics, the geographic pattern of corporate headquarters tends to be closely matched to the urban hierarchy. The headquarters of the highest-order business organizations tend to be located in the higher-order urban centers. Although the map of corporate control in the United States is less concentrated than in the United Kingdom, France, and some other European countries, the fact remains that American corporate headquarters are predominantly located in a few key centers (Figure 8.11).

Harper (1987) shows that although the leading corporate headquarters were spread over 69 different metropolitan centers in 1984, no fewer than 296 of the 367 largest business firms in the United States (81 percent) were located in just 19 centers. The relative importance of individual metropolitan centers varies according to whether number of headquarters or aggregate corporate sales is taken as a measure. As Figure 8.11 shows, some centers rate higher or lower depending on these criteria. For example, Detroit was the second most important headquarters center in terms of corporate sales controlled but ninth in terms of number of headquarters. Conversely, Dallas–Fort Worth ranked tenth in terms of sales but fifth in terms of number of leading companies headquartered there. Overall, however, headquarters of the leading U.S. com-

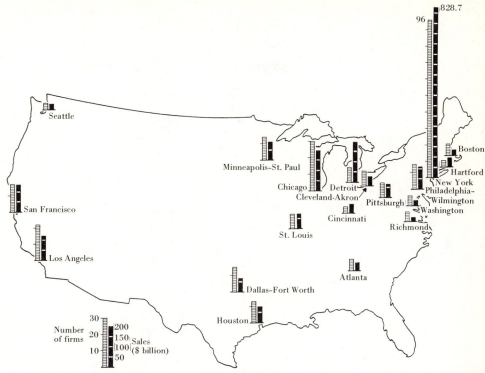

Figure 8.11 Headquarters locations of the leading U.S. corporations. *Source:* After R. A. Harper (1987), "A Functional Classification of Management Centers in the United States," *Urban Geography* 8: 540–549, table 2.

panies in 1984 were concentrated in New York, Chicago, San Francisco, Los Angeles, Philadelphia-Wilmington, and Detroit. These six metropolitan centers contained more than half of the leading companies' headquarters, accounting for almost two-thirds of their aggregate sales.

Over time, the geographic pattern of corporate control in the United States has remained remarkably stable. But changes have been occurring. At the broad regional scale, the geographic pattern has become more dispersed. This reflects mainly the rapid growth in recent years of major centers and companies in the South and West. At the scale of the metropolitan system as a whole, more stability is evident, as Stephens and Holly (1980) show for the 20-year period 1955–1975 in Figure 8.12. But at the local geographic scale, there is evidence of substantial change as many business organizations have shifted their corporate headquarters to suburban locations in the same metropolitan areas or to smaller metropolitan areas in the general vicinity. This trend has been especially apparent in and around New York City. Figure 8.12 summarizes the changes in the corporate headquarters locations of the largest 500 industrial corporations between 1955 and 1975.

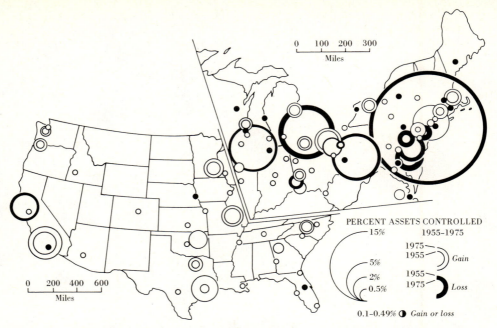

Figure 8.12 Changes in the pattern of headquarters of the largest 500 industrial corporations in the United States, 1955–1975. *Source:* J. D. Stephens and B. Holly (1980), "The Changing Patterns of Industrial Control in Metropolitan America," in S. D. Brunn and J. O. Wheeler (eds.), *The American Metropolitan System* (New York: Halstead Press), fig. 24. Reprinted by permission of Edward Arnold Ltd.

These geographic changes have been brought about by a number of processes, including the differential growth of large corporations in their existing locations. Some of the change, particularly at the local scale, is the result of the physical relocation of corporate headquarters. But only a small number of these tend to be moves to metropolitan centers in distant states. The majority are short-distance. For example, only 5 of the 33 major headquarters relocations out of New York City between 1967 and 1971 and 5 of the 13 between 1982 and 1987 were long-distance moves. The rest—almost 80 percent—were moves to New Jersey and Connecticut. However, a large part of the explanation of intermetropolitan headquarters change in the United States lies in the processes of acquisition and merger, which we have already seen to be an immensely important component of corporate change. Where a firm in one location is taken over by a firm in another location, the effect is a geographic shift in corporate control. Figure 8.13 shows the extent of such shifts in the United States between 1955 and 1968, a period that included one of the major waves of merger activity (Figure 8.6). Of the 354 companies with assets of more than $25 million acquired by the leading 200 corporations, 78 percent were acquired by firms headquartered in only six states. Of these, New York accounted for 43 percent and California for 14 percent. New York was the leading acquirer of large companies in almost every state.

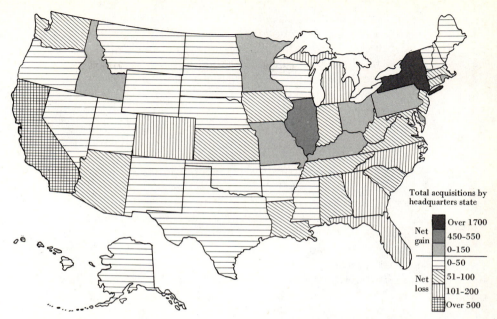

Figure 8.13 Geographic shifts in the control of industrial enterprises as a result of merger and acquisition. *Source:* P. Dicken (1976), "The Multiplant Business Enterprise and Geographical Space: Some Issues in the Study of External Control and Regional Development," *Regional Studies* 10: 401–412, fig. 1.

Corporate headquarters, therefore, tend to agglomerate in metropolitan centers that both are information-rich and also possess a high level of executive living standards. The more routine administrative functions also tend to concentrate in information-rich locations. However, because many of their contact activities can be more readily carried out indirectly via telephone, telex, fax, and mail communications, their need to locate in the highest-level metropolitan centers is less pronounced. Nevertheless, they do depend heavily on a supply of white-collar workers, good communications systems, and information flows, needs that can be met only in urban centers of some size. Thus there is a very clear association between the upper levels in the organizational hierarchy and the upper levels in the urban hierarchy.

The Geography of Corporate Research and Development   Some other corporate functions also seem to display a similar kind of geographic orientation with a strong preference for locating in large metropolitan centers. One such function is corporate research and development (R&D). We noted in Chapter 4 (see Figure 4.10) the tendency for industrial R&D activity in general to concentrate in metropolitan centers. Malecki (1979) looked specifically at the locational tendencies of corporate R&D and at changes that occurred in its geography between 1965 and 1977. The precise locational needs of corporate R&D laboratories depend on their specific type and function. Malecki distinguishes between centralized and decentralized corporate R&D. Centralized R&D tends to be large-scale team research, performed for a substantial part of total corporate business activity. Decentralized R&D facilities, in contrast, tend to carry out work for a single product division. Both types of R&D, but especially the first, require the availability of a number of factors, including access to a highly skilled technical labor force. These in turn tend to prefer the cultural, educational, and economic amenities available in large metropolitan centers. Other favored locational attributes include proximity to government research laboratories, research universities, and concentrations of appropriate types of manufacturing activity. For centralized R&D, a key locational need is geographic proximity to the firm's corporate headquarters. This is especially true for long-range or basic research that is closely linked to the firm's strategic decision making.

In Malecki's (1979) view, "The compound effect of these various attracting forces is that R&D tends to agglomerate in large urban areas where research universities, manufacturing and corporate headquarters are common" (p. 310). The specific relationships between the location of R&D facilities and corporate headquarters was clearly shown in Malecki's empirical study of the R&D laboratories of 330 of the largest United States corporations. He found that 88 percent of these firms had a joint headquarters–R&D laboratory location pattern: "This dominant locational linkage accounts, in part, for the concentration of R&D, since corporate headquarters are among the most agglomeration-oriented of corporate activities" (p. 314).

Figure 8.14 shows the changing levels of concentration of corporate R&D laboratories in the 24 leading U.S. metropolitan areas in 1965 and 1977. In both years, the New York–New Jersey area dominated the picture, but its share of the total declined from 22 percent to 16.5 percent. The biggest relative increase was in the Los Angeles metropolitan region, which gained 30 laboratories (compared with New York's loss of 14) and increased its share from 9.5 to 12.1 percent. Overall, however, the pattern corresponded quite closely to that of corporate headquarters (see Figures 8.11 and 8.12). At the same time, Malecki observed a trend toward a wider geographic distribution of corporate R&D laboratories and their location in a larger number of metropolitan areas, particularly smaller metropolitan areas, and in some nonmetropolitan areas as well. But as in the case of corporate headquarters, the overall degree of locational stability remained strong. A more recent survey by the Conference Board (1987) supports this view. In surveying the reasons for firms' choice of R&D location, the study found that the majority of new corporate laboratories were being established next to existing corporate headquarters, universities, and other research laboratories. The locational preferences of these corporate decision makers were similar to those suggested by Malecki: a local supply of scientific and technical labor, proximity to corporate headquarters, and the area's quality of living.

The Geography of Corporate Production Units   As we move down the corporate hierarchy to the level of the production units, it becomes far more

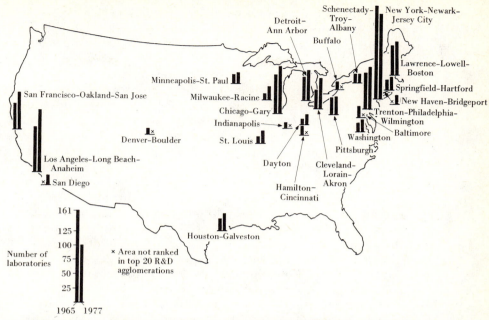

Figure 8.14 The location of corporate research and development laboratories in major metropolitan centers. *Source:* E. J. Malecki (1979), "Locational Trends in R&D by Large U.S. Corporations, 1965–1977," *Economic Geography* 55: 309–323, table 2.

difficult to generalize about their spatial tendencies. Compared with corporate headquarters, more routine administrative functions, and R&D laboratories, there is no doubt that production units tend to be more widely dispersed geographically. But no single and simple spatial tendency is common to all production activities. Whereas the locational needs of corporate headquarters, offices, and laboratories are roughly similar for all firms regardless of the industry involved, this is not true for the firm's production activities. Their locational requirements vary considerably, depending on the specific organizational and technological role they play within the organization and the geographic distribution of their specific inputs and outputs.

Hence the production units of the large business enterprise show a good deal more variability in their locational requirements—and therefore in their locational tendencies—than the other functions do. Despite the developments in production technology, which have drastically reduced the quantities of materials required for many types of manufacturing activity, the fact remains that some production units are still strongly oriented toward localized raw material sites. For some production processes, such as iron and steel manufacture, it is necessary for the separate stages in the production sequence to be located next to each other in order to save on energy costs in cooling and reheating the semifinished products of each stage in the process.

There is similar variation in the case of labor requirements. Different types of production units are sensitive to different types of labor.* Some production activities require highly skilled, highly educated labor, which tends to be concentrated in certain geographic locations, notably major metropolitan areas and a few other localities where the quality of living is high. Production units requiring labor of this kind will therefore tend to be pulled toward such locations. Of course, certain locations are especially attractive to corporate headquarters and R&D laboratories. Thus there are at least two reasons why some production units may be located in geographic proximity to headquarters and R&D: one is a functional reason arising from the need for the activities to be linked; the other is simply that different corporate activities are seeking the same types of labor.

Production units that require less-skilled labor have much greater locational choice since supplies of such labor are more widely available. This availability extends not only to other parts of the United States but also to overseas locations in such areas as Mexico and the Caribbean, Latin America, and East and Southeast Asia.† The production activities involved in such geographic dispersal, both within the United States and overseas, tend to be of two broad types. First, there are products at the mature stage of the product life cycle in which the technology has become standardized, long production runs are needed, and semskilled and unskilled labor costs are very important (an obvious example would be textiles and clothing). Second, certain parts of the production process of newer industries are also labor-intensive and amenable to the employment of such labor even through the industry as a whole is highly capital- and technology-intensive. The clearest example of this is in the electronics industry, the assembly stage of semiconductor production.

The particular geographic configuration of corporate functions has enormously important implications for the uneven development of the space economy (Chapter 6). The operation of spread and backwash effects and, indeed, the entire spatial center-periphery structure of economic activities is made even more complex by the increasingly dominant economic role of the very large multiproduct, multiplant business enterprise, with its highly specialized organizational and geographic structure. The precise way in which corporate functions are organized in space and the links that are established between them are of very great importance in "steering" growth impulses through the space economy. Norcliffe (1975) subdivides production units into activities that mainly process raw materials, those that fabricate such processed materials into finished or intermediate products, and those that integrate fabricated inputs without further processing. In dynamic terms, the locational shifts in each of these functions show distinct patterns. As we have seen, the administrative and control functions have a marked tendency to

---

* The whole issue of labor and the labor process in a geographic context is discussed in detail in Chapter 10.

† For a detailed discussion of this internationalization process, including case studies of such industries as textiles and clothing, automobiles, and electronics, see Dicken (1986).

concentrate in large agglomerations because of their information-rich and contact-intensive qualities.

Conversely, fabricating activities are tending to be moved out of the inner parts of large centers either to the edges or to medium and small-sized centers within the core or even in the periphery. The kinds of activity involved are almost invariably those of a routine and standardized nature using semiskilled often female, labor. Integrative activities, which require access to suppliers, and specialized functions such as research and development tend to remain closely associated with the core.

Such geographic segregation of functional units in multiplant business enterprises has some far-reaching spatial implications. In terms of a single multiplant firm, decisions affecting its individual units sited in different locations (to expand or contract, to close or modify, and so on) are taken at its headquarters location. This has very important repercussions for the operation of multiplier effects and for differential growth both within the core and between the core and the periphery. In this respect, Pred (1973, 1974a, 1974b) identifies three ways in which the decisions made by the multiplant enterprise are significant.

In the first place, a change in the scale of activity of a multiplant enterprise creates more complex multiplier effects than those discussed in Chapter 6. For example, suppose that a multiplant enterprise headquartered in a major metropolitan center decides to expand the output of one of its products. Wherever in the organization such expansion takes place, it will have a differential effect both on other linked units within the same organization (intraorganizational effects) and on units of other enterprises with which it has functional relationships (interorganizational effects). We already know that such linkages may be either forward or backward, depending on their input-output character. Thus, insofar as one unit's expansion creates the need for additional or new inputs, for example, multiplier effects will occur in the other units affected, either directly or indirectly. Not only do these multiplier effects have an organizational dimension, but they also have a geographic dimension. The multiplier effect of the expanded unit may be both local (other enterprise units in the same locality) and nonlocal (enterprise units located in other places).

The second way in which multiplant enterpises influence spatial change relates to their role as transmitters, processors, and receivers of specialized information, particularly that involved in "growth-inducing" innovations. Pred identifies three types of growth-inducing innovations that have effects within the urban system:

1. Innovations that provide new products or services for intermediate or final demand markets
2. Innovations that involve the use of new production processes
3. Innovations that relate to changes in the organization itself, such as new organizational structures or new planning and decision-making procedures

These three categories are, of course, not mutually exclusive; adopting one often involves adopting one or both of the others. Any one of these innovations may lead to increased employment at the location involved and, through the intra- and interorganizational multiplier, possibly at other locations. More generally, much of the flow of information in modern society is within multi-unit organizations, which means that their locational characteristics have an important influence on information flows and therefore on the kinds of decisions made. There is a tendency for many innovations to be introduced first at or very close to the enterprise's headquarters location and only later diffused to the chosen subordinate units at other locations.

The third influence of multiplant enterprises on center-periphery development relates to what Pred calls the *accumulation of operational decisions*. Quite apart from the decision to do something "new"—the growth-inducing innovations just discussed—many operational decisions, some of them highly routine, have spatial growth repercussions as they are implemented by multiplant enterprises. Each decision may not individually have a very pronounced spatial impact; however, taken as a cluster of decisions made by organizations whose control units are themselves geographically clustered, the accumulated effect of such operational decisions can be considerable.

The essence of our argument relating the behavior of large multiplant enterprises to developments within the urban system and between center and periphery is that it results in the emergence of *centers of control* with high concentrations of decision-making and innovative activity. Decisions made in these centers steer growth impulses and thus have tremendous repercussions throughout the economic system. In general, the accumulated effect of these decisions is to reinforce still further the strength of existing concentrations and to perpetuate the distinction between center and periphery.

SPATIAL REORGANIZATION WITHIN LARGE BUSINESS ORGANIZATIONS

Some General Principles

In examining the geography of the large business organization, we simply identified some general tendencies and related these, in a limited way, to the idea that different activities in the firm's value chain have different locational requirements and that these requirements can be met in certain types of geographic locations. In effect, we were aggregating across all large business organizations in order to pick out some general spatial tendencies.

At this point, we need to consider the question of spatial change and reorganization within large business organizations. Spatial change—change in the number, size, function, and geography of a firm's activities—is more than simply a question of the location decision as traditionally understood in economic geography. To repeat the point made at the beginning of this chapter, location decisions are an integral part of the broader set of investment decisions. Some of these decisions may involve the choice of a new location for a particular activity, but the majority involve the reallocation of resources and activities between existing plants.

A major difference between the large multiplant firm and the single-plant firm is that the multiplant firm has far greater potential flexibility. By definition, single-plant firms operate entirely at one geographic location. As long as they remain single-plant entities, all changes and adjustments must be accommodated at that site. Multiplant firms, by contrast, operate at several locations (in the case of very large firms, at vast numbers of locations). They not only divide their operations among these different sites in ways determined by their own organizational design and structure, but they can also reallocate functions between their plants and cross-subsidize the loss-making operations of one plant with the profit-making operations of another if this is regarded as contributing toward the firm's overall interests. Differences in potential flexibility, then, represent an important distinction between single-plant and multiplant firms. If faced, say, with local problems that interfere with production either temporarily or even permanently, the response of the single-plant firm must be either to adjust *in situ* or to find an entirely new location. Facing similar problems, the multiplant firm may be able to solve them by shifting resources to one of its other plants. In other words, multiplant firms—especially those with geographically dispersed operations—are less constrained by local economic conditions. A good deal of spatial change within the large business organization occurs as a result of acquisition and merger. Not only does this add a whole new set of activities and locations to the acquiring firm's network, but it also usually leads to the subsequent rationalization of the entire network or parts of it as the new elements are assimilated.

In the large business organizations being discussed in this chapter, some degree of spatial reorganization and rationalization is occurring virtually all the time. Sometimes this reorganization is of major magnitude and has far-reaching effects throughout the organization. More frequently, it consists of relatively minor adjustments—for example, to the speed or layout of the production process, to the volume of output, or to the pattern of flows of materials and components within the organization. The origins of organizational-spatial change, therefore, are to be found in the firm's strategic, administrative, and operational decision making—especially the strategic. Hence our detailed examination of corporate strategies in the first half of this chapter.

The specific spatial outcomes of such decisions may take a variety of forms, as Figure 8.15 demonstrates. Two broad types of spatial-organizational change can be identified: *in situ change* and *locational shift*.

In Situ Change   As noted, a major advantage of the multiplant firm is that it can make substantial adjustments without necessarily adding to or subtracting from its existing spatial network. The capacity of an existing plant can be increased to achieve economies of scale or reduced to shed surplus capacity; an existing plant's capital stock can be replaced with new technology. In such ways, the importance and even the actual function of an individual unit can be altered as the firm reallocates tasks among its existing geographically dis-

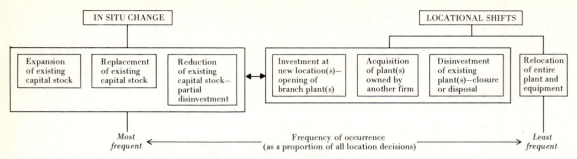

Figure 8.15 Types of spatial change within a large business organization. *Source:* P. Dicken (1986), *Global Shift: Industrial Change in a Turbulent World* (New York: Harper & Row), fig. 6.8.

persed operations. Change at any one existing plant may be either a gradual process of incremental change or a more abrupt and sudden change to its scale or function.

Locational Shift   This does involve abrupt change because it consists of either an increase or decrease in the number and location of units operated by the firm (or even, in rare cases, the complete physical relocation of an entire plant). The most common locational shifts within a firm's business operations are (1) investment at a completely new location, that is, the setting up of a branch plant; (2) disinvestment at an existing plant; and (3) acquisition of plants belonging to another firms.

Such locational shifts may well have significant repercussions on other plants within the firm. For example, a decision to establish a new branch plant in one place may be related to a reduction in scale, a change in function, or even closure of one or more plants in another place. Such locational adjustments are often associated with the introduction of new technology at different locations from those at which older technology is being replaced or with the shift of production to lower-cost locations. Similarly, the integration of acquired plants may alter the functions or the scale of existing plants. For the larger business organization, each of these processes may be occurring simultaneously in different parts of the organization and in different geographic locations. Some existing plants may be expanding, some may be contracting; some may be changing their function; new plants may be opening, others may be closing. The whole corporate adjustment process is geographically kaleidoscopic.

Even within a relatively simple structure, there are numerous possibilities for spatial reorganization. Figure 8.16 shows one such simple situation described by Healey (1984): a hypothetical firm that initially (in the prerationalization phase) operates four separate plants and produces three different products, in different combinations. Two of the plants are two-product operations, while the other two each specialize in the manufacture of a single product. Figure 8.16 illustrates four of the possible ways in which that initial structure may be changed through varying processes of spatial rationalization:

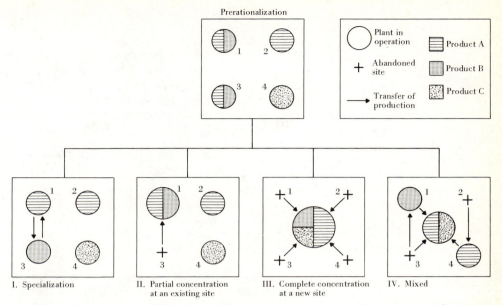

Figure 8.16 Spatial rationalization in a multiplant, multiproduct firm. *Source:* After M. J. Healey (1984), "Spatial Growth and Spatial Rationalization in Multiplant Enterprises," *Geo Journal* 9: 133–144, fig. 2. Reprinted by permission of Kluwer Academic Press.

I. *Plant specialization.* This involves the mutual transfer of product manufacture between plants 1 and 3 so that all four plants come to specialize in a single product. One advantage of such a rationalization strategy would be the achievement of economies of scale at each plant.

II. *Partial concentration at an existing site.* This involves the closure of one plant (3) and the transfer of its production to plant 1 to create an enlarged, two-product site. This may yield scale economies, especially if the two combined products are complementary; savings in administrative costs in coordinating the two plants plus the realization of the asset value of the abandoned site.

III. *Complete concentration at a new site.* This involves total disinvestment at all four of the firm's existing sites and the concentration of all production at a totally new site (a partial alternative would be to concentrate all production at just one of the firm's existing sites). The advantages of this strategy include the fresh start of a completely new plant on a new site with no inherited problems (this may have technological, plant layout, and labor supply advantages); the release of the other sites for alternative uses (including realization of the asset value through disposal); and the minimization of interruptions to the production process because the existing plants can continue to operate until the new plant is ready.

IV. *A mixed rationalization strategy.* This involves a mixture of these individual elements, including product transfer, plant closure, and new plant construction.

With the advent of the global corporation operating an international spread of activities, the whole process becomes far more complex. Figure 8.17 presents an idealized sequence of development of a hypothetical manufacturing firm that evolves from a firm with operations in only its home country to a global operator. But as Figure 8.17 suggests, this is not a linear process of simple expansion; it involves disinvestment as well as investment:

Stage I: a multiplant firm, with all its production and distribution facilities located within a single country (1). Overseas demand is served by exports.

Stage II: the beginnings of overseas expansion: (a) a licensing agreement is made with a local firm in country 4 to manufacture the firm's product for the local market; (b) a sales subsidiary is established in country 2.

Stage III: the beginnings of overseas *production*: (a) the firm acquires a domestic firm in order to enter a new product market. The acquired firm already has an overseas production plant in country 3; (b) production facilities are set up on the basis of the previously established sales subsidiary in country 2. At the same time, the firm's sales are extended into country 5 via exports.

Stage IV: substantial expansion of the firm's international activities has occurred: (a) new production plants have been built in countries 6 and 8; (b) an aggressive overseas acquisition strategy has resulted in the absorption of production facilities in countries 2, 4, 5, and 7; (c) the local firm in country 4, previously linked by the licensing agreement, has been acquired; (d) a sales subsidiary has been established in a new overseas market (country 9).

Stage V: substantial reorganization and rationalization have occurred throughout the firm's international network involving both domestic and overseas operations: (a) the five domestic plants have been reduced to two; (b) the 12 overseas plants have been reduced to seven, but with substantial alterations in their organization and functions: (i) the three plants in country 2 have been rationalized to a single large plant that serves both the overseas markets of countries 6 and 9 and also the firm's home market; (ii) a vertically integrated production system has been established in which plants in countries 4, 5, and 7 specialize in specific stages of a production process, and final assembly occurs in country 8.

Thus the locational behavior of large business organizations is complex. The important point is that location decisions are not isolated events. They arise as part of the firm's corporate and competitive strategies, which, as we

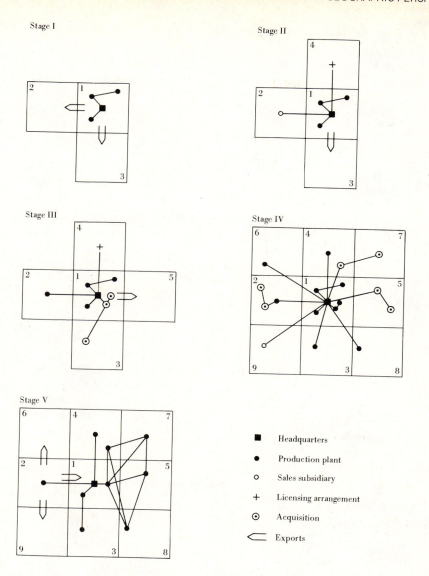

Figure 8.17 A possible sequence of firm development involving growth and rationalization. *Source:* P. Dicken (1986), *Global Shift: Industrial Change in a Turbulent World* (New York: Harper & Row), fig. 6.9.

have seen, can be divided broadly into horizontal, vertical, or diversifying strategies. Each of these strategies may in turn be pursued through either internal expansion or external combination (acquisition, merger, or colla- boration). Very often, a preference for one kind of strategy becomes adopted and established in a particular enterprise, which lends a degree of predictability to its behavior. For example, some firms show a very strong preference for acquisition strategies. The locational adjustment or rationalization of existing facilities is therefore only one of many possible kinds of locational behavior. As Table 8.1 suggests, a multiplant enterprise can implement its strategies spatially in a wide variety of ways.

The existing spatial structure of a multiplant enterprise—the number and location of its plants and the connections between them—exerts a very powerful influence on its subsequent strategic behavior. In making a strategic decision, the firm is greatly influenced by its structure. Geography plays a positive role in the behavior of business organizations. It is not merely the passive stage on which the action is played out. This influence is expressed in a number of ways. First, the geographic distribution of enterprise units (offices, production plants, warehouses, marketing units, research laboratories) is the skeleton on which the firm's task environment is built. Thus both the quantity and quality of information that the firm possesses about its external environment are at least partly a reflection of the firm's geographic structure. The more extensive that structure (the more plants it has and the more widely they are dispersed geographically), the more extensive will be its view and knowledge of the world.

A second way in which the firm's spatial structure is important is in its inertial effect. However they were originally chosen (even by some other firm subsequently taken over), existing plants represent large fixed capital investments that are not lightly abandoned. Such inertia may well be reinforced by a long-standing relationship with the community in which the plant is located or by political pressure. Also, when locational change is undertaken by a multiplant firm, its actual direction may be constrained by the prevailing geographic structure.

Table 8.1 CATEGORIES OF LOCATIONAL BEHAVIOR IN MULTIPLANT ENTERPRISES

		Locational arrangement of operations		
Scale and type of operations		Increase number of plants	No change in number	Decrease number of plants
I	Increase volume of output	1	2	3
II	No change in volume of output	4	5	6
III	Decrease volume of output	7	8	9
IV	Increase volume + new product	10	11	12
V	No change in volume + new product	13	14	15
VI	Decrease volume + new product	16	17	18
VII	Increase volume + new process	19	20	21
VIII	No change in volume + new process	22	23	24
IX	Decrease volume + new process	25	26	27

Two Case Studies: General Foods and Volkswagen

To illustrate some of these general principles in a more concrete way, let us look at two examples drawn from the world of big business in the United States. Both are examples of spatial reorganization and locational decision making within a specific strategic framework. Both are concerned with planning for new production capacity the context of a long-established network of production units.

General Foods This case study is based on detailed investigation by Whitman and Schmidt (1966). General Foods had been one of the leading industrial corporations in the United States for many years. In 1961, at the time of this case study, it ranked 36th in terms of sales. The company had grown to a very large extent by merger and acquisition since 1925 (the name General Foods was introduced in 1929) and produced a very wide range of food products (some 200), most of which were household names. By the early 1960s, General Foods consisted of more than 60 plants, warehouses, and offices in the United States, together with 17 overseas subsidiaries. Total employment was in the region of 30,000. In common with the trend in organizational structures, the company had a multidivisional structure; its eight divisions are shown in Figure 8.18.

One of the most important product divisions was Jell-O, one of the two original companies that had merged in 1925 to form the nucleus of what later

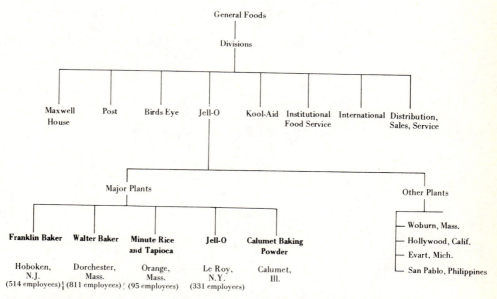

Figure 8.18 Divisional organization of General Foods. *Source:* Compiled from material in Whitman and Schmidt (1966).

became General Foods. The bulk of the Jell-O division's production of a variety of foods was concentrated in five major plants in the East and Midwest (Figure 8.19). (Some production was carried out elsewhere, but the five main plants are the focus of interest here.) Table 8.2 outlines the process of spatial reorganization carried out by Jell-O over a four-year period. The problem the company identified related to the obsolescence of the five main plants, each of which was old and technologically unsuitable under 1960s conditions. None had been selected originally by General Foods; all had been the plants of firms acquired during the 1920s. Some kind of modernization was clearly needed; the question was what kind? Two possibilities were identified. The most preferred initially was to retain and improve all five existing plants. The other was to seek one or more new locations and to consolidate production there. However, the uncertainties and possible duplication of overhead costs involved in such a strategy made it an unattractive proposition at first.

The company's search process, which was both complex and lengthy as Table 8.2 shows, can be divided fairly clearly into three distinct phases. Phase I was initiated by the establishment of a task force charged with making an exploratory investigation guided by a steering committee. The task force soon realized that improvement to the existing locations was not the best strategy; it was far better to seek new locations:

> The first step was to determine the optimum general geographic location in the United States for facilities to replace the five major Jell-O division plants. Eliminating the Far West . . . three geographical areas were considered—the South, East and Mid-West. The initial effort proved that a separate Southern plant location was not economically advantageous at the present time because it merely replaced one existing facility, and direct and transportation cost savings were not sufficient to justify the investment. It became apparent that the most significant savings were obtainable through overhead reductions by consolidation of existing Jell-O plants into fewer plants. The overhead savings that have been identified

Figure 8.19 Geographic distribution of Jell-O division plants before and after reorganization. *Source:* Compiled from material in Whitman and Schmidt (1966).

Table 8.2   STAGES IN THE SPATIAL REORGANIZATION OF THE JELL-O DIVISION OF GENERAL FOODS, 1960–1964

Identification of problem (early 1960): Need for "revamped and revitalized" production facilities to (a) maintain competitive position and (b) provide for new processes and products.

Alternative locational strategies identified
1. Expand and improve *existing* plants. This was, initially, the preferred strategy "except where other overriding advantages prevail."
2. Seek new site(s) and consolidate production there.

Search: carried out in three phases:

Phase I (early 1960–1961)
1. Establishment of task force and steering committee.
2. Early rejection of alternative 1; focus on consolidation.
3. Preliminary analysis of divisional costs, especially transportation, product allocation, construction, overhead, and shutdown and startup costs.
   Based on three geographic areas: South, East, Midwest, South rejected.
4. Six alternative consolidation plans formulated (see Table 8.3).
5. Task force studies select alternative 6: *one* new eastern plant replacing all four old eastern plants plus retention of Chicago plant. *Alternative 6 recommended.*

Phase II (July 1961–September 1962)
1. Establishment of facilities improvement group.
2. Detailed cost investigations. Confirmed likelihood of large potential savings and profit benefits.
3. Plan submitted to board. Authorized search for site.
4. Preliminary location survey of 21 locations in eight eastern states had "pointed strongly to a city in the Philadelphia-Wilmington-Baltimore area."
5. Appointment of plant location consultants. *Jell-O defined basic locational requirements:*
   (a) Plant to be within range of primary market served by four old plants, preferably within 250 miles of New York City.
   (b) Near one or more adequate ocean ports because of overseas origin of large quantities of raw materials.
   (c) Near a sugar refinery because sugar consumption more than half a million pounds per day.
   (d) In a community receptive to that type of industry.

Phase III (September 1962)
1. Consultants also recommended Philadelphia-Wilmington-Baltimore area.
2. Specific recommendation: *Dover, Del.*
3. Employment of another consulting firm to assess community of Dover as suitable location in terms of:
   (a) Attitudes and feelings of residents toward location of a General Foods plant in the city
   (b) Willingness to work in the plant
   (c) Work characteristics of the "employable people"
   (d) Status of race relations (G.F. was long-established equal-opportunity employer)
   (e) Union attitudes
   (f) Availability of skilled craft labor

September 1962: Company decides to move to Dover.
January 1964: First production lines placed in operation.
May 1965: Plant officially dedicated.

Source: Compiled from material in Whitman and Schmidt (1966).

represent approximately 80 percent of the total savings of this study. Direct and transportation savings represent the remainder. (Whitman and Schmidt, 1966, p. 16)

Thus even at that early stage and without reference to the number, let alone the specific locations, of the new plants being determined, the task force had made a commitment to a specific locational strategy. Clearly, such a consolidation strategy could take a number of spatial forms. Table 8.3 summarizes the six alternative plans put forward for consideration by the task force. Three of the plans were based on three plants and three on two plants with varying mixtures of new, expanded, or retained facilities. Comparison of the six alternatives on the basis of costs, savings, returns, and payback time showed that the most suitable strategy was to pursue plan 6, which involved replacing all four eastern plants with one new plant and retaining the Calumet (Chicago) plant in its existing state.

At this point, the task force's work was finished. However, there were several problems involved in accepting its recommended strategy. One problem related to the long-standing identification between each of the plants and the community in which it was located. Another was the potential rigidity of

> putting all of Jell-O's manufacturing eggs in one basket. The proposed consolidated plant would handle approximately two-thirds of the entire production of the division. . . . A work stoppage . . . would obviously be much more serious than such an occurrence at one of the old plants. (Whitman and Schmidt, 1966, p. 17)

Table 8.3   ALTERNATIVE PLANT CONSOLIDATION PLANS FOR THE JELL-O DIVISION OF GENERAL FOODS

Existing plant locations, 1960
Hoboken, N.J.; Dorchester, Mass.; Orange, Mass.; Le Roy, N.Y.; Calumet (Chicago), Ill.

Plan 1—2 plants:	One new plant in East
	One new plant in Midwest
Plan 2—2 plants:	One new plant in East
	Expansion of Chicago plant
Plan 3—3 plants:	One new plant in East
	Retention of Dorchester plant
	One new plant in Midwest
Plan 4—3 plants:	One new plant in East
	Retention of Dorchester plant
	Expansion of Chicago plant
Plan 5—3 plants:	One new plant in East
	Retention of Dorchester plant
	Retention of Chicago plant
Plan 6—2 plants:	One new plant in East
	Retention of Chicago plant

Source: Compiled from material in Whitman and Schmidt (1966).

Nevertheless, it was felt that the economic advantages of consolidation out-weighed such problems as these.

The next stage in the search process (phase II) was to establish a facilities improvement group to carry out detailed cost investigations. Interestingly, the group was specifically instructed to do only the work "necessary to justify or reject a recommendation for building a new eastern plant consolidating the Hoboken, Dorchester, Le Roy, and Orange operations." In other words, no further alternatives were to be sought. In fact, the group supported the task force recommendation. It had also made a preliminary location survey and regarded the Philadelphia-Wilmington-Baltimore area as most suitable. However, this recommendation was not acted on without further analysis. In phase III, location consultants were employed to carry out a detailed location survey in the East, subject to the constraints listed in Table 8.2. As it happened, the consultants also favored the Philadelphia-Wilmington-Baltimore area and ultimately recommended the city of Dover, Delaware, a community of 7,250 people.

The process did not end there. General Foods wanted to know more about Dover as a community (quite apart from the already revealed advantages of its location), especially about its attitudes to a new plant, its people, race relations, and its labor force. Accordingly, another firm of consultants was hired to "backstop" the location consultants' report with a detailed community survey. Their investigations were satisfactory, so in September 1962, the General Foods board decided to locate the company's new plant in Dover. Two years later, the plant began limited production and was fully operational by mid-1965, some five years after the original investigations began. As a result, Jell-O's spatial structure was transformed from a five-plant structure to a two-plant structure (Figure 8.19).

Volkswagen   The General Food case study demonstrated something of the complexity of the strategic investment process in the context of the need to reorganize spatially a set of outdated plants. A different kind of complexity is illustrated in our second case study, that of the Volkswagen (VW) assembly plant in the United States. The basis of this example is geographer Gunter Krumme's detailed 1981 analysis of the VW decision, supplemented by more recent material that extends Krumme's study, although not quite in the direction foreseen at the time of his work.

The additional complexity inherent in the VW case derives from three sources. First, it is an international investment decision involving the es-tablishment of a plant by a West German automobile manufacturer in a foreign country, the United States. Second, the organizational structure of VW is unusual in the sense that the key decisions on all major resource allocations are made by a supervisory board whose membership includes workers' represen-tatives and also representatives of the state government (Lower Saxony) and the West German federal government. The third source of complexity arises from a combination of the first two. A decision to build an assembly plant outside the firm's home country creates domestic pressures and opposition because of the concern that it involves "exporting jobs" and displacing exports.

Table 8.4 summarizes the major phases in VW's strategic decision-making process. The initial stimulus to consider establishing an assembly plant in the United States had arisen as early as the mid-1950s when VW had actually purchased a plant from Studebaker in New Brunswick, New Jersey, with the intention of building the VW Beetle there. Nothing came of this, the plant was sold, and not until the late 1960s did the possibility of a U.S. plant reemerge at VW. Between 1968 and 1970, VW's share of the U.S. market was at a peak of nearly 6 percent. But in the following eight years, this market share fell to 2.3 percent. Among the reasons for this were the growing success of Japanese car manufacturers in penetrating the U.S. small-car market and the increasing exchange value of the deutschmark against the dollar, which made imports from West Germany more expensive for U.S. customers. In addition, production costs in West Germany had been increasing relative to those in the United States and exceeded U.S. levels by 1977. For example, in 1960, the hourly labor cost in the United States was $2.67 and in West Germany $0.85. By 1977, the position had been reversed: U.S. hourly labor costs were $7.60, while West German labor costs were $7.73 (Krumme, 1981).

For all of these reasons, the view emerged among the VW top management that it was essential, for defensive purposes, to establish a production plant in the United States to serve the U.S. market directly rather than by exports. However, not all members of the VW supervisory board agreed with this view. That institution was, as Krumme observed, a "highly politicized body" subject to powerful political and social pressures from labor, regional, and national interest groups. The most obvious problem was that the plant responsible for manufacturing cars for export to the United States—Emden in northwestern Germany—was in a depressed economic region. This created implications for local and regional economic policy, implications reinforced by the fear of a ripple effect on other VW plants elsewhere in West Germany. The federal government was also concerned about the export and balance of payments repercussions of shifting production overseas.

The existence of such a plethora of interest groups explains the protracted predecision phase (phase I in Table 8.4).

> To most observers, this decision-making structure, together with VW's weak financial position, was at the heart of the delays in reaching a decision about a United States assembly plant during this time. In such an atmosphere, a decision which would create jobs outside Germany could hardly have been expected. As the crisis continued and worsened, the logic of the argument that the American market would be vital for VW and that only lower U.S. production costs would save it became stronger on economic grounds but weaker on social and political grounds. (Krumme, 1981, p. 347)

Eventually, after much delay—and the resignation of the chief executive, who had been closely identified with the U.S. project—the VW board formally decided to build an assembly plant in the United States and authorized the search for a location and site.

The next three phases shown in Table 8.4 took much less time. Once the decision had been made to proceed, progress, though not without its problems, was relatively rapid. The board gave its formal go-ahead in April 1976; two years later, the first U.S.-built VW Rabbit was produced. During that two-year period, however, a number of hurdles had to be overcome. VW had already collected some data on potential sites at an earlier stage. It then built an extremely thorough and detailed locational search on this foundation. VW's announcement of its U.S. plans generated a frenzy of activity among most U.S. states and many local communities, each attempting to outbid its rivals and seduce VW. Krumme quotes several examples, including that of the city of Baltimore, Maryland, which placed a full-page advertisement in the *Wall Street Journal* in the form of an open letter to VW's chief executive. Part of the city's case included a series of reasons why VW should not seriously consider three named alternatives (New Stanton, Pennsylvania; Brook Park, Ohio; Detroit, Michigan). The letter concluded that Baltimore " 'wants you so much so, we'll let you write your own terms. . . . Baltimore is willing to go all the way with VW' " (Krumme, 1981, p. 351).

Table 8.5 lists the incentives offered to VW by the state of Pennsylvania, which became VW's favored option—specifically a site in New Stanton in Westmoreland County. The carrot for Pennsylvania was a major boost ($80 million) in tax revenues, up to 15,000 direct and indirect jobs (6,000 at the plant itself), and an addition to the area's payroll of between $140 and $165 million. After some delays, the formal agreement between VW and the state of Pennsylvania was signed in September 1976.

The implementation phase took a further 18 months. Problems of obtaining a supply of body stampings were resolved when VW acquired the American Motors stamping plant at South Charleston, West Virginia. The state government got around the fact that VW was likely to contravene Environmental Protection Agency pollution regulations by reducing its own pollution levels to bring down the state average. Pennsylvania state employment agencies helped VW in its massive recruitment campaign and in its efforts to cream the best prospects from the more than 40,000 applicants. Finally, on April 10, 1978, the first VW Rabbit was produced on the New Stanton assembly line. After a shaky start, the plant achieved profitability and reached a peak output of 1,000 cars a day. VW announced that it planned to expand production at New Stanton and eventually to build a second assembly plant in the United States.

Unfortunately, the expansion phase was relatively short-lived. Within five years, the fortunes of VW's U.S. plant had changed dramatically. The plan to build a second plant was abandoned in 1983. The South Charleston body stamping plant was closed. In 1985, output at New Stanton was cut back to one shift per day. By 1987, the plant was operating at half capacity and employment was down to 2,500. Finally, in November 1987, VW announced that the plant would close completely in 1988. The causes of such a final decision were the major shift that had occurred in the U.S. car market and changes in VW's overall production strategy worldwide. The market segment served by the New Stanton plant was being squeezed by Japanese producers and by more competitive U.S. producers. The lower end of the market was being served by VW through its imports of the Brazilian-built Fox. The upper market segment was still being served by imports from West Germany. As VW's chief executive

Table 8.4   STAGES IN THE DECISION-MAKING PROCESS FOR VOLKSWAGEN'S AUTOMOBILE PRODUCTION PLANT IN THE UNITED STATES

Reasons for considering a production plant in the United States

1. Decline in VW's market position in the United States
2. Increasing convergence of production cost structures between West Germany and the United States

Alternative locational strategies considered

Retain all production for the U.S. market in West Germany and continue to serve US via exports. This option was especially strongly argued on social and political, rather than economic, grounds.

Decision phases

Phase I: Predecision (late 1960s–April 23, 1976)
1. Much procrastination by VW board as the U.S. project was subject to major internal disagreements.
2. By 1974, it was clear that VW's position in the United States was continuing to deteriorate. Executive manager commissioned a feasibility survey in the United States. Formally discussed by VW board, September 1974.
3. Increased controversy within VW and in West Germany concerning likely domestic job losses if U.S. project went ahead. Executive manager resigned.
4. Between 1974 and 1976, VW's position in the U.S. market worsened.
5. On April 23, 1976, the VW board decided to build an assembly plant in the United States. Authorized the search for a location and site.

Phase II: Location and site selection (April 23, 1976–May 2, 1976)
1. VW built on its earlier feasibility study to screen alternative locations and sites. Extremely detailed and thorough locational search undertaken in the United States.
2. Intense competitive bidding from individual states and communities for the VW plant.
3. On May 2, 1976, VW announced that it would negotiate formally with the state of Pennsylvania for the development of a site at New Stanton, in Westmoreland County. This was identified as the company's first-choice site.

Phase III: State negotiation (May 2, 1976–September 15, 1976)
1. The state of Pennsylvania had put forward an extremely generous incentives package (see Table 8.5).
2. The benefits to the state and local community were estimated to be $80 million in additional tax revenues over five years, 6,000 new jobs at the plant itself and a further 7,000–9,000 indirect jobs, and $140–$165 million in additional payroll to the area.

Source: Krumme (1981), miscellaneous press reports.

3. There were several delays in signing an agreement because of difficulties in arranging the long-term financing of the facilities at New Stanton.
4. The state government was not able to deliver on all its initial promises.
5. On September 15, 1976, master agreement signed between VW and Pennsylvania.

Phase IV: Implementation (September 15, 1976–April 10, 1978)
1. VW now displayed great urgency in getting the plant into production after the previous lengthy delays.
2. There was a problem of obtaining body stampings for the new plant. Eventually, VW purchased the American Motors plant at South Charleston, West Virginia.
3. VW also faced the problem of potential infringement of the Environmental Protection Agency's pollution regulations. The state of Pennsylvania adjusted its own pollution emissions to solve VW's problem.
4. State employment agencies assisted VW's massive recruitment drive. More than 40,000 applications were received.
5. On April 10, 1978, the first VW Rabbit was produced on the New Stanton assembly line.

Phase V: Production expansion (April 10, 1978–1983)
1. After early difficulties, the plant became profitable and reached a peak output of 1,000 cars per day.
2. The use of components made in the United States gradually increased; a number of West German component manufacturers established U.S. plants (though not solely to serve VW). However, VW had agreed to maintain substantial sourcing of some components from West Germany.
3. VW planned to expand production at New Stanton and to build a second assembly plant in the United States.

Phase VI: Contraction and closure (1983–1988)
1. In 1983, the plan to build a second assembly plant in the United States was abandoned.
2. The body stamping plant at South Charleston was closed.
3. In 1985, output at the New Stanton plant was cut and work reduced to one shift per day.
4. In 1987, the New Stanton plant was operating at half its planned capacity. Employment was down to 2,500.
5. In November 1987, VW announced its plans to close the New Stanton plant by 1988.

Table 8.5 INCENTIVES EXTENDED TO VOLKSWAGEN BY THE STATE OF PENNSYLVANIA

1. A $40 million loan by the Pennsylvania Industrial Development Authority to (nonprofit) Greater Greensburg Industrial Development Corporation to be used for land and plant purchase and renovation, to be repaid by Volkswagen over 30 years at an interest rate of 1.75 percent over the first 20 years and 8.5 percent over the last 10 years. Purchase price paid to Chrysler was reported to be $28 million. (Chrysler estimated the cost of completing the plant to be about $100 million.)

2. Highway improvements (through a $26.9 million bond issue) and a rail spur into the plant (through a $6.7 million bond issue); both needed and received legislative approval.

3. Originally, two large state pension funds for public employees had offered to lend VW $135 million over 15 years at 9 percent interest. The interest rate, however, was slightly higher than had been promised by Pennsylvania as part of its original financing package. Volkswagen eventually accepted only a loan of $6 million on these conditions (and financed the remainder through the private capital market).

4. Tax concessions were offered. VW would pay under the original plan 5 percent of its taxes during the first three years, 50 percent for two more years, and 100 percent thereafter. Under the revised plan, 5 percent for two years after production begins, 50 percent for another three years, and 100 percent thereafter. According to a county official (*Wall Street Journal,* July 7, 1976), the revised plan would give the VW corporation a $6 million tax break over $6\frac{1}{2}$ years.

5. A very intense program, using federal and other funds, to train workers for employment at the Volkswagen plant. The *Wall Street Journal* reported (June 1, 1976) that ''critics of New Stanton location have asserted that the immediate area lacks the pool of skilled labor offered by other sites, such as the Cleveland area.''

Source: After Krumme (1981), table 9.5.

stated: "The market segment has changed, the competitors have changed, and the fragmentation and segmentation of markets has changed" (*Financial Times,* November 23, 1987; p. 18). No clearer statement could be made of the intimate relationship between corporate competitive strategy and location decision making.

LINKAGES REVISITED: ANOTHER LOOK AT INTERFIRM RELATIONSHIPS

There is a long tradition of geographic linkage studies; we referred to some of them in our discussion of agglomeration economies in Chapter 5. Recent research has tended to be highly critical of much of this geographic work.* In particular, most geographic linkage studies have abstracted linkage from its broader context, the unequal power relationship that exist between firms occupying different positions in the economic system (in the center and pe-

---

* For example, McDermott and Taylor (1982); Scott (1983a, 1983b, 1984); Taylor and Thrift (1982).

ripheral segments, respectively). Many geographic linkage studies can also be criticized for separating out the dimension of geographic space to the exclusion of other dimensions. Linkages have been looked at in purely spatial terms, and the emphasis has been on the role of geographic proximity.

Interfirm relationships are far more complex than many geographers have appreciated. First, the concept of the segmented econmy, discussed in Chapter 7, tells us that the modern economy consists not merely of different types of business organization but also that organizations in different segments of the economy have different degrees of power over resources and over their relationships with other firms. Second, our discussions of organizational integration of business activities based on the extent to which a firm internalizes or externalizes its various transactions (Chapter 7) tell us that the boundary between the firm and its external environment is subject to change over time. Such changes are the direct result of the kinds of competitive strategies we have been examining in this chapter. Particularly important in this respect are the strategies of related diversification and the extent to which a firm chooses to increase or decrease its degree of vertical integration, that is, to perform or not to perform particular activities in its value chain.

At the most basic level, each business has to choose which activities to perform itself in-house and which to procure from other firms (commonly known as the *make-or-buy decision*). This apparently simple alternative is less simple than it sounds. Sheard (1983) has pointed out that "between the two extremes of internal organization (in-house production) and free market transactions lies a spectrum of intermediate interfirm arrangements" (p. 51). We briefly discussed one such intermediate arrangement earlier in this chapter: the formal collaborative venture between independent firms. Another especially important arrangement for organizing interfirm transactions is the *subcontracting relationship*. This form of economic organization has been attracting more and more attention among geographers.*

Holmes (1986) defined a subcontracting relationship as follows:

> A situation where the firm offering the subcontract requires another independent enterprise to undertake the production or carry out the processing of a material, component, part or subassembly for it according to the specifications or plans provided by the firm offering the contract. Thus, subcontracting differs from the mere purchase of ready-made parts and components from suppliers in that there is an actual contract between the two participating firms setting out the specifications for the order. (p. 84)

Subcontracting, then, is a kind of halfway house between complete internalization within the firm and arm's length transactions on the open market. Generally, the firm placing the order is known as the principal firm; the firm carrying out the order is known as the subcontractor.

The precise advantage of subcontracting to the principal firm depends very

---

* See Dicken (1986); Holmes (1986); Sayer (1986); Sheard (1983).

much on the kind of subcontracting involved (see Table 8.6). In general, however, it is one way in which the principal firm may avoid investing in new or expanded plant. Subcontracting also offers a degree of flexibility: it is easier to change subcontractors than it is to close down or reduce the firm's own fixed capacity. At the same time, by entering into a contractual agreement, the principal firm gains a certain amount of control over the operation. It is also one way of externalizing some of the risks of certain operations, which are, in effect, passed on to the subcontractor. Indeed, small subcontracting firms have been described as performing a "shock-absorbing" role for large firms.

Subcontractors tend to be both expandable and expendable, particularly when they are small firms in an unequal power relationship with large firms. There may be further problems from the subcontractor's viewpoint when the work being carried out for a particular customer is a large proportion of the subcontractor's total output. In effect, the subcontractor becomes part of a vertically organized and integrated operation without the full benefits of such an arrangement. As such, its freedom to move into new products or new markets may be limited. The problem is greatest for the small subcontractor when the principal firm specifies the product or process in detail and the subcontractor depends on the principal for product and process development. On the other hand, small firms may well gain substantially from their sub-

---

Table 8.6   ELEMENTS OF THE SUBCONTRACTING RELATIONSHIP

1. Technical Aspects of Production
   (a) Subcontracting processes ⎫
   (b) Subcontracting components ⎬ (industrial subcontracting)
   (c) Subcontractng whole products (commercial subcontracting)
2. Nature of the Principal Firm
   (a) Producer firm (both industrial and commercial subcontracting)
   (b) Retailing or wholesaling firm (commercial subcontracting)
3. Type of Subcontracting (Motivation of Principal Firm)
   (a) Speciality subcontracting
   (b) Cost-saving subcontracting
   (c) Complementary or intermittent subcontracting
4. Types of Relationship Between Principal and Subcontractor
   (a) Time period involved: long-term, short-term, single-batch
   (b) Principal may provide some or all materials or components
   (c) Principal may provide detailed design or specification
   (d) Principal may provide financing
   (e) Principal may provide machinery and equipment
   (f) Principal may provide technical or general assistance and advice
   (g) Principal is invariably responsible for all marketing arrangements
5. Geographic Scale Involved
   (a) Within-border (domestic) subcontracting
   (b) Cross-border (international) subcontracting

Source: P. Dicken (1986), *Global Shift* (New York: Harper & Row), table 6.1.

contracting role. Most important is the otherwise unattainable access gained to particular markets via brand names, continuity of orders (in some cases over a long period of time), injection of capital in the form of equipment, and access to technology. Subcontracting operations are especially important for small firms. Many firms actually start their lives as subcontractors to larger firms, and it is certainly an important channel through which small entrepreneurial firms can operate. Such observations have led to the view that large and small firms are related in a particular kind of unequal power relationship in which large firms dominate. There is a good deal of truth in this, but it is not the whole story. Subcontracting relationships are not confined solely to ones in which large firms dominate the small. In some industries—for example, aerospace and automobiles—very large firms act as subcontractors to other large firms.

Table 8.6 sets out the basic elements of the subcontracting process. It shows that subcontracting occurs in both the industrial and the commercial sectors of the economy and that it can cover not only processes and components but also complete products. Commercial subcontracting involves the manufacture of a finished product to the principal's specifications. The subcontractor plays no part in marketing the product, which is generally sold under the principal's brand name and through its distribution channels. Industrial subcontracting can be divided into three types according to the motivation of the principal firm. Speciality subcontracting involves the carrying out of specialized functions that the principal chooses not to perform itself but for which the subcontractor has special skills and equipment. Cost-saving subcontracting is self-explanatory: it is based on differentials in production costs between principal and subcontractor for certain processes or products. Complementary or intermittent subcontracting is a means adopted by the principal firm to cope with occasional surges in demand without expanding its own production capacity. In effect, the subcontractor is used as extra capacity, often for a limited period or even for a single operation. The actual relationship between principal and subcontractor can also take a variety of forms, as Table 8.6 indicates. The length of time of the agreement may be long or short. The principal's involvement may vary in terms of finance, technology, design, and provision of materials and equipment. Invariably, however, the principal is solely responsible for marketing the finished products or for arranging further assembly or processing.

Geographically, the scale at which the subcontracting relationship operates has increased dramatically. Until very recently, its geographic range was very restricted. Subcontracting occurred solely within national boundaries, both parties being located in the same country and, not infrequently, in the same locality. Such nationally bounded subcontracting systems are characteristic of all industrialized economies. One of the most highly developed systems of all exists in Japan, where each large firm tends to be surrounded by a constellation of small and medium-sized subcontractors. Figure 8.20 shows the immensely complex subcontracting system that exists in the Japanese automobile industry. Sheard (1983) estimated that the production system of any one auto maker averaged 171 first-layer, 4,700 second-layer, and 31,600 third-layer

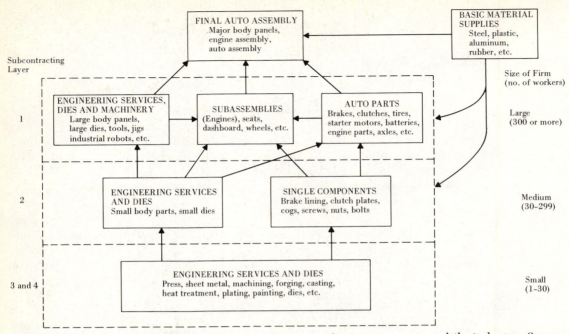

Figure 8.20 The subcontracting system in the Japanese automobile industry. *Source:* After P. Sheard (1983), "Auto Production Systems in Japan: Organizational and Locational Features," *Australian Geographical Studies* 21: 49–68, fig. 2.

subcontractors (p. 56). This kind of highly intensive subcontracting system depends, to a very large extent, on geographic proximity between principal firm and subcontractors. It is the epitome of the local linkage network so often studied by geographers (see Chapter 5), but with the added (and often overlooked) element of a particular system of power relationships between the participants. So it is not mere geographic proximity alone that is important but also the fact that these firms are locked together in a tightly organized hierarchical and vertical system of control and organization. However, a major development among Western firms in the past two or three decades has been the extension of subcontracting across national boundaries: the development of international subcontracting. In such a global operation, an American firm, for example, may set up a subcontracting relationship with a firm in South Korea, Taiwan, or Singapore to perform a particular part of its production operations (see Dicken, 1986).

The relationship—especially the geographic relationship—between a firm and its subcontractors depends fundamentally on the firm's specific production and procurement strategy as well as on the technologies of transportation and communication. The revolution in transportation and communications technology, together with developments in the production process itself that have allowed some processes to be split up into discrete stages, has created the potential for firms to establish subcontracting and procurement networks over vast geographic distances. This, indeed, is what many American and some European firms have done, especially in industries like garment manufacture, electronics, automobiles, and many other consumer goods. Relatively low transportation costs, plus the ability to control and coordinate the operation of a long-distance subcontracting system, allowed firms to take advantage of very low labor costs in developing countries.

But the use of long-distance subcontractors (and of other sources of supply) also involves the firm in holding large inventories of materials and components in order to ensure against interruptions in supply or faulty components. Some writers have termed this the *just-in-case system*. Sayer (1986) writes:

> Both functionally and geographically the relationship between firms and their suppliers tends to be distant; how the suppliers operate is of no concern to the purchaser, provided the price is right, and this lack of inter-firm contact, together with the infrequent nature of deliveries, allows suppliers to locate at often considerable distances away from the purchaser, if by doing so they can minimize labor costs and other expenses. (pp. 47–48)

In recent years, however, there has been a move toward a very different kind of procurement system: the *just-in-time system*.* It is particularly highly developed in Japan but is becoming more widely adopted in North America and Western Europe as well. The just-in-time system, like the just-in-case system, is much more than merely a system of procurement of supplies. It is part of the broader system of the organization of production adopted by the firm. The essence of the just-in-time system is that work is done only when needed and in the necessary quantity at the necessary time. Very small inventories, approaching zero, are held by the firm. Supplies of the needed materials and components are delivered just in time to be used in the production process. These characteristics "require that orders to the firm's suppliers and subcontractors [be] small and frequent, indeed deliveries may be made several times a day and hence proximity to suppliers is essential" (Sayer, 1986, p. 55). The layers of subcontractors in the Japanese automobile industry shown in Figure 8.20 are all located within a very small geographic area: a highly localized geographic agglomeration.

This raises the specific locational question of whether an increasing adoption of just-in-time methods by American firms will lead to a degree of geographic recentralization of production in relatively localized areas. Estall (1985) seems fairly sure that it will. In his view:

> In countries like the USA, where multi-sourcing has become common, where supply lines to assembly plants have often become extended over many hundreds of kilometers and the holding of large stocks at assembly locations has become normal practice, important consequences seem bound to follow. . . . Insofar as the system is adopted in the USA, therefore, it will tend to restore the significance

---

* The just-in-time system has a fast-growing literature in the management field. Within geography, the most comprehensive account is provided by Sayer (1986). See also Estall (1985).

of the economies of geographical concentration in relevant industries, after a period in which some have thought them to be becoming less significant. (pp. 130–131)

So far, there is only limited empirical evidence of this, most notably in the automobile industry, although Sayer cites examples from the U.S. electronics industry as well. But it is all too easy to inflate one or two individual cases to universal generalizations. As in so many instances in a rapidly changing economic environment, the exhortation must be to await further developments.

The geography of interfirm linkages, then, is but a part of the much broader context of the strategic behavior of business organizations in a volatile and uncertain environment. As far as the procurement of materials and components is concerned, the organization has a number of options. It can opt for *single sourcing* to gain the economies of scale of production but with the concomitant risk of putting all its procurement eggs in one supply basket. Alternatively, it can opt for *dual* and *multiple sourcing*, spreading its orders more widely. In either case, the firm has the further option of procuring its inputs internally, through a strategy of vertical integration or of obtaining them from external sources. Here, again, there is a range of choices. One is to buy at arm's length on the open market. Another is to establish a collaborative venture with a more or less equal firm. Yet another is to develop a subcontracting relationship with an independent company while retaining control over design and marketing.

In all of these various possibilities, the role of geography is not easily generalizable; different geographic arrangements are likely to develop in different circumstances. In terms of physical distance, geography may be very important indeed, as the just-in-time system implies. But it may also be very unimportant, as the development of international sourcing and international subcontracting in a just-in-case system suggests. But geography may well be important in a different—and more traditional—sense: as place, rather than simply as space. Places have particular attributes or qualities that make them attractive or otherwise as locations for investment or as sources of inputs, particularly labor. In Chapters 7 and 8, we have treated labor as a commodity like any other. In fact, it is very much more. This will become clear as we explore a rather different theoretical perspective on location in space in the following chapters.

FURTHER READING

Ansoff, H. I. (1987). *Corporate Strategy* (rev. ed.). Harmondsworth, England: Penguin.

Chandler, A. D., Jr. (1962). *Strategy and Structure: Chapters in the History of the Industrial Enterprise*. Cambridge, Mass.: MIT Press.

Chandler, A. D., Jr., and F. Redlich (1961). "Recent Developments in American Business Administration and Their Conceptualization," *Business History Review* 35:1–27.

Dicken, P. (1976). "The Multiplant Enterprise and Geographical Space: Some Issues in the Study of External Control and Regional Development," *Regional Studies* 10:401–412.

Holmes, J. (1986). "The Organizational and Locational Structure of Production Subcontracting." In A. J. Scott and M. Storper (eds.), *Production, Work and Territory: The Geographical Anatomy of Industrial Capitalism.* London: Allen & Unwin, chap. 5.

Johnson, G., and K. Scholes (1988). *Exploring Corporate Strategy* (2d ed.). Hemel Hempstead, England: Prentice-Hall, chaps. 6–10.

Krumme, G. (1981). "Making It Abroad: The Evolution of Volkswagen's North American Production Plans." In F. E. I. Hamilton and G. J. R. Linge (eds.), *Spatial Analysis, Industry and the Industrial Environment: Vol. 2. International Industrial Systems.* Chichester, England: Wiley, chap. 9.

Malecki, E. J. (1979). "Locational Trends in R&D by Large U.S. Corporations, 1965–1977," *Economic Geography* 55:309–323.

Porter, M. E. (1985). *Competitive Advantage.* New York: Free Press, chaps. 2–4.

Stephens, J. D., and B. P. Holly (1980). "The Changing Patterns of Industrial Control in Metropolitan America." In S. D. Brunn and J. O. Wheeler (eds.), *The American Metropolitan System.* New York: Halstead, chap. 11.

Chapter
9

Insights from the Social and Economic Theories of Marx

By now, we have established beyond doubt that location in space and the development of spatial patterns cannot be examined by restricting ourselves solely to geography and the exploration of spatial causes. It has also become clear that while the neoclassical models of Part One teach us a great deal about model building and method, they can offer little help in the task of understanding the complexity of the contemporary world, whose firms, consumers, markets, and networks bear little resemblance to flat plains of equal fertility uniformly populated by innate profit and utility maximizers. In trying to grapple with the real world, we have to accept that neoclassical models cannot take us very far.

Economic geographers began to adopt this view in the late 1960s. In response some launched themselves into more descriptive pursuits, trying to map and measure what was going on. Others looked into behavioral theories of the firm for new inspiration (which we have just done in Chapters 7 and 8). Another small but influential group examined more general Marxist theories of economy and society (which we are about to do). All, however, agreed that although new theories were badly needed, it was becoming increasingly difficult to see them as coming narrowly from within geography or economics alone.

Consider our present position in this book. We have established that the firm can no longer be treated as a black box. We need to open up and explore the decision-making processes that go on inside it. We need to know how investment decisions are arrived at and put into operation; how product lines are selected, installed, expanded, closed down; how workers are selected, placed, used, valued; how management is organized, distributed, made accountable; and so on. Given the game theory examples of Chapter 7, we also have to consider the actions and reactions of competitors, the responses of trades unions and governments. On top of this are the demands of bankers, the pressures generated by technological change, and the effects of cyclic changes in the economy. Any actor in the processes must be explored as part of an interconnected web of economic, social, and political relationships.

By contrast, people and firms appeared to have a rather restricted part to play in what went on in the society described by the simplified models of Part One. They produced. They consumed. They made rational choices as to how space was to be used or traversed. They invested. They traded. They stayed put or relocated. No one would deny that these are elemental things that people do, but we were able to say little about the broader social context in which these activities were performed. Nor were we able, except by simplification to one or two variables, to say very much about the real complexities of the web of social and economic relationships in which such behavior was effected.

What seems to be missing is some concept of the ways in which some individuals', groups', or firms' choices preclude or enhance the choices available to others. We all know how much life confronts us not with free but *structured* choices—some that we can choose and some that we cannot, however rational our judgment. We are equally aware how many decisions represent trade-offs not simply in terms of rational calculation of opportunity costs but because of the webs of interpersonal, intergroup relationships in which we find ourselves.

In short, what is lacking in the approach of Part One is a sense of society as well as a sense of history. The concepts that have to be addressed are of ever-changing conflicts and compromises, evolving structures of power and dependency, arrays of the possible and the proscribed, and the accumulated power of social structures. Clearly, at the extreme, this would take us all the way to demanding a theory for the "meaning of life," but we have to set ourselves more limited aims. What seems to be required is a realization that actors whose behavior conditions outcomes in the geographic world we seek to understand operate in a context that cannot sensibly be abstracted from the workings of society as a whole. Decision, action, and outcome reflect as much the structured interrelations that make up society as the conscious choices and decisions of the firms, workers, consumers, and governments that are the stock-in-trade of orthodox economics. It is axiomatic, from this viewpoint, that all are not equally free to choose. The giant transnational enterprises described in Chapters 7 and 8 have considerable (though by no means absolute) power to make things happen by comparison with the small "laggard" firm. More fundamentally, capital (firms; banks; owners of industrial, financial, and property capital) has more "relative freedom" in this respect than labor (people who work for a wage). Indeed, this may be regarded as the primary asymmetry in access to power and relative freedom in the society we have to try to understand. But we shall come back to this in due course. For the moment, the point is that the adoption of approaches that stress the inseparability of economy from society and focus on systems of largely asymmetrical social relationships rather than narrower economic "rules" has a profound effect on how economic geography is to be viewed.

Perhaps the first key insight from this perspective is that there can be no universal theory of economic behavior applicable to all societies at all times. Where societies differ, social contexts differ, and outcomes cannot be determined outside such contexts. Historical circumstances serve to influence social outcomes, and these outcomes in turn condition the unfolding processes of social development. Insert *geographic* for *historical* in the preceding sentence and it becomes clear that what we have been narrowly defining as economic geography becomes understandable only as part of the unfolding of social relations of production in general and further that this unfolding is part of an ongoing dynamic process in which the parts of the system mold each other and the development of the system as a whole.

Taking this line, academic divisions of labor that recognize economic geography as distinct from social or political geography have, of course, to be reexamined. On top of this, there are some important methodological considerations. Where in Part One we could use the orthodox economists' approach of exploring complex interrelations by the "other things being equal" method, this abstracted from the multifaceted and complex relationships that allow us to explore the social relationships of production in their fully connected form. Above all, an approach that does not accept the existence of some universal theory invariant across time and space and instead focuses on space-time variation has to take a dynamic viewpoint. It is no longer adequate to make assumptions about equilibrium as a static condition; change and flow must be seen as a norm to be grappled with.

Such issues have come to dominate the debate in economic geography in recent years. For the most part, they have emerged from the gradual buildup of a radical, largely Marxist school of thought within the discipline. The traditional subject matter of economic geography—the location of industry, the geography of employment, and urban and regional economic development—has been reexamined from a perspective that views it as part of the general development of capitalism and its social relations. Some aspects of this have of course served to challenge the very roots of the discipline and have provoked more than their fair share of acrimony. Even analysts who have sought to reject this incursion as unwelcome in purely theoretical terms have, however, been forced to make an effort to come to terms with it. Resistance to grappling with Marx has powerful political and ideological roots rather than narrowly academic ones. As a theoretical critique that polarized the postwar world into attackers and defenders of capitalism and fostered the potentially disastrous cleavage between Communism and the Western democracies, Marxism raises passions unconnected with the insight contained in its underlying theory.

Not all of the acrimony and dispute has been between the "orthodox" and "radical" camps, however. Debates have raged within the emerging Marxist school itself. Attempting to address them here is rather like entering a mine field under crossfire, and there would be good reason on these grounds alone for seeking to duck the issues. Another good reason is that the body of ideas we are seeking to elucidate are complex by their very nature and have an unfortunate tradition of being described in relatively obscure language. These ideas have, however, had a powerful impact on geographers of all branches and have leaked piecemeal into the teaching syllabus, and it is necessary to review the Marxian experience on its merits insofar as it provides new and worthwhile insights to help us cope with a complex world. Whether we accept it or reject it, we must first grapple with it and seek an informed understanding of the logic and structure of Marx's theory.

STARTING POINTS: THE CAPITALIST MODE OF PRODUCTION

According to David Harvey (1983):

> Geographical knowledge deals with the description and analysis of the spatial distribution of those conditions (either naturally occuring or humanly created) that form the material basis for the reproduction of social life. It also tries to understand the relations between such conditions and the qualities of social life achieved under a given mode of production. (p. 189)

From the very beginning of the book, we have been talking about the economy and the way it operates as if there was only one possible kind of economy. We have accepted without question or comment that some individuals and groups have rights to acquire and benefit from privately owned property. It has been the norm that some people or the firms that represent their interests hire the labor services of other people (the employers) while others sell their services on the labor market (the workers). We have also accepted as an axiom that the purpose of the economic system is the maximization of profits by the first group and the maximization of utility by all consumers (both employers and workers) subject to efficient utilization of resources. Our easy assumption of these axioms has been based on the fact that we all knew what we were talking about. Capitalism in its fully developed form has only been in existence for a few hundred years but has come to dominate the world economy; more important, it has become so powerful a concept that most people today accept the rules and structures of it as universal.

Karl Marx, writing from a vantage point closer to the beginning of the capitalist era, saw things differently. For him, capitalism was simply the newest of a historically evolving series of what he called *modes of production* through which people sought to solve what, in Heilbroner's term, we described at the beginning of the book as the "economic problem." Far from being some universal system destined to dominate the world for all time (no doubt many saw feudalism as equally "universal"), capitalism is seen as a particular solution rooted in its time and place, one that itself is in a constant state of evolution and development.

Figure 9.1 shows the essential elements of a mode of production as envisaged by Marx. There were two elements: the *forces of production* and the *relations of production*.* The forces of production are already familiar to us. They are the tools, physical resources, and labor inputs of the production process. The relations of production derive from Marx's specific viewpoint on the nature of economic systems. The distribution of economic power and the ownership of society's productive forces are, for Marx, critical dimensions of any mode of production, and it is not sufficient simply to have theories for the efficient allocation of resources and the determination of prices.

In the specific case of capitalism, Marx perceived the relations of production as fundamentally based on capitalists' economic ownership (effective control) of the forces of production while workers "owned" no more than their capacity to perform useful work in a system of wage labor. Another key feature of the capitalist mode of production is that relations between people, as well as the relations of people to the natural world, are configured around the production of *commodities*—goods exchanged at values determined in the marketplace and not simply by their intrinsic usefulness. A third key feature of the capitalist mode is that the object of production is the accumulation of surpluses (profits) by capitalists.

This description offers no more than a caricature of Marxian insight into the nature of the capitalist mode of production, but our purpose here is simply to introduce and define terms that have increasingly found their way into the geographic literature. There are essential differences between the orthodox approaches of the earlier discussions in this book and the perspectives we are now exploring. In particular, we took it for granted that the basic features of contemporary economies implied that people would be drawn into relationships with each other in the processes of production, exchange, and consumption. Now these relationships come to the fore as issues for which we need theories. They are key features that have a powerful effect on the kinds of outcomes we, as geographers, must concern ourselves with.

---

* It is not our purpose here to probe deeply into precise definitions and meanings for the elements Marx so carefully specified in his theory. Hunt and Sherman (1986) set out the elements of Marx's social and economic theory at a level appropriate for readers of this book. For formal definitions of terms and a deeper review of each concept, see Bottomore (1983). By far the best simple guide to Marxian economics is Fine (1984).

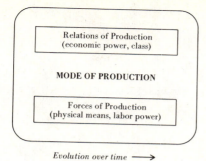

Evolution over time ⟶   **Figure 9.1** A mode of production.

THE LABOR THEORY OF VALUE, LABOR POWER, AND THE LABOR PROCESS

In Chapter 4, we talked about the well-known orthodox economic concept of *factors of production*. Recall that these were land, labor, and capital, with technology added in some cases as a fourth discrete factor. Each was seen to be a force to achieve some form of marketable output, and this view is common to both orthodox and Marxian approaches. Entrepreneurs, as we called them in that chapter, hired each of these factors in the anticipation that they would bring some form of return on this act of investment. The object was, of course, to enhance the original investment through some accumulation of further reward as profit, rent, or interest. Entrepreneurs were seen to be hiring the productive capabilities of the factor with a view toward putting it to profitable use. The Marxist viewpoint seeks to go behind the essential allocation problem, which is addressed in the neoclassical analysis. As we have pointed out, it asks different questions, not purely economic ones but social, political, and moral questions as well. There is a tendency, from orthodox perspectives, to accept on face value the obvious fact that combining land, capital, and labor in production is the normal way of producing useful commodities. However, Marx, beginning from a different perspective known as the *labor theory of value*, addressed the issue of the production of commodities from a wholly different standpoint. Since he saw labor as the ultimate source of the intrinsic value of commodities in exchange, then simply to see it as one of the elements in the "trinity" of the production function is to devalue it. Labor is also, of course, people, and to see it as a factor of production as we did in Chapter 4 takes a particularly abstract view of it.

The Labor Theory of Value

The Marxian alternative to the orthodox, utility-based theories we addressed earlier is known as the labor theory of value. In traditional theory, production is a venture in which people contribute freely and equally. People are rewarded appropriately for their contribution as workers, as owners of land, and as owners of capital, tools, and resources. In the labor theory of value, the starting proposition is that all value derives from the application of human labor to the forces of nature. In the capitalist mode of production, what is significant is that this value is for the most part expropriated by one class, the capitalists, with labor receiving only such returns as are necessary to reproduce itself. This is the primary characteristic of production and social relations in capitalism.

The labor theory of value is a source of considerable dispute both between orthodox and radical economists and within the radical camp. Indeed, Roemer (1987) goes so far as to describe it as "simply wrong," but not in ways that destroy the basis for radical economic theory. Trying to address it here in the simple terms necessary to our project presents great difficulties, but we seek to capture the essence of it by means of a parable, based on Hunt and Sherman's (1986) masterly exposition of the theory.

Imagine a tropical island where food and shelter are available as the free bounty of nature and where the only necessary productive activity is catching fish. These are caught by hand, with every member of the 100-strong population spending six hours a day on the process. The fish, once caught, are distributed equally among the inhabitants, and any surplus is taken by boat to a second island, where the inhabitants operate a similar system for the production of sweet potatoes. If we were to assume for the sake of simplicity that these agriculturalists spent six hours a day in the sweet potato cultivation process, then by the labor theory of value they would find their products exchanging with fish at parity. The barter price would be the same. As geographers, we might be sensitive to the fact that interisland transport is not without its labor and time costs, and to keep the notion of price parity, we might have to insist that the islanders agreed to take it in turns to ship the goods to each other. Were, however, sweet potato cultivation to take only three hours a day to achieve the same perceived value as fishing, the fish–sweet potato price ratio would stand at 2 : 1—a true reflection of the differences in labor embodied in their production. Tackling the more orthodox question of what happens if for some external reason the demand for fish or sweet potatoes rises, the labor theory of value suggests that this will serve to alter the amount of each commodity produced but in the final analysis would not change the long-run price difference based on the ratio of labor inputs.

Going back to our fishing community, suppose that it dawns on some of the fish-producing islanders that by collecting natural hemp, which grows freely, rope could be made and then nets to catch fish more efficiently. This technological and social innovation is given a try by allocating 50 people to the hemp-rope-nets activity and 50 to fishing. Just from this example, we can dispose of one of the knottier problems of the labor theory of value. The net producers are actually in the business of trying to catch fish—their nets are "partly caught fish," if you like, and the nets, as tools, embody the labor used to make them. What, in the traditional production function, might then have been labeled "capital"—tools, machines in a physical form as the means of production—can now be seen as "partly made commodities" and the products of previous labor ("dead" labor, as Marx called it). So by this form of analysis, land (cleared, cultivated, made accessible) and capital (products of labor on capital goods) cannot be simply pulled apart from the factor of labor. Labor infuses all three.

Back now to our example, the results of production using nets increase the fishermen's catch by a factor of 2. The society now has some new choices: it can raise its consumption of fish for the same amount of effort or reduce fishing and netmaking to three hours per day and find other useful or rewarding things to do with the extra time. Notice that if nothing similar has happened to the sweet potato producers, the price ratio of the new high-yield fishing will shift to reflect the reduced labor content of the same amount of product. The sweet potato producers are now working twice as hard as their fishing neighbors and may themselves look for some way of applying laborsaving methods to release them from the burden of work.

Suppose, however, that the original inventors of the netmaking technique or some other self-appointed leaders decide to institute a system of private property and claim ownership of the newly acquired hemp, rope, and nets. Ten of them have sufficient power to achieve this, leaving the remaining 90 free to choose to use the others' nets or not. Private hand fishing is also now decreed to be poaching and punished accordingly. These new moves derive from an emerging set of social relations and are not from the nature of the forces of production available to the community. The prospect of "net technology" provided an opportunity for some change but did not determine any particular outcome. This is a point we have been stressing all along—that the social relations of production have a fundamental impact on emerging economy and society (and, therefore, on its geography) and cannot be effectively separated from the material forces of production.

Back to the island. The new social order has choices of the allocation of resources and the product of labor. Fifty people may now continue to catch fish as employees (for fish wages), ten can be put to work to repair and renew nets, which gives them a longer life. The remaining 30 can now be put to work at the behest of the net owners to find other sources of profitable revenue or to be servants and produce luxuries for their masters. There was nothing inevitable about this particular set of social relations opposing owners and workers, masters and servants, but the opportunities of the new technology were there to be exploited in a variety of ways.

Under the new system, 50 fishermen plus 10 net repairers could be used to meet basic production needs. The remaining 40 represent surplus labor, and depending on the evolution of the new social order, their labor could be used in a variety of ways. If strict inheritance laws are established, laws to protect private property are enacted, and a policy is implemented for the maintenance of law and order that includes not only offenses against the person but offenses against private property, the system is set to evolve in what might be described as capitalist form.

From such a parable, then, the elements of the labor theory of value can be revealed, showing how value is assumed to be created and disbursed as the system evolves. In this way, both the social and the production relationships are revealed in a system that from orthodox perspectives might simply be described in terms of the impact of technical change (the net) on fish production where there is a shift from labor to more capital-intensive production and a resultant rise in productivity, profit, and investable capital.

Labor Power and the Labor Process

Marx argues that a particular feature of the capitalist mode of production is that labor (the ultimate source of all value) is regarded as a commodity bought and sold in exchange transactions in a marketplace. To account for this, Marx argues that what is hired by capitalists (entrepreneurs) is the workers' "capacity or ability to work," their *labor power*. This is, in essence, a commodity that workers sell and employers buy. The employer is conceded the right to decide how labor power should be applied to production. People under capitalism, then, participate in a particular form of the *labor process* (see Figure 9.2*a*): they perform some "purposeful activity in transforming nature to some useful form." A distinction is made between the *objects of labor*—whatever is to be changed: land to be tilled, metal to be forged, timber to be cut—and the *instruments of labor*, the tools. The labor process described in this way could apply to any mode of production; what distinguishes capitalism is that the labor process is based on the provision of labor for a wage by some members of society and is hired by others who allocate it as a force for production and who control its use in production. The particular form of the capitalist labor process is set out in Figure 9.2*b*. The relations of production as well as the forces of production are clearly identified. The particular form of the labor process sees the work itself (purposive productive activity), the objects of labor, and the instruments of labor all under the economic ownership (control) of the capitalists and dedicated to the production of commodities for exchange. Capitalist commodity production is essentially a value-creating process as well as a labor process, and from Marx's point of view, the extraction of value from labor in the form of a surplus to drive accumulation characterizes capitalism.

SURPLUS VALUE AND THE ORIGINS OF PROFIT

One of the most important aspects of the labor theory of value is the insight that it provides on the underlying social and economic processes that see value created, circulated, and accumulated. Going back to the fish and sweet potato producers of our earlier example, we were able to show how simple commodity production works in dynamic terms. Figure 9.3*a* shows how that system operated. The clear purpose of production and exchange in this case was to satisfy people's needs for particular combinations of commodities. As the diagram shows, the starting and end points of what would be a continual circuit of exchange are these commodities, which are in demand. Money sits in between as a convenient medium for making trade easier in exchanging different goods.

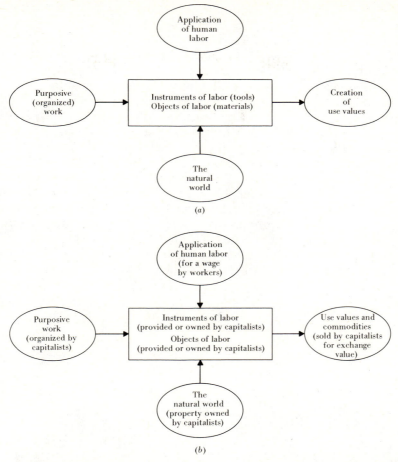

(a)

(b)

Figure 9.2 (*a*) The labor process in general terms; (*b*) The capitalist labor process.

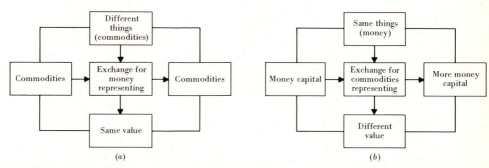

(a)

(b)

Figure 9.3 (*a*) Simple commodity exchange; (*b*) Capitalist commodity exchange.

However, when we introduce a form of social relations of the kind that characterize capitalism, the circuit of money and commodities looks very different (Figure 9.3*b*). Here, the starting point is some kind of money capital. The production of commodities now becomes the facilitating means to a different kind of objective, the accumulation of more money (capital). From this we can derive Marx's view of capital as not so much an object as a *process*—as "value in motion" or "value in the process of expansion." This helps us to distinguish capital in motion from the various forms it takes as part of the process. These may be machines, bank accounts, shares, or currency as a medium of exchange. We shall come back to this idea of circulation in more detail later. The labor theory of value provides one means of exploring how money can expand in this process to provide more at the end of each cycle than was put in at the beginning. This lies in the concept of *surplus value*. Figure 9.4 shows diagrammatically how surplus value emerges in the process of capitalist commodity production. We start with the money capital applied to the purchase of two other forms of capital. The first of these is *constant* capital, the value of which represents the historic (or "dead") labor embodied in its creation. This is made up of the machines, tools, and raw materials (the objects

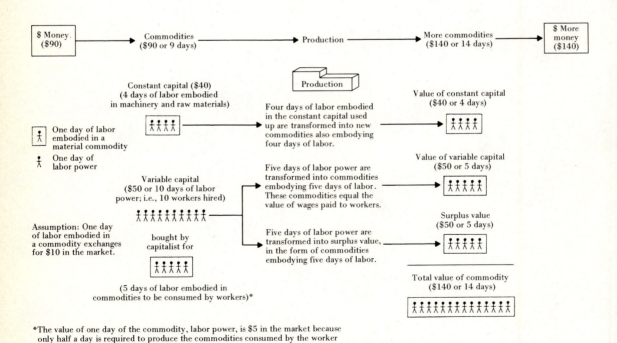

*The value of one day of the commodity, labor power, is $5 in the market because only half a day is required to produce the commodities consumed by the worker

Figure 9.4 The labor theory of value: a graphic illustration. *Source:* After E. K. Hunt and H. J. Sherman (1986), *Economics: An Introduction to Traditional and Radical Views* (5th ed.) (New York: Harper & Row), fig. 18.1. Reprinted by permission.

and instruments of production) that go into the production process itself. The second is *variable* capital, the labor power purchased from workers for a wage and applied in the labor process. Both are applied in the process of production, which converts one set of commodities into another capable of attracting exchange value in the marketplace. In this process, the values attached to the constant and variable capital inputs are themselves transformed.

While constant capital is assumed to be transformed in ways that pass all of its original value ($40 in the diagram) into the new commodities (hence the term *constant*), variable capital transformation in production has two elements. First, the value equivalent to the original cost of the labor power purchased at the outset ($50 in the diagram) and, second, surplus value. The latter arises from the additional value that accrues from the exchange of the new commodities ($50 in the diagram) that is not returned to its source—labor —but is expropriated by capitalists. This surplus value, from the perspective of the labor theory of value, provides for the expansion of capital at the end of the circulation process and is the basis for the *accumulation* that provides the motive force for capitalism. Couched in terms of the social relations of production rather than the circulation of capital, it is identified as the process of *exploitation*.

Capital and Labor

From the orthodox perspective, profit, as the driving force of capitalism, is seen as a reward to the capitalist (entrepreneur, employer) for thrift and enterprise. Such members of society, by denying themselves consumption (or as Marshall added later, by waiting rather than consuming directly) are entitled both to accrue investable capital (savings, inheritance) and to use it to seek profit and further accumulation by wisely allocating it to the acquisition of factors of production and the pursuit of market opportunity. From such a perspective, members of society receive in return what they contribute to social well-being in general. Since capitalists contribute more, it is appropriate that they receive these "just" rewards for thrift and enterprise. Workers similarly receive, in wages, rewards commensurate with what they contribute "freely" to society's wealth. The rule for capitalists is therefore the maximization of profits in the free competitive marketplace. This ensures also that scarce resources are used efficiently. With regard to labor, as Chapter 4 showed, efficient allocation and effective return mean that workers will continue to be hired by entrepreneurs to the point at which the marginal revenue gained is equal to the marginal cost of employing them. This ensures that wages are set to reflect market valuations of products and that no worker is employed whose labor does not contribute to "fair returns."

Starting again from entirely different questions and probing issues of justice and morality rather than simply efficiency and resource allocation, the logic based on the labor theory of value holds the same process against wholly different criteria. To begin with, the capitalist mode of production is seen as being founded on a fundamental antagonism between labor and capital. One minority class, the capitalists, have rights to own and control the means of production—raw materials, tools, machines, land—that are essential to the production of most use values and all exchange values. Opposed to them, of necessity, are the class of wage laborers, who, having no rights of access to the means of production, are thus dependent on the capitalist. They must sell their labor power on terms dictated by employers in relation to the exchange value of labor as a commodity.

In the historical context of his time and place, it was this essentially class-based feature of society that Marx set out to reveal from logical analysis. He was thus exploring the social relationships that infused what we have more narrowly described as an economic system. Marx was concerned to know, for example, in the case of any economic activity—simply digging a ditch perhaps—not just how many man-days are required at what cost (the economic parameters) but who is digging, for whom, with whose equipment, on whose land, for what purpose, with what end in view, and with what view of the nature of the project. This was not through idle curiosity. His point was that the simple economic variables—cost, time, resources—are predicated on these social conditions and cannot be divorced from their social context and the meanings attached to the whole enterprise. This was the point of the parable of the fishing community. Once again we see that the questions of relationships, interdependencies, and social meanings provide a basic context for the interpretation of economic issues.

SOCIAL RELATIONS OF PRODUCTION

The central distinguishing feature of capitalism is, then, that it pivots around the social relations between capital and labor, employers and employees, investors and wage earners. These are essentially relations of economic power, within which control lies with capital, employers, and investors. The "right of managers to manage" is a largely accepted and uncontroversial state of affairs in the societies explored in this book.

While this central class relationship lies at the core of the capitalist mode of production, other key relations exist that emphasize its social, rather than simply economic, dimension. There is, for example, the relationship between people and nature. Particular commodities—food, clothing, shelter—have certain material qualities that make them intrinsically valuable for the satisfaction of human needs and wants. These Marx distinguished as *use values*. The fact, however, that in common parlance we speak of the "exploitation" of natural resources gives a strong clue to contemporary attitudes to and relations with nature. In a system dominated by the market, in which production is primarily for exchange, the natural world provides a basic source of commodities to be used in the marketplace in the search for profit or *exchange values*. In this form of the exploitation of nature, the means of production and labor power are applied in accordance with levels of expected return on invested capital. This is how the natural world is "valued," and the relations of people not just with each other but also with the natural world are set within this framework. There are, of course, other dimensions of the relationship between humanity and nature in capitalism, and environmental issues are taking on increasing importance. The point being made in Marxist theory is, however, that the *central* relationship between people and the natural world in capitalism is a commodity relationship based on the search for profit for its own sake.

What about people-people relations? We have seen one aspect of this in the class relationship, but by standing in relation to each other through the system of commodity production for the market, the Marxist analysis suggests that everybody in capitalist society is drawn into a particular set of social relations with their fellows. In a much quoted phrase from the preface to his "Contribution to the Critique of Political Economy," Marx (1859) described the general situation in the following terms:

> In the social reproduction of their life men enter into definite relations that are indispensable and independent of their will, relations of production that correspond to a definite stage of development of their material and productive forces. The sum total of these relations of production constitutes the economic structure, the real basis on which rises a legal and political superstructure.

The question of whether the economic structure gives rise to the legal and political superstructure of society in a simple one-way relation has exercised many commentators. The point for our purposes, however, is that one cannot abstract from these relations of production in trying to understand any society. This has motivated a number of geographers in recent years to abandon orthodox neoclassical models in the search for theories that explore the social relations of production from a spatial perspective.

Divisions of Labor

Given the picture we have sketched out of capitalism as a system based on the production, consumption, and exchange of commodities in which certain "mode-specific" relations of production have grown up, radical perspectives on divisions of labor are of a different order from those that emerge from more orthodox literature. We have already encountered one classic description of the division of labor in Chapter 5. This was Adam Smith's famous example of the pinmaker. It was a very specific case of the division of labor, one that has had a powerful effect on how the concept has come to be described in orthodox economics textbooks. Remember that here the division of labor into a series of specialized technical tasks contributed to enhanced productivity and efficiency. Wire drawers and straighteners, cutters and sharpeners, by specializing in branches of production rather than undertaking the entire pinmaking process themselves, were seen to apply their labor power more effectively. As a result, productivity (output per person per hour or day or per unit wage) increased, providing a greater return so long as the market was large enough to absorb the extra product. Few commentators dispute this proposition in its own terms, and the evolution of industrial society since Adam Smith's time bears witness to the emergence of ever more sophisticated divisions of labor.

Marx, however, wanted to explore a rather wider range of questions about the division of labor in capitalist society than those addressed by Adam Smith. As we have seen, he was concerned not only with questions of efficiency but also with questions of equity and social justice. The detail or technical division of labor in this production process, which is revealed in the case of the pin-maker, has much to tell us about the impact the organization of the labor process has on productive efficiency with a given level of technology. However, as Braverman (1974) helps us to understand and as we shall look at more closely in Chapter 10, other questions remain. For example, who organizes the process of production? In whose interests is this done? What principles determine the allocation of particular individuals and groups to the necessary tasks? How is mental labor applied in production as opposed to manual labor, and how does this affect the particular form of the labor process? How are reward systems worked out, and do they relate only to productivity or to other factors that are not economically determined? It is not that orthodox theorists deemed such questions trivial. Indeed, whole sciences of industrial relations and human resource management have grown up specifically to address them. What is so important about the insight offered by Marxist writers is that these questions cannot be partitioned into issues for management scientists alone. They are central to the ways in which the economic world as a whole operates, and without some sense of these relations of production, our understanding is incomplete.

So far, we have limited our discussion of divisions of labor to those operating at the level of the plant, the firm, or of the industry—that is, divisions of labor in production. At this level, it is relatively easy to see how a particular labor process draws people into relationships in ways determined by the nature of the production processes. However, at a much broader level, these economically regulated relations bring about divisions of labor around which key aspects of people's lives are configured. We can describe this as the division of labor in society (see Massey, 1984). It is a feature of the capitalist mode of production that people are related to each other in a system of commodity exchange not only by their workplace roles but also by virtue of the goods they produce and the broader social roles they perform—by their position within the social division of labor. They are linked by objects (commodities) that they sell, produce, and consume and by the market through which these commodities are traded for exchange value.

Expressed more formally, what people produce through the social division of labor are described as *social use values* (valuable or exchangeable things for society in general). These are produced by people working "together" but through the divisions of labor, commodities, and the market. It is fundamental to this process of producing and exchanging commodities in the context of the market that such a division of labor occurs. At the level of the individual production process in plants, in offices, or on farms, there is a detail or technical division of labor as we have described in the example of the pinmaker.

Critically important to capitalist society as a whole, however, is the division of labor that exists between plants, offices, sectors, regions, and nations as people perform their various functions in an integrated capitalist economy. Such a system, which we rarely see as other than the purported norm, produces particular (and from our point of view geographically interesting) relationships between each of the elements. Social use values are created at the level of society as an integrated whole and traded through the system of exchange. In this, people's relations with each other are mediated through the marketplace rather than through personal or social contact. They are "brought together" by the demands of the production system of which they form a part, and their personal relationships are powerfully molded by their functional roles within the exchange-driven system of commodity production.

To Marx, these special features characterized the historically specific mode of production, which in his later writings he labeled *capitalism*. What we have been describing in this chapter is the same "picture" we explored in Part One using orthodox, utility-based concepts. The terminology in this chapter, however, reflects a different theoretical perspective, one dedicated primarily toward an exploration of the nature of the relationships between people in particular societies.

THE DYNAMICS OF CAPITALIST PRODUCTION

Other insights that have accompanied the emergence of Marxist approaches to economic geography influence the way that we attempt to deal with changes over time—with the dynamics of economic and social systems. Recall that our approach in Part One was to set up ideal models of the working of the market system under certain assumptions and then to attempt to bring them closer to reality by adding in other influences, which were treated as deviations from the norm. By this method, for example, the huge concentrations of economic power in the hands of the giant transnational companies might be seen to be deviations from some normal condition of perfect competition in which it is assumed that no firms are large enough to influence the way the market operates. From the point of view we are now exploring, however, this would be unnecessary.

For radical theory, the "normal" condition of operation of the capitalist (or any other) mode of production lies in the way it works, not in particular ideal states to which it may naturally tend to move. As we shall see, Marx was able to anticipate the evolution of capitalism in the direction of the giant enterprise, even though that evolution had barely begun. The concentration of capital in the hands of giant oligopolists, which we described in Chapters 7 and 8, does not, from this perspective, have to be seen as a deviation but as an inevitable part of the evolution of a dynamic system in which constant processes of change are taking place. This sort of statement would probably be largely uncontroversial, but Marx and his followers set out some theoretical propositions as to why such a tendency is inevitable in the capitalist system, and these have been a subject for hot dispute ever since. Nevertheless, the notion that we are dealing with a dynamically evolving, constantly learning and changing system where simple rules (perhaps latent within it) tell us very little and where deviations from the simple models become impossible to handle has a great deal to commend it.

Once again, of course, the system we have described from a Marxian perspective is the same one that we were exploring in the first part of the book. There, however, it was taken as unchallengeable that the purpose of capitalist production is to accumulate further investable funds through interest, rent, and profit. We were, of course, not concerned at that stage with questions of morality, justice, and equity but simply with the nature of the workings of capitalism, accepting (or at least not seeking to address) questions about its legitimacy.

The Circuit of Industrial Capital

In volume 2 of *Capital*, Marx extends his analysis of capitalism to explore the way in which a whole economic structure can be described in terms such as those we have just outlined. The focal point of this was an overview of the *circuit of industrial capital*. We can use this concept to explore some of the more complex aspects of capitalism as a fully integrated social and economic system before we go on to look at how the system evolves over time. Figure 9.5 is a familiar one in the literature. Industrial capital appears in three forms. The first of these is money capital (M); the second, productive capital (P); and the third, commodity capital (C). Each element in this fully connected system

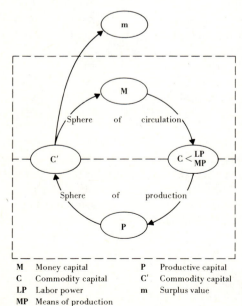

M	Money capital	**P**	Productive capital
C	Commodity capital	**C′**	Commodity capital
LP	Labor power	**m**	Surplus value
MP	Means of production		

Figure 9.5 The circuit of industrial capital.

depends on the others for the circuit to be complete. Money capital (M) is converted through the application of a labor process involving labor power and the means of production into productive capital (P). The product of this is further converted into commodity capital (C), from which money capital (M) is again derived on the realization of value from exchange in the market. Through the process we described earlier, money capital derived at the end of the circuit is greater than at the beginning. This is because of the creation of surplus value (m) in the process of exploitation. Surplus value as expanded capital is thus seen to provide the motor force for the continual reproduction of the system and for the process of accumulation.

One of the features of Figure 9.5 is that it enables us to distinguish the sphere of production (the lower half) from the sphere of circulation (the upper half). It does not matter where we begin the circuit of capital since it is a complete circuit. For example, if we take the circuit of money capital, we can describe an activity whose prime purpose is handling money with a view to expanded returns. Production becomes a secondary (but necessary) concern, as are the commodities that derive from it. Breaking in at another point, we can describe a circuit whose focus is production. Making profits from the production process itself is the primary aim, but this cannot be achieved without the sphere of circulation and the elements of money and commodity capital. Finally, there is the circuit of commodity capital where trading in commodities for gain is the central objective. This cannot, however, be separated from the circuits of production and circulation and is part of the fully connected system with productive and money capital.

The power of this view of the capitalist economy is that it reveals it as a system of integrated flows driven by a continuing dynamic process. Figure 9.6 shows how the different circuits "lock together" in a continuum, each supporting and continuing the other in a process of "autoexpansion." Such a view is not so revolutionary to us these days, but at the time of its emergence in the nineteenth century, the concept had great influence, and much of it has found its way into more orthodox economic thinking. Joseph Schumpeter, for example, one of the great theorists on innovation and technical change, acknowledged his debt to Marxian literature while opposing the general thrust of its theory. Similarly, John Maynard Keynes, the father of state economic management and much of modern macroeconomics, owed a debt to Marx for some key insights into the dynamics of the capitalist system.

In particular, the well-being of the system and of those who invest in it depends on an ability to realize their returns on capital. Capital put out in one form or another must see "fair" or at least average return through the circuits we have described or else a degree of breakdown begins to occur for the firm, the sector, or the economy at large. Where such circuits begin to break down at the level of the system as a whole, conditions of crisis emerge. Marx's own view was that such crises in a variety of forms were endemic to capitalism and that the system would eventually be destroyed by the contradictions inherent in it. Orthodox economists and Marx's critics would, however, point out that the very concept of capitalism as a dynamic process implies that iron laws are

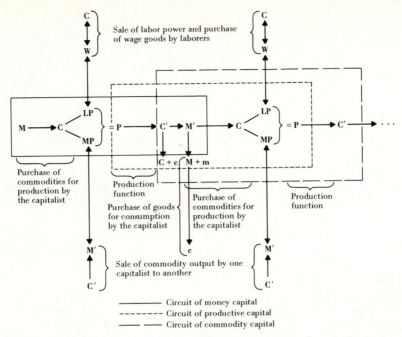

Figure 9.6 Interlocking circuits and the expansion of capital. *Source:* D. W. Harvey (1982), *Limits to Capital* (Oxford: Blackwell), fig. 2.1. Reprinted by permission.

unlikely to prevail and that the system would seek adjustments and change to cope with its crisis-prone nature. History so far bears witness to the ability of capitalism to resist final crisis. Nevertheless, as we shall go on to show, crisis, or perhaps less apocalyptically, recession, does have a central place in the dynamics of capitalist production.

Concentration and Centralization

In Chapter 7, we made the point that modern capitalist forms of business enterprise, competitive structures, and geographic ranges of activity are very different from those envisaged in the neoclassical models of the perfectly competitive small-enterprise economy. Indeed, one of the difficulties in trying to evolve neoclassical theories to address more contemporary reality is that they are largely unable to cope with oligopolistic competitive structures dominated by giant enterprises with a global span of activity. Marxian theory does, however, have something to say about the likelihood that capitalist economies will give rise to such features, even though, historically speaking, the earliest development of the theory predates the growth of the giant firm in the last quarter of the nineteenth century. On the basis of the logic inherent in his theoretical propositions, Marx identifies concentration and centralization as key structural features of evolving capitalism. The increasing scale of production and the technical benefits that accrue through economies of scale are seen both to promote and to require an increasing concentration of capital (sheer scale of enterprise), largely through the process of accumulation itself. At the same time, and we saw the evidence for this in Chapter 8, concentration tends to be accelerated by a process of centralization of capital in which the larger enterprise gobbles up the smaller either through outcompeting it or by merger and acquisition. Given the attention paid to this in Chapters 7 and 8, we have no need to repeat the details here. What Marxian approaches see in this phenomenon is, at base, the need to extend more and more control over a turbulent marketplace. This becomes a self-generating drive as capitalist enterprises become physically larger and have greater scales of fixed assets to defend and justify by economic returns.

However, it is misleading to think of concentration and centralization as a one-way street. Harvey (1982), for example, reminds us of Marx's belief that such a process was not unidirectional but involved processes of both centralization and decentralization simultaneously as new ventures were spun off and some activities were discarded. This, of course, echoes closely the description of current events set out in Chapter 8 and the work of Williamson (1975) in exploring these centralizing and decentralizing tendencies over time. In many ways, it is perhaps the awareness of these impending developments in the Marxian literature and the inability of neoclassical theories to cope adequately with them that has led to this particular aspect of radical theory being most widely taken up by economic geographers in recent years. As always, however, in the Marxian literature, simple borrowing of one element of the general theory tends to ignore the interrelatedness of the elements in the entire theoretical construct, and borrowing the terms or a relevant sentence or two describing concentration or centralization leaves the borrower in a theoretical no-man's-land. Nevertheless, the emergence of monopoly (or oligopoly) forms of capitalist competition of centralization and concentration was predicted in a body of writing that predated the more extreme instances of these events, and it is understandable that the events should have drawn people to explore a theory that showed some awareness of them.

Time and Cycles: The Role of Crisis

Continuing to focus on the dynamics of capitalism as Marx and his followers described them, we come to another aspect that neoclassical theories cannot help us to understand (since they were not designed for that purpose). Looking at capitalism in the last quarter of the twentieth century, the notion of crisis and the work of theorists who have addressed it seems quite reasonable. It would not have captured attention in boomtime, but the depression years between the wars forced it then, as now, onto the agenda.

It was, of course, central to the Marxian thesis that capitalism as a system was prone to crises both cyclic and structural and that, being based on a conflictual opposition of capital and labor, such crises would serve to exacerbate the class struggle endemic to the system. Among orthodox economists, the cyclic nature of the capitalist system is also widely acknowledged, but the use of the descriptor *crisis* is resisted as having a particular meaning derived from the context in which radical theory sets it.

The notion of crisis has, however, made a strong impact on the social sciences in general and economic geography in particular through the related concept of restructuring, which we shall deal with in Chapter 10. Like so many terms that have become part of the intellectual currency of modern debate, *restructuring* has a meaning within the Marxian corpus of thought that is interpretable only in the light of the elements of the whole theory. Accepting that crisis is perceived to be a feature of a system whose circulation and realization fall out of synchronization and in which class struggle will, from time to time, damage the thrust of the drive to accumulate, readjustment is considered to be inevitable for survival. From this viewpoint, restructuring as a value-laden concept may be defined as a dynamic process by which profitability is restored to capitalist societies. When set within the broader body of Marxian theory, then, restructuring is an inevitable stage in the evolution of capitalism as a response to crises that the structure and dynamics of the system itself throw up. The clear purpose of such restructuring from this viewpoint is to restore accumulation, and in terms of the logic set out earlier, this is likely to involve a restoration of the extraction of surplus value from the labor process. Defining the labor process broadly as involving not just wages in relation to output but the whole bundle of possibilities that surround the creation and extraction of value through labor (see Braverman, 1974; Massey and Meegan, 1978, 1982) restructuring pervades the entire system. It will frequently include these activities:

> *Intensification*—the extraction of greater labor time from a given work day

> *Rationalization*—the removal of means of production and their workers on which returns are deemed unsatisfactory

> *New technology*—the application of new methods and machines that raise levels of output for the same application of labor time

It may also include various measures to achieve a "deskilling" of the labor process, separating mental labor from manual labor as a means of giving management greater control over the value chain and the extraction of the surplus. In addition, however, restructuring may not simply act within the sphere of production but may also affect the reproduction of the labor force. Examples of the latter might include finding ways to have the state bear the costs of training workers or employing housebound women for parts of production in some form of outwork system. Clearly, *restructuring, concentration* and *centralization*, cannot simply be borrowed from Marxian theory unless its nuances in the context of that theory are clearly understood.

Like most of the concepts dealt with here, restructuring can be addressed from a more orthodox standpoint (as it was in Chapter 8). It is generally used in this case to denote reconfiguration, reorganization, and repositioning of a business to restore it to profitable returns. Indeed, the methods used may be identical to those described in the Marxian literature—giving managers the "right to manage," ensuring that workers deliver a "fair day's work for a fair day's pay," scientifically organizing production to avoid "wasteful" downtime, making sure a "fair return" is achieved on capital invested, and so on. Once again, the concrete processes observed are the same; where orthodox and radical theories differ is in ascribing different meanings to those processes.

CONCLUSION: ALTERNATIVE PERSPECTIVES TO LOCATION IN SPACE FROM RADICAL THEORY

In striving to explore Marxian literature, economic geographers have opened a Pandora's box of complications. While this has brought a flood of new ideas and insights, it has also brought a dramatic increase in the complexity of the processes to be taken into account and some thorny problems of how to engage with the deeper philosophical and political debates that Marx and his followers were primarily addressing. Perhaps the first clear lesson that raising the veil on Marxian theory brought with it was that location itself becomes important in entirely new ways and that some traditional views of locational decision making are so incomplete as to be of only marginal importance when the fully interconnected economic system is explored through spatial lenses. It is to these issues that we now turn to examine the ways in which radical theory has contributed to evolving theory in the economic geographic literature.

FURTHER READING

Aglietta, M. (1979). *A Theory of Capitalist Regulation.* London: New Left Books.

Bottomore, T. (1983). *A Dictionary of Marxist Thought.* Oxford: Blackwell.

Braverman, H. (1974). *Labor and Monopoly Capitalism: The Degradation of Work in the Twentieth Century.* New York: Basic Books.

Fine, B. (1984). *Marx's Capital* (2d ed.). London: Macmillan.

Green, F., and B. Sutcliffe (1987). *The Profit System: The Economics of Capitalism.* London: Penguin.

Harvey, D. W. (1982). *The Limits to Capital.* Oxford: Blackwell.

Hunt, E. K., and H. J. Sherman (1990). *Economics: An Introduction to Traditional and Radical Views* (6th ed.). New York: Harper & Row.

Chapter
10

Social Relations and the Geography of Production

If, on the basis of Chapter 9, we are to take the view that the essence of capitalism is its focus on accumulation through the "production of commodities by means of commodities" and that social relationships within it are closely conditioned by this mode of production, a number of important things follow. Perhaps the most important concerns our attitude to location in space. In earlier chapters, we have seen space largely as a resource or as a costly medium to be allowed for in the locational decision. In the von Thünen agricultural model of Chapter 1 and the Weber model of industrial location in Chapter 3, space was occupied by entrepreneurs (capitalists) in accordance with its natural properties such as site, position, physical attributes of fertility, and mineral resources. These properties (of which we singled out relative location) gave place its value (rent) for the conduct of certain forms of commodity production. Capitalists competed to acquire the productive services of particular locations in space, and the neoclassical model of allocative efficiency under perfect competition saw to it that sites were occupied by the most rewarding and efficient uses.

In this sense, space, place, and site were commodities traded at their exchange values on a free market, and the combination of marginal cost and marginal return ensured that so long as the market was unconstrained, places would be adopted (or abandoned) according to economic logic. Only places able to outcompete others in attracting a given form of commodity production and those able to outbid rival claims could be assigned appropriate roles in the geography of production. Thus economic geography, as the simplified models showed, was the product of a system of efficient bidding by activities for different places. Economic geography ultimately reflected optimal production allocation over space. This is what Christaller, Lösch, von Thünen, and Weber set out under the restrictive assumptions of their models. That the real system of commodity production reflected at least some of these theoretical principles was also demonstrated in Part One. We were able to add the time dimension to this approach in Chapter 6, showing how places (once the stable equilibrium assumption is discarded) come to perform changing roles in the production process over time—being selected, abandoned, and reselected as their economic geography unfolds historically.

But, of course, we were viewing these processes through a particular theoretical lens. The questions we were asking were the classic ones about how alternative uses are allocated to different places. As our brief excursion through Marxist theory showed, the answers we were getting were contingent on the particular mode of production we were examining at a particular moment in its development. We would not necessarily expect to ask the same questions or get the same answers from an examination of feudal economies or socialist ones as we did for capitalism. In particular, we were asking questions about the relationships of objects or things. Places were identified as "portions of space," "territories," "sites," and we were seeing through a lens that allocated objects—factories, farms, railways, crops, houses, offices—to places in accordance with logical rules derived from neoclassical theory. We were also seeing space as something relatively passive. Geographic patterns projected the result of the distribution of economic processes over space, rather like a canvas on which the outcomes of locational decisions were painted. Insofar as space had a role to play, it was often conceived as having some kind of a "distorting" effect, pulling central places, factories, and farms into shapes that reflected the costs of buying, using, or traversing land.

We must now change lenses to reflect the input of the Marxist insights to geography discussed in Chapter 9. First of all, we must ask different questions: questions about people, societies, social relationships—about how space, place, and territory affect and reflect not just the forces of production but also the relations of production. Whether or not we decide to adopt a formal labor theory of value and to view the process of accumulation as the expropriation of surplus product by capitalists from labor, there is considerable merit in recognizing that the space economy functions by means of social processes between people and not simply by models that assign land uses to places like pegs to appropriate holes.

At the same time, this approach demands a change in the way we conceptualize space. We need to remove the separation between space and social process. Place and space are to be seen as active forces, with the character of places not simply offering distortion but playing a central role in the way social processes work themselves out. From this perspective, it is too limiting to consider space simply a platform on which the furniture of the economy is placed and from time to time moved about. This, then, is a key insight that the radical approach brings. Space and place in the world we inhabit reflect not only the product of decisions about allocation and efficiency but also, through a kaleidoscope of continual change, the ways in which people see, interpret, give meaning to, and use the portions of space with which they come in contact.

As we have seen, people interpret, act, and react to space in the context of the complex sets of social relationships around which their lives are configured. From a Marxian perspective, the most fundamental ones are the social relations of production, which we discussed in Chapter 9. From this point of view, the key social structure would be class, and the fundamental distinction would be between capitalists and workers. Broader, more pluralistic perceptions might see people positioned in systems of social relations—by gender, ethnicity, social group, occupational status, age, and so on. The important point is the recognition that the social context and its systems of social relationships cannot be ignored if we are to have a complete view of the causal processes that influence the structure and dynamics of spatial form. Equally important, space and place have an active role to play in the working out of social processes. Not only are social processes constructed over space, but space itself is "socially constructed." Doreen Massey (1984) puts it most clearly:

> If the social is inextricably spatial and the spatial impossible to divorce from its social construction and content, it follows not only that social processes should be analysed as taking place spatially but also that what have been thought of as spatial patterns can be conceptualised in terms of social processes (p. 67)

From these preliminary observations, then, let us have another look at "location in space," widening the scope to take account of social processes in general but retaining our essential focus on the sphere of production.

KEY THEMES IN RADICAL LITERATURE ON SPATIAL STRUCTURES OF PRODUCTION

Having established that, from the Marxist perspective, an analysis of the geography of production must take account of both broader social processes and the interactive role of space, we can now try to put some flesh on these bare conceptual bones. How have geographers used these insights in their work, and what have been the outcomes for our understanding of the mechanisms that produce particular spatial effects? In addressing this question, we have to acknowledge at the outset that the room at our disposal in this chapter is far from adequate to review the product of what, for at least a decade, has been one of the most exciting and dynamic fields in human geography. All we can do is review briefly some of the key areas of work; you may want to follow through with your own search of the literature.

The organizing theme for this chapter is the series of approaches (often set out as particular "debates") that Marxist insights have spawned. We begin with the organizing theme that has been variously labeled the structural or restructuring approach.* This emerged particularly in British geography in the middle years of the 1970s and continues to be a rich source of discussion and research. It arose primarily both from a dissatisfaction with the explanatory powers of the approaches we reviewed earlier in the book and from events in the real world as economic recession began to grip many of the older industrial regions of the United Kingdom. New questions were thrown up by conditions where the most important locational decision was becoming the decision to close down and to rationalize production units and not the decision to choose sites for new plants. Existing theory in economic geography was simply unable to cope with this, so a search for new insights began.

A key element in the restructuring approach that has its own particular history and character is the *labor process debate*. This emerged directly from Marxist theory and focuses particularly on the means by which capitalists seek to extract value from the application of human labor in the process of producing goods for the marketplace. It focuses both on the technical aspects of the division of labor in production and on the social relations that surround it. Central to this debate is the work of labor process theorists like Henry Braverman (1974). The scale of analysis is, for the most part, set at the level of the technical division of labor inside the plant or production unit. It explores the ways in which labor is organized according to the dictates of management for the purpose of achieving profitable production.

In conceptual terms, the restructuring and labor process debates are not really separable in what is, after all, a whole. We shall, however, treat them here in separate sections for ease of exposition. Crosscutting and central to both debates in ways that have had a powerful effect on geographic thinking is the concept of the *spatial division of labor*. We shall use this as a unifying theme and as a summary device.

More recently, the literature of radical economic geography has been reverberating with another key debate, the flexibility debate, with which we will end this chapter. It explores whether we are witnessing the emergence of a new regime of accumulation following the breakdown of the Fordist era, which sustained the advanced capitalist economies for much of the twentieth century. The debate focuses on flexible accumulation as a potential successor regime, and we shall explore its dimensions.

---

* Literature on this is plentiful: see, in particular, Massey, 1978, 1984; Massey and Meegan, 1982; Massey and Allen, 1984; Scott and Storper, 1986; Peet, 1987.

THE RESTRUCTURING APPROACH

Alternative Perspectives on the Locational Decision

From the perspective adopted in Chapter 9, we can start by reaffirming that the driving force for capitalism is accumulation—the search for profit through the creation and realization of exchange values (Figure 9.5). In the circuit of productive capital, money capital (M) is transformed into productive capital (P) for the generation of commodity capital (C) and realized in the form of expanded value in money capital (m) as profit. The approaches that the neoclassical models adopted concentrated on only one part of this process of capital accumulation. Productive capital (P), once put in place, has the physical form of a fixed asset—a factory, farm, or warehouse and its associated equipment, for example. This represents the "economic furniture" whose allocation was the center of attention in traditional economic geography.

Radical approaches, however, which attempt to understand the spatial structure of production at the level of the system as a whole, rather than through a single element with geographic connotations, start at the beginning with (M), the commitment of money capital, as an investment decision and follow it through in the entire process of accumulation. What this means is that location theory has to be seen as part of investment theory and investment theory as part of a general theory of capitalist accumulation (see Walker and Storper, 1981). Put another way, locational choice becomes investment choice, and investment choice depends on the overall analysis of the prospects for accumulation. Of course, this was also true in the simplified models of the first part of the book. What is different here is that the investment choice itself is conditioned by the full range of commercial, social, and political circumstances and not simply by closely constrained economic assumptions dedicated to the exploration of the distorting effects of location in space. The restructuring approach seeks to cut into the causality of spatial events at the level of the capitalist system as a whole. Marxist political economy is the most common vehicle for the dissection of capitalism as the relevant mode of production.

Let us take a brief example to make these differences clearer. Suppose we are considering the locational decision surrounding the choice of a site for a new automobile plant. Traditional approaches in industrial location theory from Weber onward would take as given that the natural basis for production is the search for profit in a system of production dedicated to the needs of the marketplace. Prior and essentially aspatial decisions of whether to invest in the first place (given, say, the economic and sociocultural conditions of the time) or what to invest in (given, say, the returns to be expected from constructing another plant in the auto industry as opposed to diversifying into another business) would also be taken for granted. The focus of the location theorists' attention would be on where to choose. The chain of causality would start at this point rather than at a level that explores the role of space in the working of the system as a whole. The skills of the economic geographer would then be concentrated on judging, either after the event (causal analysis) or before it (prescription), the parameters of the choice of location by entrepreneurs who had already decided to invest in an auto plant.

In contrast, the restructuring approach would seek to cut into the causal chain at the wider level of the full circuit of production and would be concerned with social, political, and moral issues (political economy) as well as more narrowly economic perspectives. It would begin with the commitment of the money capital itself. What factors were involved in releasing the investment program in the first place? Against what criteria? Why in the particular business of the auto industry? What makes the approach radical (the management strategy approaches of Chapters 7 and 8 ask these same questions) is that the restructuring approach also considers who made such decisions, how they were empowered to do so, in what social and political (class) context, in whose interests, and with what effects on social relations. Only at that point does the location decision, in the narrow sense in which we used it earlier, come into the frame of reference, and even then, as we shall see, the questions asked by radical approaches are significantly different.

As to geographic outcomes, the restructuring approach acknowledges that at certain periods there may be very significant spatial results from general decisions in a particular sector or in the economy as a whole not to invest at all. Where, for example, a particular industry has been localized in a city or region and a decision is made at the sector level not to reinvest, continuing depreciation may go on to prime plants for later rationalization. No initial decision has necessarily to be made to choose a site. The choice becomes an outfall of the prior distribution of the capital stock and the intrinsically nonspatial decision to cease investment in a line of business. From this viewpoint, periods of restructuring may be marked in the industrial economic landscape by "missing strata"—with significant gaps in the sequence of layers of productive capital that make up the visible geography of the built form.

To address such real issues at a time of major recession in the economies of Western world in the mid-1970s and the early 1980s, the restructuring approach attracted growing popularity among economic geographers of all kinds, both radical and not so radical. At this time, the events overtaking space economies in many key regions of Europe and North America involved closure, rationalization, and retrenchment (see Bluestone and Harrison, 1982; Martin and Rowthorne, 1986; Massey and Meegan, 1985). The location theory of site selection at a time of expansion, investment, and growth seemed to have little to say to a generation confronted with a need to know more about the causes of decline and emerging urban and regional deprivation. The search for causality promoted a breaking out from narrowly geographic explanations. Attention shifted from theory set at the microscale in the context of free market competition with a strong concentration on geographic factors of location to approaches that looked either to the broader traditions of political economy or, as we did in Chapters 7 and 8, more toward theories of decision making among corporate enterprises and to management strategies. Advocates of both approaches would recognize that explanations for geographically significant events demand perspectives that take the theorist far beyond the narrow confines of the approaches we discussed in the first part of the book and would accept that causal chains for local events frequently begin with factors that operate at the scale of the capitalist world economy as a whole.

What have we learned from our introduction to the restructuring approach?

1. Location is but one (and often a minor) element in the complete equation that confronts a capitalist enterprise in making decisions that have significant impact on the evolution of what we earlier described as the space economy.
2. Location theory is, in essence, a subset of investment theory with the investment decision preceding the location decision and laying down the key parameters subsequently used in making a locational choice.
3. Investment is rarely a discrete event, and the decision to invest in a new plant demanding a new site is relatively rare. Most investment decisions deal with ongoing situations—whether to invest in new plant and equipment, whether to continue to support an existing business or product line, whether to apply new technology to an existing system, whether to scrap an existing production facility or discontinue a planned program.
4. Investment has an upside and a downside, and geographically significant events are produced by both.
5. Above all, the investment decision (and therefore the locationally significant decision) is a dynamic ongoing process. Theory must deal with a process in motion—a state of dynamic evolution in a mode of production dominated by the necessity for expanded accumulation and based on the capitalist labor process.
6. Geographic outcomes both reflect the constellation of social relations in capitalism and, at the same time, represent an active force conditioning the evolution of that constellation of social relations.

Dynamic Perspectives: Waves of Development and Rounds of Investment

It is central to the restructuring approach that the essence of causality in capitalism is the process of accumulation. As we showed earlier, this process contains an inbuilt tendency toward cycles of expansion and contraction, boom and slump. The *long* or *Kondratiev wave* has been identified as a recurring cycle spanning roughly 50 years with an upswing of approximately 25 years' duration followed by a downswing of the same length (Marshall, 1987; Hall, 1987; Freeman et al., 1982; see Box 10.1). Shorter-amplitude business cycles have been identified, with roughly ten years from peak to peak. This cyclic (long- and short-wave) tendency within capitalism thus features a progression of waves of development followed by waves of restructuring as the system seeks to restore itself to profitability. The essence of the restructuring approach is therefore that the causal processes are driven by this constant need for capitalism to find ways of maintaining the drive to accumulate.

Having established that location theory is a subset of investment theory

Box 10-1 Long or Kondratiev Waves

Recent years have witnessed a popular revival of the notion of long or Kondratiev waves. These are roughly 50-year cycles of change that can be discerned in the historical evolution of the capitalist system. First noticed by a Russian economist (Kondratiev, 1925), these waves are characterized by upswings of roughly 25 years followed by downturns of approximately the same length. For example, key troughs in economic development appear to have occurred in the 1830s, 1870s, 1930s, and 1980s, with the peaks of upturns in the 1850s, 1900s, and 1960s. Edwards and colleagues (1975), Hall (1985), and Marshall (1987) review the long-wave phenomenon in considerable detail (see Figure 1).

Much of the literature on long waves has become associated with the debate on the role of technical change in economic development; Schumpeter (1939) in particular regarded innovations as powerful energizing forces in driving upswings of capitalist development. Later work along the same lines by Mensch (1979) sought the motor force for the upswings in the bunching of particular innovations—steam power, electricity, the internal combustion engine, and so on—at the bottom of each trough subsequently powering the breakout from recession. This view has been strongly criticized by Freeman and colleagues (1982), whose work is based on the "technology web," a system of mutually reinforcing technical innovations and social structures that render already available technologies capable of generating virtuous circles of growth and development. Mandel (1980), also a contributor to this literature, reviews the long-wave phenomenon along strictly Marxian lines. He sees historical developments in capitalism as reflecting phases of intensive accumulation followed by periods of crisis and breakdown. In these, the emergence of realization crises and the rising power of labor at the top of the upswing demand a social and economic restructuring of the system to restore accumulation.

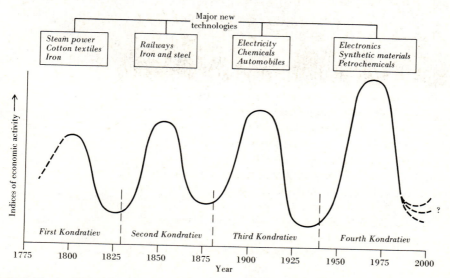

Figure 1 The long or Kondratiev wave. *Source:* P. Dicken (1986), *Global Shift* (New York: Harper & Row), fig. 2.2. © 1986 P. Dicken. Reproduced by permission of Harper & Row, Publishers, Inc.

and that locational outcomes are predicated on investment decisions, it is a short step to recognize that where investment waves and cycles exist in the macroeconomy, they will produce spatial effects in the form of the built environment. Figure 10.1 shows, for example, the existence of these waveforms in the evolution of the built environment in the United States and Britain. Such changes in the scale and spatial distribution of production will in turn flow through to have substantial effects on the nature of social processes. These rounds or layers of production, as Massey (1984) labels them, provide us with a geologic metaphor revealing that places can be visualized as deriving key aspects of their physical and social characteristics from the succession of roles they have played historically in the evolution of capitalist production.

For example, strong and powerful roles at the controlling centers of the economy will tend to leave legacies of the fine buildings, the state-of-the-art transport and communications structures of the day, the factories and warehouses of the period, together with the homes and community facilities required to service the system.

Figure 10.2 shows for a single element of the built form—housing—how these past waves of investment have left a legacy in the contours of housing age in Minneapolis. These appear like tree rings of different widths representing the housing investment climate of the time as well as the particular technologies of housebuilding and travel that set the context. Clearly, the real background conditions for the shape of the housing surface for Minneapolis are far more complex, but we use the example here simply to fix the notion that waves of investment leave their spatial footprints in the built form.

By contrast with strong roles in economy and society, weak and dependent ones may leave much less to show, but if they follow previous periods of success and development, the downside features of decay and social deprivation may leave powerful marks in the physical landscape. Such roles also stamp the social character of places, their cultural forms, the nature of their class structures, and the shape and hierarchical ordering of civil society.* The powerful role inevitably has its social and cultural effects, but so, in ways that lack of physical change may fail to reveal, does the weak and dependent. So much of what we understand by the concept of place and the identities of cities and regions is derived from these historical forces. They emanate from the evolution of their current physical and spatial form.†

---

* These are also reflected in the landscape of the built form, urban design, and architecture in the shapes and places that reflect social value systems. A growing literature is exploring this; see, for example, Meinig (1979), Cosgrove (1989), and Soja (1989).

† The fullest exploration of this aspect of the restructuring process is to be found in Marxist approaches to urban problems. Harvey (1973, 1978) and Castells (1977) feel that evolving urban forms reflect the class struggles central to capitalism as well as the physical manifestations of accumulation in the built form of the city.

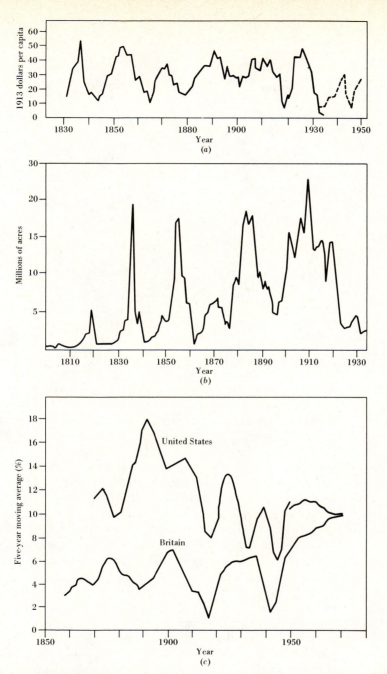

Figure 10.1 Waveforms in the built environment in the United States and Britain: (*a*) U.S. building activity per capita; (*b*) Sale of U.S. public land; (*c*) Investment in the built environment in relation to GNP (U.S.) and GDP (U.K.). *Source:* D. W. Harvey (1989), *The Urban Experience* (Oxford, Blackwell), figs. 6 and 7. Reprinted by permission.

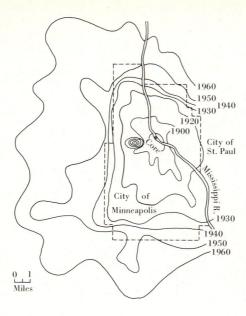

Figure 10.2 The legacy of waves of investment: housing age contours in Minneapolis. *Source:* J. S. Adams (1970), "Residential Structure of Midwest Cities," *Annals of the Association of American Geographers* 60: 37–62, fig. 16. Reprinted with permission.

Waves of Investment: Some Examples

Let us look at a few examples to illustrate more clearly what we mean by these relatively abstract ideas. We start with an example of the favorable conditions that existed for U.S. companies during the postwar reconstruction period in Europe. Such circumstances persuaded them to extend their investment programs and to position themselves favorably for a revival in the European market for consumer goods. The "whether to invest" criterion was met by a positive general response, the "what to invest in" decision pointed firmly to foodstuffs and consumer durables, and the "where to invest" decision targeted the United Kingdom as a particularly favorable location. The outcome, which had great geographic significance for the places subject to its distinctive effects, was a round of investment in the parts of Britain favored at the time by government investment incentives. Local results saw northern England and Scotland in particular capture a considerable number of new U.S.-owned subsidiary and branch plants.

One particular area to benefit from this wave of inward investment, as Figure 10.3 shows, was northwestern England. A declining port, coal, and cotton economy whose traditional industries were struggling to recover in the postwar period saw a massive influx of plants with U.S. parent companies. These investment projects brought with them new production technologies and new working conditions. They favored greenfield sites in developing areas, often on the margins of the more traditional industrial towns, and many of them tended to draw in women workers rather than men. This particular round of investment thus had in own particular technological capabilities, its

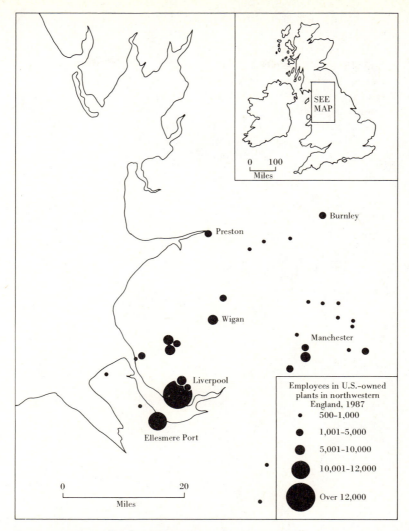

Figure 10.3 Rounds of investment: U.S. post–World War II investment in Northwestern England. *Source:* North West Industrial Databank.

own kinds of labor requirements, its own kinds of working practices and labor relations, and its own special geography. It was based on an emerging internationalization of capitalist production and marked the beginning of a long period of global integration in the world economy.

More recently, a new wave of inward investment has turned its attention to the United States itself with a massive influx of Japanese auto companies to the Midwest. As Rubenstein (1989) shows, the opening of six Japanese-owned automobile assembly plants has significantly altered the industrial landscape of the United States. Although the new plants have selected the nation's old manufacturing belt, the incoming Japanese firms have generally not selected communities traditionally associated with automobile production. Figure 10.4 shows the spatial pattern of the choices: Honda (Ohio), Nissan (Tennessee), Mazda (Michigan), Toyota (Kentucky), Mitsubishi (Illinois), and Subaru-Isuzu (Indiana), mostly located in areas not occupied by domestic automakers. As with the earlier example of U.S. investment in Britain, new industrial spaces have been occupied by the incomers, bringing with them new labor requirements and forms of work and substantially altering the lives of the communities into which they have implanted themselves.

"Layered" on more traditional technologies of production and working

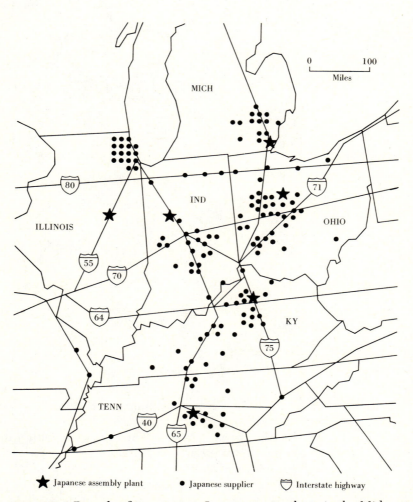

Figure 10.4 Rounds of investment: Japanese auto plants in the Midwest. *Source:* J. M. Rubenstein (1990), "The Impact of Japanese Investment in the United States," in C. M. Law, ed., *Motor of Change* (London: Routledge), fig. 5. Reprinted with permission.

requirements, the injection of U.S. direct investment in Britain and Japanese investment in the United States brought to the communities that captured them new physical and social conditions that then evolved in different ways. The American or Japanese dimension changed the nature of such places, by contrast with those that missed out on a share of this wave of development. A change occurred in the spatial organization of production in Britain and the United States predicated on global shifts in the pattern of industrial investment and brought to older industrial communities in specifically designated parts of Britain and to some cities in the Midwest a new round of investment that sharply changed their physical and social character. Figure 10.5 shows how such global forces may be transmitted through the decisions of companies of various kinds, through their decisions to open, expand, or close plants to produce outcomes "on the ground." It also serves to show how explanations for localized events often have to be sought in terms of global processes (Dicken, 1986; Thrift, 1985; Lloyd and Shutt, 1983).

No search for causality that did not look at (1) the particular stage of development through which the capitalist world economy was passing, (2) the changing role of U.S. and Japanese companies in that economy, and (3) in the particular case of Britain the critical role of state regional development policies could begin adequately to understand what had happened to bring about changes in geography and society in particular industrial towns. That such changes altered the identities of places is undoubted. Corporate managers charged with bringing new technologies and working practices to these areas would also doubtless confirm that local conditions had to be taken into account in ways that modified the nature of the investment and labor management systems put in place.*

Anatomies of Job Loss

Having looked at these essentially upside effects, now let us reverse the lens and look at the downside—the impact of *disinvestment*. One of the most significant early pieces of work on the restructuring approach was conducted by Doreen Massey and Richard Meegan (1978, 1982). They set out to examine from a geographer's viewpoint the effects on industrial cities in Britain of a major attempt by the U.K. government to reorganize the production of a whole range of products under the broad heading of electrical engineering in the early 1970s. Once again, a world view proved essential to an understanding of why such a program was thought necessary. First of all, there had been significant technological changes in the nature of production of heavy electrical engineering equipment, electrical consumer goods, and computers during the 1960s. Second, in organizational terms, the same period had seen the emer-

---

* Interest in this issue has been growing in recent years; studies by Massey (1985), Morgan and Sayer (1985, 1987), and Morgan (1986) have examined both local restructuring under the auspices of incoming firms and the adjustment processes by which incomers adapt to local conditions.

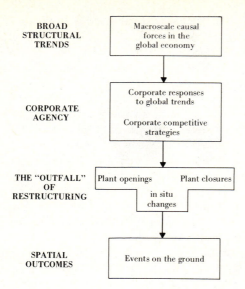

BROAD STRUCTURAL TRENDS — Macroscale causal forces in the global economy

CORPORATE AGENCY — Corporate responses to global trends / Corporate competitive strategies

THE "OUTFALL" OF RESTRUCTURING — Plant openings / Plant closures / in situ changes

SPATIAL OUTCOMES — Events on the ground

Figure 10.5 Causal forces and spatial events.

gence of a globally configured system of production with the key roles being played by major oligopolistic companies. These and other features of the new system set the conditions for change.

United Kingdom—owned companies, of which there were several, remained too fragmented and too small to compete effectively in this new large-scale, global marketplace. In response to this, the U.K. government of the day set out to reorganize production to achieve a rationalization and consolidation of the industry around a small number of companies strong enough to engage in the new competitive process. The inevitable result was a process of induced concentration and centralization, giving growth in some companies and significant reductions in the competitive position of others. Causality here, then, lay with responses to the emergence of a global reconfiguration of production, to new technological and market opportunities, and to more oligopolistic structures of competition. Once again, the state was a key actor.

What Massey and Meegan (1978, 1982) attempted to do was to explore this causality in the context of more general, Marxist-informed theories about the evolution of the capitalist economy and go on to explore how this led to events on the ground. They were interested both in spatially significant changes in the social relations of production and in the industrial geography of cities and regions in the United Kingdom. By contrast with our earlier example of the U.S. branch plants in the postwar period, geographic changes in this case were brought about as a result not of overt locational choices but of the spatial outfall of a restructuring process. This was conceived as a process with systemwide causes but local outcomes. Plants selected for rationalization were, for example, chosen by their position within the system of production in the electrical goods industry and by the position of their controlling companies within the system of corporate competition. The narrowly spatial aspects of their location were simply part of this general package of causal factors.

Geographic outcomes, as they emerged, were those of in situ survival or closure and rationalization. Little construction of new plants was involved, but even in the plants chosen for survival, the application of new technologies or the reshaping of working practices brought about changes. These were captured not so much in visible physical structures as through significant changes in the relations of production within plants and between communities. In the electrical engineering study and in a more widely drawn study derived from it in *The Anatomy of Job Loss*, Massey and Meegan (1982) identified changed working practices to deliver more productivity as resulting from the strategies we referred to in Chapter 9 as common methods for coping with the periodic crises of capitalism: the intensification of work and the implementation of new technologies to change the nature of the technical division of labor. The rationalization of certain lines of production was also evident, sometimes leading to obvious physical closure but in many cases to less visible internal changes in the activities carried out. All of these altered the balance of employment from place to place. Table 10.1 shows how each of the three restructuring strategies was applied with different weight across whole sectors of the U.K. economy during the recession years, bringing about changes in the distribution of employment and producing what we shall come to identify later as a new form of the spatial division of labor.

Table 10.1   IMPORTANT FORMS OF PRODUCTION REORGANIZATION, 1968–1973

Rationalization	Intensification	Investment and Technical Change
Grain milling	Cycles	Crackers
Iron castings	Textile finishing	Sugar
Metalworking machine tools	Leather	Brewing and malting
Electrical machinery	Men's and boys' tailored outerwear	Coke ovens
Insulated wires and cables		General chemicals
Locomotives and track equipment, railroad cars	Footwear	Synthetic resins, etc.
Weaving of cotton, linen, and synthetic fibers	Printing and publishing	Fertilizers
Woolen and worsted		Iron and steel, steel tubes
Paper and board		Aluminum
		Miscellaneous base metals
		Textile machinery
		Scientific and Industrial instruments and systems
		Cutlery
		Jute
		Carpets
		Bricks and refractory goods

Source: After Massey and Meegan (1982).

Taking a much wider canvas, Bluestone and Harrison (1982) produced a book with the deliberately alarmist title *The Deindustrialization of America*. They were driven, like Massey and Meegan, to respond to events unfolding around them at the time. These were the closure of plants and the virtual destruction of industrial communities, both large and small. The thrust of the argument developed in the book is the essential contradiction that exists between the requirements of economic growth and the damage that appeared to have to be done to restructure the nation's industrial base. The economist Joseph Schumpeter (1939) had suggested that part of the process of restoring capitalist economies to profitability after the downswing of the long wave were what he called "gales of creative destruction." We referred to them briefly in Chapter 6.

Bluestone and Harrison (1982) reported just such a gale blowing through communities as far apart as Detroit, Michigan, and Anaconda, Montana, provoking a generalized process of deindustrialization. At the regional level, however, they were able to show that the gale blew with differential force. There were significant variations both in scale and in the causal forces behind it. For instance, every state in the Northeast saw private industry destroy more jobs through closures than were created in openings. In the Northwest and the Sunbelt, by contrast, startups exceeded shutdowns. Closures were, however, the most powerful events everywhere. "Boomtown versus busttown" was seen to be a particularly spatial outcome of the process they were describing. At the urban level, the gale was significantly altering the economic and social fabric of cities, favoring some and damaging others on a scale running from expanding Houston to reviving Boston to declining Youngstown. The core of Bluestone and Harrison's argument followed a restructuring approach with the need to restore the drive to accumulate, producing, through its spatially distributed effects, a major reworking of the roles of U.S. cities and regions in the geographic distribution of production.

Summary: Dynamic Processes of Restructuring

Generalizing the processes that we have been describing in both the upside and downside examples, we can see the essence of the restructuring approach. First of all, causality is ascribed at the level of the system as a whole. Within this, Marxist political economy sees certain structural characteristics as necessary features of the capitalist mode of production. The primacy of the drive to accumulate; the preeminence of the relations of production; the dynamic tendency to concentration, centralization, and internationalization as evolutionary forces; the importance of crises and phases of restructuring to restore the conditions for profitable production—all are seen as fundamental causal forces. From the conceptual viewpoint adopted by the restructuring approach, they have to be understood before proceeding to the study of changes in the spatial distribution of production and the spatial division of labor.

For given, socially constructed portions of space—towns, cities, regions, or nations—the built form is recognized as the imprint of the rounds of invest-

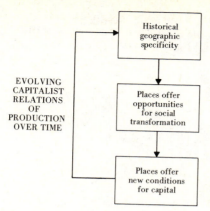

Figure 10.6 Cyclic processes in sociospatial evolution.

ment that marked a succession of roles in the spatial distribution of the fixed capital stock. In parallel, their social formations are understood as the product of this same succession of roles in the spatial division of labor. In all this, space and place are not seen as inert forms, simply a screen on which these images are projected. Particular places, with their social and cultural attributes and pools of work skills, are visualized as presenting capital with a bundle of opportunities and constraints to be adopted or rejected in subsequent development stages. So, in the terms of Figure 10.6, the process of sociospatial evolution takes place on capital's terms and is cyclic. The specific historical and geographic legacies of past rounds of investment offer differential opportunities for future waves. Against this, places are adopted or abandoned in the process of change. That these opportunities and constraints are themselves the product of a past history of capitalist exploitation closes the circle on a dynamically evolving process that itself expresses in spatial form the ongoing nature of capitalist relations.

THE LABOR PROCESS DEBATE

Remember that from the viewpoint of Marxist social and economic theory outlined in Chapter 9, the key source of value in capitalist society was considered to be the application of human labor to the products of the natural world. We described this as the labor process, the "productive activity which is the basic driving mechanism of capitalism" (Scott, 1982, p. 186). At base, the labor process represents the fundamental set of social and economic relationships through which society reproduces itself day by day. Capital controls both the broad process of investment and the day-to-day workings of the system. Indeed, the capitalist relation of the dominance by capital not only of the labor process but also of the broad fabric of life is central to the Marxist thesis.

Back in Chapter 4, we explored labor as a factor of production from the orthodox perspective to show how variations in the spatial distribution of production could emerge primarily from differences in labor cost. Labor from this viewpoint was simply a commodity to be offered up at a market price, the going wage. The choice to hire labor services lay with capital or, in the terminology of that chapter, entrepreneurs. The capitalist relation was accepted as the normal conditon, as was the notion that variations in investment and production over space would respond primarily to differences in the cost of labor as a factor of production.

For Marxist analysts, however, this whole argument about differential costs of labor is based on a deeper one about the relations of production and, in particular, the control of labor. This lies at the heart of the labor process debate. It was Marx himself who saw the introduction of new technologies as a critical determinant of the particular forms and shapes of labor process observable in capitalism. The introduction of machinery, in his view, enhanced the power of capital over labor by reducing skilled workers to machine minders and by making labor more uniform in its characteristics. This enabled workers to be more readily substitutable for each other. In this way, the traditional craft divisions of labor, which gave a degree of differential bargaining power to workers, would be broken down. These, for Marx, would be replaced by what he describes as more "natural" differences in the "qualities" of workers, such as age and sex (Marx, 1967, p. 545). Though he did not specifically mention it, perhaps we could, for our purposes, add location.

The theme of control has been taken up again recently, after a long period when the technological determinism implicit in Marx's view was subject to criticism. Henry Braverman's influential book *Labor and Monopoly Capital* (1974) put the labor process back on the agenda. The proposition was still one of the deskilling of workers by capital, but the focus was on managerial control through the introduction of scientific techniques of management rather than the direct influence of technology. Braverman saw these new management techniques as attempting to separate the mental from the manual component of work—removing the power of conceptualization and decision from the shop floor and reducing workers to machine minders in a working environment closely controlled by scientific management. A central component in the work of both Marx and Braverman is the need for capital to be concerned not simply with costs of labor but with the more generalized struggle for control of the production process.* This has considerable importance for our attempts to understand the ways in which the social relations of production reproduce themselves over space. For instance, from place to place, labor may prove to be more or less tractable to the needs of capital. As a result, issues of control and tractability may have a considerable bearing on the geographic structure of

---

* More recently, radical writers on the labor process have become critical of the deskilling assumptions implicit in Marx and Braverman, seeing the struggle between capital and labor in the workplace as by no means always producing an inevitable result but being "persistently incomplete." Variations are seen to exist constantly in the degree to which capital can subjugate the labor under its control (Wood, 1982; Friedmann, 1977).

production in ways not simply reducible to the cost differences we explored in Chapter 4. It was as a proxy for this that the orthodox literature frequently made use of the "degree of unionization" variable in its model.

Another important issue to be taken into account is that it is often critical for the lifetime of a particular round of capital investment that labor continue to be reproduced in ways that can maintain the use of the capital stock. Such "reproduction" may be seen in terms that include demography or the availability of skills. It may also depend crucially on the stable reproduction of attitudes, stable industrial relations, and supportive local institutions. This point is summarized in Figure 10.6. More than simple tractability, the reproducibility of labor may represent a significant variable in investment choice and in restructuring strategy. This represents another critical way in which places—towns, cities, regions, nations—may differ from each other as potential sites for productive enterprise. In short, to borrow Massey and Allen's (1984) echoing phrase, "geography matters." Taken to a greater extreme, Peet (1987) attempts to make the point about control, tractability, and reproducibility by superimposing the observed growth of employment in manufacturing in the United States on his state-based map of "class struggle" (Figure 10.7).

We have now arrived at a point in the analysis of factors that may lead regions or places to be attractive or not for new rounds of investment where we can distinguish not only cost but also control and reproduction as significant for the attraction of waves of development or rounds of investment.* What we have been able to do is to examine the ways in which, throughout the evolution of the capitalist system, the spatial characteristics that we have been concerned with from the very beginning of the book are laid down. The radical approach sees them in very different terms as a product of broad social transformations and as a result of the succession of roles played by particular places in the spatial distribution of production. We have therefore been able to shed some light on the role played by space in the reproduction of the social relations of capitalism. What we can do now is to explore in closer detail the nature of these roles in the production process, regarding them as the expression of a changing spatial division of labor.

---

* Ranging much more widely, it is possible to see that these same conditions for the regulation of the labor process are very important to capitalism in general during the course of its development. This approach identifies a number of regimes of accumulation—manufacture, machinofacture, Fordism, and neo-Fordism—through which capitalism is seen to have passed (see Aglietta, 1979). Each of these had its own particular characteristics in terms of labor process, systems for circulation and realization of capital, and general social relations of production. From a more geographic point of view in the terms we are dealing with in this chapter, each also had particular spatial form, characterized through the ways in which the system of production was allocated spatially and feature the social processes of the time that made places what they were (see Dunford, 1977 and Dunford & Perrons 1983). Diane Perrons (1981) brought this idea most accessibly before geographers in her study of Ireland. She showed how Ireland's role in the spatial division of labor changed over time in association with the regimes of accumulation, rendering its present position understandable only in terms of its integration into the neo-Fordist world capitalist economy. (Fordism and its successors are defined and discussed at the end of this chapter.)

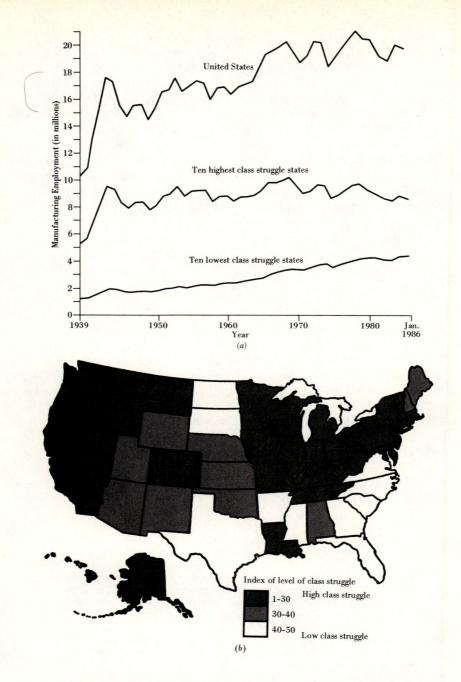

Figure 10.7 (*a*) Employment change in manufacturing, 1939–1986; (*b*) geography of class struggle in the 1970s. *Source:* R. Peet (1987), *International Capitalism and Industrial Restructuring*, figs. 4.1 and 4.2. Reprinted with permission.

SPATIAL STRUCTURES OF PRODUCTION AND SPATIAL DIVISIONS OF LABOR

Figure 10.8 helps us to keep track of the story so far. We began by exploring the restructuring approach. This cut into the problem of exploring location in space and the geography of spatial differentiation at the level of the system as a whole. As we learned in Chapter 9, the capitalist system at this stage of its development features a number of broad structural characteristics that define its contemporary form. First, it tends, as we saw in Chapters 7 and 8, to be dominated by large enterprises with concentrated economic power on which increasing control has become centralized through the merger and acquisition process—concentration and centralization. Second, the system as a whole is international. Indeed, it is global in its scope for action, with capital becoming almost hypermobile in its ability to seek out opportunities for profit wherever in the world they exist. The cyclic instability to which capitalism has been historically prone remains a feature of its evolutionary path, with both long waves and business cycles historically identifiable.

In exploring the labor process, we have made the point that production is, in essence, a social process and that to maintain itself, capital must reproduce itself through the maintenance of its system of social relations. Critical to this are both the issue of control of the labor process and the conditions under which labor is reproduced. The so-called labor factor is therefore of far wider significance to this approach than is represented by its narrow treatment as a variable cost input in the simplified models of the first part of the book. The

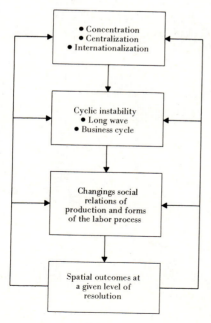

Figure 10.8 A top-down view of the restructuring approach.

essence of the system of capitalist relations is, from this theoretical perspective, the dominance of capital over labor, a condition that is considered far from automatic and complete in modern versions of the labor process debate. A more or less active struggle for control is waged in the production process. The significance of this for our exploration of spatial divisions of labor and the spatial structure of production is that variety exists in the social relations of capitalism from time to time and place to place and that these variations may have a very signficant impact on the rate of accumulation.

The social processes that we have identified in capitalism take place over space, whether they are seen in terms of the macrostructures or they are concerned with the microfeatures of day-to-day living. They produce spatial outcomes identifiable at whatever levels of resolution are set for the analysis. We have tried to show this graphically in Figure 10.9. In their turn, these spatial outcomes, whether in the form of the built environment or in social transformations, mold the relations of production. In this way capital is presented with new conditions from which to make subsequent choices. New choices make for new transformations, and so the process runs on as a continuous process. This, in a nutshell, is the centerpiece around which contemporary radical approaches to economic geography are set. The project is essentially incomplete, but it has provided the discipline with an innovative wave of ideas and insights.

Spatial Divisions of Labor

One of the most developed perspectives in this area has been through the work of Doreen Massey (1984), and perhaps the most fully developed exposition of the concept of spatial divisions of labor has been through the exploration of the ways in which the relations of production in the multifunctional firms we

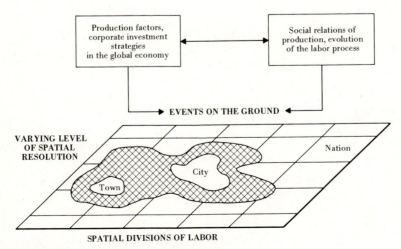

Figure 10.9 Causal processes and levels of resolution.

discussed in Chapters 7 and 8 have been "stretched out" over space. Rather than pursue the issue in abstract terms, let us illustrate what we mean by using an extended version of the hierarchy of levels of production we used in Chapter 7 (see Figure 10.10).

We examined the spatial allocation of the production activities of the multiplant firm by using a simplified three-stage functional hierarchy. Level 1 was the controlling headquarters, level 2 the administrative function, and level 3 the production activities. Figure 10.10 shows these functions in a rather different way. In the past and before transport and communications technologies permitted their geographic separation, all three levels in the hierarchy would have been in the same place (Figure 10.10a). The result would have been regional complexes of relatively small, autonomous firms in similar industries. This would have given rise to what we can describe as a sectoral-spatial division of labor.

Sectoral-Spatial Divisions of Labor

Spatial variations in the organization of production at this stage would then have been based on particular kinds of products. Cities and regions, as the old geography books would have described them, would have been cotton and coal regions or shipbuilding and steel towns. The ownership of capital and the locus of control would have been internal to such places, with the capitalist perhaps easily identifiable as the man in the tall hat living in the large house at the top of the hill. In terms of social structure, a wide range of classes, social

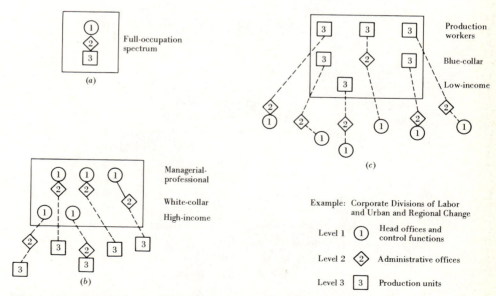

Figure 10.10 Sectoral-spatial and spatial-hierarchical divisions of labor: (a) Traditional sectoral-spatial; (b) Center of control; (c) Peripheral region.

groups, and occupations would be represented. Local communities (depending on the sectoral base) would offer a partial reflection of the class structure of society as a whole. Places would be recognized as mill towns, coal towns, or factory towns, with their geographic character socially constructed from this role. A mill or coal town is what it is, then, not just by virtue of its spatial form but also by the role it performs in the spatial division of labor and by the way this role is played out in the local community.

Spatial-Hierarchical Divisions of Labor

Once it became possible, through technological changes and changing organizational structures for business, to separate each hierarchical level of activity and allocate it as a specialized facility to selected places (we discussed the reasons for this in Chapter 7), this had a profound effect on the spatial organization of the social relations of production. By way of example, Figure 10.10*b* shows one of the possibilities that can emerge from this. For instance, some very special places may attract predominantly the headquarters level of organizations together with a smaller number of administrative functions. Often, as we showed in Chapter 7, these will be centers rich in information—capital cities, centers of government, commercial and banking centers. As we learned from our exploration of agglomeration effects in Chapter 5, the concentration of these high-level functions will have strong pulls on others through the existence of the benefits derived from clustering. The process is cumulative; beyond a certain point, it may be difficult for certain businesses *not* to be there. We might call such centers, mostly represented by the great world cities, *centers of control.**

Centers of Control

Stating the case more formally, relations of economic ownership and control will be concentrated in such centers, giving them a specialized role in the structure of society (see Table 10.2). Such places will be assigned a controlling role in the organization of the production system and the spatial division of labor, and spatial patterns will reflect this dominant position within the relations of production. Such places, for example, will often represent the key focal nodes in the circulation of capital, those from which investments emanate and back to which interest, rent, and profits flow for recycling.

Places assigned pivotal roles in the spatial structure of production may, of course, exist at a variety of levels. From the top of the hierarchy to the bottom, they may range from global centers for the recycling of circulating capital to the national, regional, and local centers. With progression downward, the degree of purchase of each kind of center on the economic ownership and control of

---

* John Friedmann (1986) has developed a series of hypotheses about such world cities that is set within the conceptual framework discussed in this chapter.

Table 10.2 CENTERS OF CONTROL

1. Key role in circulation and realization of wealth
2. Focal point: exporting investment funds, importing profit, interest, rent
3. Tendency to focus on money forms of capital: banking, insurance, finance
4. Occupational bias toward professional, white-collar, high-income groups
5. "Virtuous circles" of increasing incomes, high multiplier effects
6. Support large service group of workers in unstable, secondary occupations

production falls, and with it the balance of dominance and dependency shifts. This is much the same idea that we expressed from a very different conceptual viewpoint in the central place theory models at the beginning of the book.

In terms of their social characteristics, centers that capture the higher-level ownership and control functions of capital (company headquarters, banks, finance houses, property companies) will be characterized by occupations at the top of the hierarchy, the white-collar skilled-professional jobs. They will have the highest potential incomes and purchasing power and, as a result, will also tend to draw to themselves "virtuous circles" of linkage multipliers. To service this aspect of their economies, they may also draw in large numbers of workers attached to lower-grade service tasks with low skill needs and low pay. Although this is oversimplified as an outline, the example is designed less to describe real places than to show the particular forms of social relationships that exist both between and within centers in such hierarchies. The key role of such centers in the relations of production is control and dominance, and this is reproduced in their function within the spatial division of labor.

Peripheral Regions

Turning to the second example in our simplified review of the ways in which the capacity to stretch out the organization of production over space can construct different kinds of places (Figure 10.10c), we can explore areas that attract a predominance of level 3 production activities. These are regions in which the key role in social relations and the spatial division of labor is the creation of value in the process of production. They are dominated by production activities in manufacturing, agriculture, or mining, for example. Within the more detailed distribution of the spatial system of production, such regions may attract branch plants in which control and ownership functions are essentially absent because they are located elsewhere (in the centers of control).* These are essentially places at which roles in the technical division of labor are based on the organization of production in the factory, on the farm, or in the mine. The particular labor process in operation under these conditions will strongly

---

* This feature has been long a subject of interest in the orthodox literature of economic geography as the issue of external control (see, for example, Watts, 1981).

condition the social relations of production within such communities. In the social division of labor, they produce goods to be traded in the wider marketplace (see Table 10.3).

If we take as an example places where the role in the spatial division of labor is dominated by the large, integrated, continuous processing plant, shift working patterns will condition the cadence of life, gender divisions of labor in the household, and much of the fabric of social contact both at work and at home. Many traditional mining areas now being recolonized by modern manufacturing establishments employ women and demand a reconstitution of worker attitudes and household divisions of labor. As a result, social relations of production may be forced to undergo profound changes and with them the general social life of communities. A feature of all such regions is that workers are drawn to compete with those in other such regions to attract future rounds of investment capital. Thus while a key relation in the one direction is a dependency on those who control the capital allocation process (spatially expressed in the concept of the centers of control), a second key relation is of competition with other workers in bidding for continuing investment (spatially expressed in intercity or interregional rivalries).

Socially, regions dominated by level 3 functions tend to be characterized by blue-collar production workers, often with a strong segmentation by skill or occupational group. The upper echelons of the occupational stratum may be missing, making them culturally working-class towns or factory suburbs. The level of incomes may be such that income multiplier effects may be limited and the service sector of employment attenuated. As a result of both the generalized tendency for the system as a whole to undergo varying degrees of cyclic instability and the effects of interregional competition, such places may have only a tentative hold on continued development. Historical geography is full of examples of places with formerly powerful roles in the spatial organization of production reduced to ghost towns, depressed cities, or deprived regions.

Spatial Divisions of Labor: More Sophisticated Approaches

Doreen Massey (1984) takes the exploration of what she calls "whole bundles" of hierarchies in the process of production and their allocation across space much further than we can here. Places are not only ordered by their position

---

Table 10.3 PERIPHERAL REGIONS

1. Key role in the creation of value through characteristics of resource endowments and attributes of labor pools
2. Compete with each other regionally and internationally to attract capital
3. Transfers of value from periphery to core; no guarantee that return investment will continue
4. Subject to finance capital process: whether, what, where to invest? Will they be chosen?
5. Characterized by waves of investment and phases of restructuring

within the headquarters-branch hierarchy in the way we have just explored. They are also ordered by the particular roles that individual plants play in the process of production itself. Some places, for example, attract the special degree of autonomy that goes with capturing a major share of corporate R&D activities (we looked briefly at this in Chapter 8). Others may be specialized in the more skilled functions that surround the manufacture of quality components, tapping the qualities available in the local skill pool. Screwdriver assembly plants will form another kind of production allocated to places— most often these days after considerable lobbying by national and local political interests. There are places where small firms thrive and those where no small-firm tradition exists. In some, the plants are clones of others within a corporate hierarchy. In others, the plants form part of a spatially separated but functionally continuous production system.

Increasingly, geographers are using this kind of conceptual framework to help them understand what is going on in real cities and regions. Significantly, geographic allocation is seen as having an active potential for management since different control strategies can be effected in different places. Space thus becomes more than just a platform for outcomes but a medium that can be used through the allocation of structures of production to enhance the rate of accumulation. Massey (1984) provides examples of three identifiable types of spatial structures, called *part-process, cloning,* and *concentrated,* reflecting the ways in which different forms of production structures and their contingent social relations can be "stretched out" over space to achieve cost, control, and reproduction advantages.

Tackling the same issue from a different direction, Gordon Clark (1981) examines how a hypothetical company, Data Inc., effects an internal spatial division of labor to enhance its profitability and to ensure its own control of potentially contagious forces for the destabilization of labor relations. This is achieved by choosing "a smaller, older, and slacker labor market" in a peripheral location for its production activities and by separating its management and R&D from manufacturing functions. The mental and conceptual activities of company management and R&D, which are crucial to its continued ability to compete, require an entirely different form of the labor process from that in the production plant. Good working conditions, fringe benefits, favorable wages, and secure contracts of employment are needed both to retain the much-sought-after work force and to draw their enthusiasm in the interests of the company. By contrast, the production workers need to be closely controlled, with little freedom to depart from established routines. Productivity requirements in a competitive marketplace mean that wages plus benefits must be kept to a minimum and a less secure form of contract is beneficial to the company to absorb the ups and downs of production.

Having both kinds of operations colocated at the same site makes these incompatible regimes difficult to manage since the obvious variation in conditions between the two groups of workers is likely to be a basis for disaffection. One strategy for achieving control of this bifurcated labor process is therefore the "stretching out" of the organization of production and the active use of space as a strategy for accumulation. Such an analysis of the reasons for locational decisions to decentralize production, while it can be reduced to effective costs, offers a richer interpretation and reinforces for us the power of the concept of the spatial division of labor as a means of understanding geography's importance in the social relations of production.

From yet another perspective, Allen Scott (1982) explores the same phenomenon of bifurcation of labor processes in the context of spatial divisions of labor in the city. Having already reviewed intraurban industrial location theory and reset it in terms of a theory of production in capitalist society (essentially a form of the restructuring approach), Scott goes on to explore the ways in which, as a firm grows, a split hierarchy of management and production appears. Set at the intraurban level of resolution, Figure 10.11 shows how a spatial division of labor can emerge at a smaller scale. The management and control functions gradually become concentrated at the central city location while the production functions become decentralized. Once again, what we see as the geography of industrial location in the city can be interpreted from the perspective of an emerging spatial division of labor focused around the separation of functions of ownership and control from the production process itself.

Summary: Spatial Divisions of Labor

We have seen that each wave of capital accumulation, each round of investment, allocates spaces and places to different functions within the social rela-

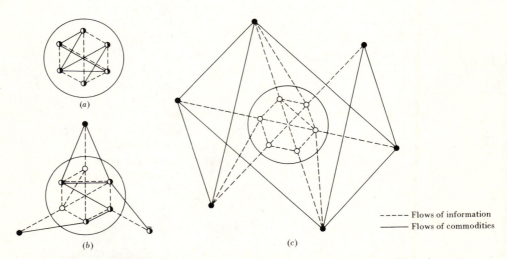

Figure 10.11 Functional bifurcation and spatial divisions of labor in the urban setting: (*a*) *T* = 1, clustering in the central city; (*b*) *T* = 2, incipient decentralization of production, (*c*) *T* = 3, effective decentralization of production and centralization of control. *Source:* A. J. Scott (1982), "Locational Patterns and Dynamics of Industrial Activity in the Modern Metropolis," *Urban Studies* 19: 11–142, fig. 2. Reprinted by permission.

tions of production. Some take controlling roles, some subordinate ones. Where places are identified with the functions of capital, they achieve control. Where they are closely associated with labor, they are subordinate.

Within this broad framework, left-radical theorists like Massey point to the variety of roles played by places over time. Some remain at the focal, controlling points, while others have an ephemeral existence, being drawn in and discarded perhaps to rise again at some propitious future time. But place is not passive. The legacy of past roles and the forms of social relations that they engendered set the possibilities for the future. Centers of control may be able to defend their positions forever, while Amazonian rain forests, militant coalfields, and obsolescent inner cities may have truncated lives.

The overall message is that capitalism moves on. Change is the norm, and places play evolving roles as times change. Through this dynamic process, then, Marxist analysis sees in the spatial structure of social relations a window on the asymetric struggle for power between capital and labor. Through this perspective, what we saw in Part One as economic geography moves into the wider sphere of *political economy.*

Retaining the notion of continuous dynamic change, we turn finally to the most recent of the debates in the literature. This addresses the process of change itself and asks whether, in the contemporary world, we are witnessing a revolutionary shift in the nature of capitalism itself and of the spatial structures of production that characterize it.

THE END OF FORDISM?: THE FLEXIBILITY DEBATE

As with the other debates we have looked at, the march of events has generated a new debate in recent years in another area of the radical literature. This one has sufficient range and depth to draw into it almost all shades of academic endeavor in the sphere of economic geography. A number of labels have emerged to describe it, but common to them all is the word *flexibility.* A second feature is the idea that capitalism itself is moving from an "organized" to a "disorganized" form (Offe, 1985; Lash and Urry, 1987). Historically, a thread of debate also views advanced capitalist societies as having moved from a Fordist to a "post-Fordist" stage (Harvey, 1987, 1988). All of them convey a sense that change is under way and that some sort of "phase shift" is taking place (Lloyd, 1989). The flexibility debate implies, for some, no more than a reworking of old structures into different but less rigid forms. Others see flexibility itself as the essence of a new regime of *flexible accumulation,* which is replacing the Fordist regime at a time of crisis for capitalism (Harvey, 1988).

Certainly the oil shocks and inflation of the mid-1970s and the world recession of the 1980s represented significant crises for the capitalist world economy and ushered in major rounds of deindustrialization and restructuring. However, the combination of these with a breakdown of the postwar political consensus and the emergence of Reaganomics and Thatcherism began to foster a view that something genuinely new was under way. Precisely what such a new era will turn out to be and to what extent it will produce significantly different social and spatial relations from those of the past are the subject of a flood of literature at the present time. We can do no more than skim the debate, picking up the main features that have emerged.

The Crisis of Fordism

Fordism is a relatively stable regime of accumulation in the evolution of capitalism under which compatibility is strong among the particular technologies in use, the form of the production and labor process, and the nature of consumption in society as a whole. In general, Fordism is identified as the dominant form of advanced capitalism in the period since the 1920s—the period of mass production of standardized consumer products. It is characterized by vertically integrated plants using shift systems and scientific management techniques (often called Taylorist methods after their best-known advocate). In spatial terms, Fordism is closely identified with the large suburban plant employing thousands of workers and, as we saw earlier, with a hierarchy of production units geographically separated as headquarters, branch assembly plants, and administrative offices. Indeed, much of the geography of spatial structures of production discussed earlier was based on Fordism. Inside the plant, Fordism represented the very essence of Braverman's (1974) thesis of deskilling, with management seeking to separate mental (management and supervisory) tasks from manual (shop floor) jobs in a bid to raise productivity and achieve greater control. A feature of the Fordist era was a balance between the enhanced ability of the plants to produce relatively cheap mass outputs and the willingness and ability of the population at large to consume large enough quantities of them to see profits realized out of substantial investments. Central to the Fordist system was the role of the state. By managing the economy, by providing a supporting infrastructure for the health, education, and training of workers, by regulating the labor market, and by sustaining levels of consumption through the operation of social welfare policies, the strong, centrally involved state was fundamental to the success of the system as a whole.

The paradox lay in the fact that the factors that underpinned the success of the system also sowed the seeds of its downfall. The economies of scale available to the large integrated plant became tempered by the fact that a strong measure of bargaining power was delivered to work forces and unions in the large plants. Strikes and growing crises of labor relations were instrumental in having firms seek new locations both nationally and internationally. Crises of underconsumption were an ever-present threat to a system so highly geared to the mass market, and the demands on the state to manage the economy on Keynesian lines drew it into an increasingly ambiguous role. In smoothing the downturns while maintaining social welfare policies and damping the upturns by taxation to offset inflation, the state expanded as a force in the economy and became open to accusations of interfering with the free flow of market forces. As a response, supply-side economics and monetarism rose as radical forces at end of the 1970s. These were reinforced with "right to work" legislation and a new drive to "let the market decide." Dressed in their full political regalia, these "new realist" forces entered the stage in the United States as Reaganomics and in the United Kingdom as what became labeled Thatcherism. Much of what we have just said verges on caricature. To understand the emerging flexibility debate, however, it is essential to have a broad overview of the economic and political context.* Among theorists unconvinced that Fordism is monolithic and a thing of the past are the French regulationists, from whose work came the notion of regimes of accumulation. For example, long before the current debate, Aglietta (1979) had already suggested that Fordism as a regime was undergoing change. He introduced the term *neo-Fordism* to describe what he saw as a remolding of the original regime, pointing in particular to the internationalization of Fordism. The vertical structures of production were becoming stretched out over global space to both foster and take advantage of a new industrial division of labor (see Frobel et al., 1980; Scott, 1987). Others (Lipietz, 1987; Leborgne and Lipietz, 1988) specifically attack the idea that Fordism ever revealed a clearly definable, uniform pattern in its organizational forms. They argue that much of what the flexibility theorists are seeing was already present in preexisting Fordist structures.

The Emergence of Flexible Specialization

By contrast with those who see no evidence of a clean break in the evolution of capitalism, the proponents of the *flexible specialization* thesis claim that we are entering a genuinely new ball game. Sabel (1982), and Piore and Sabel (1984) believe that we are experiencing something genuinely new, as they signal in the title of their book *The Second Industrial Divide*. More important, as the subtitle of this same book—*Possibilities for Prosperity*—suggests, theirs is an optimistic vision to counter the gloom of the restructuring and crisis debates. Such optimism arises, in part, from a view that with a range of new opportunities opening up for small firms and for different kinds of industrial communities, it is possible to move to nonconflictual forms of bargaining among those engaged in the production process. This offers a direct challenge to the Marxist thesis of class conflict—a challenge that has, not unexpectedly, been taken up with vigor in the literature.

Flexible specialization is, first and foremost, a system of organization perceived to be emerging through a process of fragmentation. In the case of production, this is associated with the breakdown of the classic vertically integrated Fordist firm. By a process of deintegration, firms are, as we saw in

---

* The notion of caricature is a part of the debate itself, and some theorists (see Amin, 1989) argue that the newfound flexibility that many seem to be witnessing is of flexibility only in relation to a generalized and overly rigid description of Fordism such as the one we have just set out. In short, one view of the flexibility debate is that it is little more than a reaction to a caricature and that no such clear-cut unidimensional system as the one we have just described ever existed.

Chapter 8, increasingly turning to subcontracting and outside purchases for goods and services that they have hitherto sought in-house. Through decentralization, they have been disaggregating themselves into clusters of quasi-independent profit centers with the headquarters role limited to the provision of investment capital and strategic management. By devolution, many firms have sought to move themselves away from direct production or service provision by the use of franchising and leasing arrangements (see Shutt and Whittington, 1986; Rainnie, 1984). There has also been a strong recent tendency toward divestment, with companies seeking to focus their skills and activities on core businesses, generating a wave of management buyouts and the sale of subsidiaries.* The effect of many of these changes is perceived to be an extension of the opportunities available to small and medium-sized enterprises, giving potentially strong competitive advantages to communities and cultural settings able to nurture and support them.

Central to the debate about flexible specialization is the role played by the new technologies, particularly those resulting from convergence between telecommunications and computing. The possibilities offered by these information technologies for coordination and control and their ability to achieve virtually costless transmission over space is introducing a revolutionary era perhaps on the scale of the first introduction of the railways. We can offer only a few examples here. As we saw in Chapter 8, just-in-time systems for inventory and component input management have potentially profound effects on the role of subcontractors in the production system. They demand new, more exacting quality standards and give companies like the major automobile assemblers wider, more flexible choices of supply configurations. At the same time, they encourage greater spatial proximity between plant and subcontractor. New capabilities for design instructions and product modifications to be transmitted faithfully and almost costlessly over global space give entirely new possibilities for the spreading of disintegrated, decentralized, and devolved forms of production all over the world. These allow for the particular comparative advantages of localized communities and labor pools to be tapped wherever in the world they may be. New systems for the manufacturing process itself with computer-aided design and computer-assisted manufacturing (CAD/CAM), flexible manufacturing systems (FMS), and automated manufacturing technology (AMT) also have the power to revolutionize the allocation of manufacturing operations between sizes and types of firms. For example, smaller business units can use these techniques to string together runs of small, high-quality batch production to begin to capture some of the economies of scale available to the large unit. At the other end of the scale, the large units, by breaking their production down into manufacturing cells and by using information technology to connect internal "islands of automation" within the

---

* Not all of these tendencies are described as part of a generalized process of deconcentration. Shutt and Whittington (1986), for example, see them as essentially furthering the interests of the major corporate enterprises by a movement from direct to indirect forms of control.

plant, can begin to emulate the flexibility of smaller operations. The already vast and rapidly growing literature on the impact of technical change on the process of manufacturing makes it clear that the possibilities for the achievement of greater flexibility across the board are very considerable indeed. What is in dispute in the debate as it currently stands is the extent to which the possibilities are achievable in reality.

The third major dimension of the flexibility debate concerns labor. Clearly, the technical possibilities provide many alternatives for the form of the labor process, and great popular attention is being given to questions of "flexible manning." With the drive to reduce the power of the unions and to break out of the demarcation disputes that have tended to blight many of the older industries, a good deal of attention has been paid to questions of flexibility in the use of labor (Atkinson, 1984, 1985; Morgan and Sayer, 1985). Firms are seen to be trying to achieve three types of flexibility:

1. *Functional flexibility*—applying workers to a greater variety of tasks; multiskilling
2. *Numeric flexibility*—adjusting the quantities of labor employed to fit more closely the pattern of shifts in product or service demand
3. *Financial flexibility*—adjusting wage rates for the job to the demands of the competitive situation.

Much of this emerges in real terms as the drive to bring inputs of labor more closely into harmony with the needs of production. In reality, this emerges as employing more and more workers on part-time, short-term, and temporary contracts. Less of the newfound flexibility seems to involve genuine multiskilling and new working practices among full-time workers (MacInnes, 1987). It also seems clear that up to the present, much of the impact of flexible working strategies has been within the more labor-intensive service sectors, particularly among women (Christopherson, 1989; Hakim, 1987). Indeed, many people who challenge the optimism of the flexible specialization debate point to the emergence of increasing dualism in the labor market, with much of the flexibility being achieved at the expense of women, minorities, and the least skilled, who become the victims of an increasing polarization (Bluestone and Harrison, 1986; Harvey, 1987). In the more idealized perspectives on flexible specialisation, however, part of the argument is that enhanced opportunities for small firms, family-based forms in the organization of labor, and the collaborative possibilities that may exist in certain community settings may achieve many of these flexibilities in an essentially harmonious rather than exploitive way (Sabel and Piore, 1984).

The new drive to achieve some form of flexibility on the production side of the economy is also bound up with a number of changes perceived to be taking place on the demand side. Markets appear to be fragmenting, and the mass markets of Fordism seem to be breaking down. Attacking the niche market has, it seems, become proportionately more important, and firms are now seeking economies of scope in addition to economies of scale. At the same time, there is considerable market volatility, and the need for flexible responses to sudden demand shifts is at a premium. Here, again, it assumed that the smaller firm—or at least the more flexible small-firm style—is essential to competitive success. The movement of significant market segments toward higher value-added products is also assumed to be having some powerful effects. Insofar as it influences the clothing, fashion, and style industries, this trend is a central plank in the perceived effect of flexible specialization in reviving craft and artisan industries by reintegrating them into the world marketplace. It should be quite clear from this brief overview that the thrust of the flexible specialization argument is that broad trends in the evolution of the capitalist system are tending to favor both the smaller enterprise form of organization and, by extension, the particular social settings or milieux that provide appropriate seedbed conditions and supportive political, social, and economic infrastructures. Cynics argue that there is more than a hint of utopianism in this and that what is really afoot is a complex reworking of economic organization in capitalism with little effective change in the relations of power and the nature of the social relations of production (see Amin, 1989). Between the two lie the unconvinced, who see change but no fundamental transformation (Gertler, 1988).

Space and the Issue of Flexible Specialization

In these debates, it is clear that geography and geographers have a key role to play. For example, particular geographic examples provided a base of evidence for the flexible specialization thesis. Emergent industrial communities in central Italian provinces like Emilia Romagna formed a crucial seedbed for Sabel and Piore's (1984) speculations on the new small-firm, small-community form of production organization. The work of Italian writers like Bagnasco (1981), Brusco (1982, 1986), and Garofoli (1984) examining regional case studies of the knitwear, textile, shoe, ceramic, and machine tool industries in small towns and industrial districts sounded echoes of the earlier work on agglomeration economies that we reviewed in Chapter 6. There was a strong convergence with work being undertaken on the new high-tech industries of Silicon Valley and the emergent Sunbelt by such writers as Scott and Angel (1987). The counterurbanization debate in the United States and Great Britain served also to reveal the apparent ability of the smaller rural or exurban community to show greater economic robustness in the face of the disindustrialization process (Dicken and Lloyd, 1981; Fothergill and Gudgin, 1982; Bluestone and Harrison, 1986). Paradoxically, there also emerged evidence of the renewed buoyancy of the great metropolitan regions, but a buoyancy based chiefly on small-scale networks of labor-intensive producers within a shell provided by the rapid expansion of professional business services (Scott, 1988; Scott and Cooke, 1988). What seemed to be emerging were new industrial regions or, as Scott (1988, 1989) calls them, *new industrial spaces*. Are we therefore witnessing a period of revolutionary change in the space economy? Or are we seeing a return to the principles of agglomeration economies that Alfred Weber brought to our attention from a period that predated the Fordist experience?

Whatever the resolution of a debate that is only just warming up, one thing is already abundantly clear. The properties of space and place—socially constructed and not simply reducible to factor endowments—are, if anything, becoming more important as capitalism restructures and reforms itself. The most forceful proponents of flexible specialization point to the growing importance of the small firm, the urbanizing countryside, and the revived inner-city craft quarter as new spatial forms. Those more concerned with the role of high technology see the emergence of R&D clusters, Silicon Valleys, and new forms of "technopolis" based on advances in information technology. All these new "industrial spaces," down to the level of individual localities, are available to be selected by capital, which literally has the whole world to choose from.

By contrast, some theorists focus on the neo-Fordist drive toward global integration and the massive horizontal stretching out of the hierarchical structures associated with monopoly capitalism. New spatial divisions of labor will continue to reflect the upside and downside effects of restructuring that we discussed at the beginning of this chapter. For spaces and places not selected for the positive effects either of the new specializations or of the upside benefits of restructuring, optimistic scenarios of less conflictual forms of bargaining are less significant than the specter of increasingly uneven development.

From this perspective on what we started out by calling economic geography, we are seeing the spatial reflection of unequal and asymetric power structures in capitalist society as a whole. What we see on the ground is conditioned by the way society itself works. It works, of course, to produce unevenness (spatial differentiation, we would have called it earlier). This is not some distortion from an idealized norm but rather a true measure of the way the system actually works. From time to time, some places, spaces, and people benefit and some suffer, but the process moves on. Uneven development emerges both as a result of past processes and as the terrain over which future strategies will be selected. In a very fundamental sense, geography matters.

FURTHER READING

Bluestone, B., and B. Harrison (1982). *The Deindustrialization of America: Plant Closings, Community Abandonment, and the Dismantling of Basic Industry.* New York: Basic Books.

Green, F., and B. Sutcliffe (1987). *The Profit System: The Economics of Capitalism.* London: Penguin.

Gregory, D., and J. Urry (eds.) (1985). *Social Relations and Spatial Structures.* London: Macmillan.

Lash, S., and J. Urry (1987). *The End of Organized Capitalism.* Cambridge: Polity Press.

Marshall, M. (1987). *Long Waves of Regional Development.* London: Macmillan.

Martin, R., and R. Rowthorne (eds.) (1986). *The Geography of Deindustrialization.* London: Macmillan.

Massey, D. (1984). *Spatial Divisions of Labour: Social Structures and the Geography of Production*. London: Macmillan.

Massey, D., and J. Allen (eds.) (1988). *Restructuring Britain: The Economy in Question*. London: Sage.

Massey, D., and R. Meegan (1982). *The Anatomy of Job Loss: The How, Why and Where of Employment Decline*. London: Methuen.

Massey, D., and R. Meegan (1985). *Politics and Method: Contrasting Studies in Economic Geography*. London: Methuen.

Piore, M., and C. Sabel (1984). *The Second Industrial Divide*. New York: Basic Books.

Scott, A. J., and M. Storper (eds.) (1986). *Production, Work, Territory: A Geographical Anatomy of Industrial Capitalism*. Winchester, Mass.: Allen & Unwin.

Postscript

This third edition of *Location in Space* has taken us a long way in understanding the nature of the space economy. It is a measure of the loss of innocence that knowledge brings that from the apparent certainties of location on a flat plain of equal fertility we were to come in the end to a statement that geography matters but that there are very wide divergences of view on why it matters and how its influence is transmitted.

The first edition of *Location in Space* was published in 1972 at a time when geography as a discipline was proud of its newly won status as a the "science of space." The quantitative revolution was in full swing, and neoclassical economic theory provided the ideal vehicle for a modeling of spatial form. The contingent circumstances of the time were growth and expansion. The mood was a holdover from the 1960s—buoyant, vibrant, self-confident. The practical topics of interest to geographers were how to estimate the income and employment impact of new steel complexes, plan for population expansion, and capture a share of international direct investment.

By the time the second edition of was published in 1977, the mood had changed. The self-confidence had given away to serious doubts about the nature of the economic geographer's mission. Recession had arrived, the cores of the great cities were in disarray, and the manufacturing belt had become the "rust" belt, giving way to counterurbanization and the growth of the Sunbelt. Misgivings were growing about the validity of tried and tested theories. Plant closures, deprivation, and rising unemployment seemed to find no place in rigorous orthodox models (although input-output multiplier analysis could easily be run backward to estimate decline). At the same time, the sophistication of some modeling approaches had become so great as to make them largely inaccessible to all but a few experts. The rise of the giant corporations was also proving a particular problem for economic geographers. Interest was increasing in this feature of the economic system, but theoretical "handles" were lacking. The second edition of *Location in Space* did include a widely read chapter on the large organization, reflecting the shifting locus of interest.

Between the second and this third edition of the book, the changes that were slowly appearing in the 1970s accelerated. These changes arose both from the contingent circumstances of the world at large (with the appearance of Reaganomics and Thatcherism) and from the academic realm. As a response, many economic geographers abandoned, altogether the neoclassical insights of Part One of the book—throwing, as some would argue, the baby out with the bath water. Some took up a position that focused on the corporate enterprise as the key agent in promoting significant changes in the global economy. In some cases, this was pursued from an essentially orthodox view; in others, the focus lay on the Marxist concept of monopoly capitalism. A growing number moved toward the political economy approach, abandoning entirely the concept of a narrowly economic geography for theories that sought critical insights at the level of the capitalist system as a whole.

We live in interesting times! We have set out the array of current theoretical approaches available to the economic geographer, but these are far outweighed by the extraordinary complexity of the world that confronts us. To achieve an understanding, we must cut its variety down to manageable but faithful proportions. This is what each of the theoretical perspectives we have reviewed has set out to do. We believe that each still has its merits, and it is necessary to resist the desperate view that the world outside cannot be understood. The most difficult problem to handle has been the pace of change. The neoclassical models of Part One largely assumed it away, but there is no escaping it. All of the approaches in Part Two had to have a means of addressing it, and it is necessary to take the view that change is the norm and our approaches must have a view of dynamics. This is, of course, not a problem for the Marxian theories in Part Two, since change is central to this way of looking at social evolution. Our exploration of the flexibility debate seems to suggest that the pace of change is accelerating still more and that, perhaps, well-known and understood structures are even now breaking down in front of us. New technologies are driving a revolution in economic and social (and therefore spatial) forms and is producing new futures as yet only dimly perceived. Change, then, is the force to be reckoned with, and the challenge is for us to improve our understanding of the dynamics of the systems we are dealing with.

But the old problems remain—poverty, deprivation, the rape of the environment. With all our newfound skills and conceptualizations, we still confront a terrain of such startlingly unequal opportunities that we must strive to understand the underpinnings of spatial unevenness. From the political economy perspectives of Chapter 10 emerges the view that it is naive to expect any amelioration of the condition since it is, in essence, inequality that provides the muscle to drive the invisible hand of the market. Yet blank acceptance of the logic of capital seems so negative, so defeating. However much the utopian strain of the flexible specialization debate seems no more than straws in the wind, there is a need continually to explore the new possibilities opening up in the capitalist system to seek, as Lösch put it, "to improve our sorry reality."

The geographer's task remains extremely important. What we have done in this third edition of *Location in Space* is to look over our shoulder at evolving approaches that over the span of two decades have served to guide our understanding of space and place in economy and society. Each approach, we submit, has something to offer. Each throws light on the complexity of our world. Whether or not it is possible to combine parts of each distinctive theoretical perspective into an eclectic theorization is moot. All in all, the record is of enormous endeavor, debate, and progress. What has been revealed is a learning process, a journey of exploration that continues and will continue.

Bibliography

Abler, R., J. S. Adams, and P. R. Gould (1971). *Spatial Organization: The Geographer's View of the World.* Englewood Cliffs, N.J.: Prentice-Hall.

Aglietta M. (1979) *A Theory of Capitalist Regulation.* London: New Left Books.

Alexander, J. W., S. E. Brown, and R. E. Dahlberg (1958). "Freight Rates: Selected Aspects of Uniform and Nodal Regions." *Economic Geography* 34: 1–18.

Alonso, W. (1960). "A Theory of the Urban Land Market," *Papers and Proceedings of the Regional Science Association* 6: 149–157.

Alonso, W. (1964). *Location and Land Use: Toward a General Theory of Land Rent.* Cambridge, Mass.: Harvard University Press.

Alonso, W. (1967). "A Reformulation of Classical Location Theory and Its Relation to Rent Theory," *Papers and Proceedings of the Regional Science Association* 19: 23–44.

Amin, A. (1989). "Flexible Specialization and Small firms in Italy: Myths and Realities," *Antipode* 21, 1: 13–34.

Ansoff, H. I. (1987). *Corporate Strategy* (rev. ed.). Harmondsworth, England: Penguin.

Appleton, J. H. (1963). "The Efficacy of the Great Australian Divide as a Barrier to Railway Communications," *Transactions of the Institute of British Geographers* 33: 101–122.

403

Atkinson, J. (1984). "Manpower Strategies for Flexible Organizations," *Personnel Management*, August: 28–31.

Atkinson, J. (1985). "The Changing Corporation," in D. Clutterbuck (ed.), *New Patterns of Work*. Aldershot: Gower.

Averitt, R. T. (1968). *The Dual Economy: The Dynamics of American Industry Structure*. New York: Norton.

Bagnasco, A. (1981). "Labour Market, Class Structure and Regional Formations in Italy," *International Journal of Urban and Regional Research* 5: 40–44.

Bailly, A., D. Maillat, and M. Rey (1987). *Nouvelles articulations des systèmes de production et rôle des services: Une analyse comparative internationale et interrégionale*. Lausanne: CEAT.

Bain, J. S. (1954). "Economies of Scale, Concentration and the Condition of Entry in 20 Manufacturing Industries," *American Economic Review* 44: 15–39.

Bain, J. S. (1968). *Industrial Organization* (2d ed.). New York: Wiley.

Bannock, G. (1971). *The Juggernauts: The Age of the Big Corporation*. Indianapolis: Bobbs-Merrill.

Barber, R. J. (1970). *The American Corporation*. London: MacGibbon Kee.

Barloon, M. J. (1965). "Interrelationship of the Changing Structure of American Transportation and Changes in Industrial Location," *Land Economics* 41: 169–182.

Barnum, H. G. (1966). *Market Centers and Hinterlands in Baden-Württemberg*. University of Chicago, Department of Geography Research Paper No. 103.

Beavon, K. S. O., and A. S. Mabin (1975). "The Lösch System of Market Areas: Derivation and Extension," *Geographical Analysis* 7: 131–151.

Beckmann, M. (1955a). "The Economics of Location," *Kyklos* 8: 416–421.

Beckmann, M. (1955b). "Some Reflections on Lösch's Theory of Location," *Papers and Proceedings of the Regional Science Association* 1: N2–N8.

Beckmann, M. (1958). "City Hierarchies and the Distribution of City Size," *Economic Development and Cultural Change* 6: 243–248.

Benson, I., and J. Lloyd (1983). *New Technology and Industrial Change*. London: Kogan Page.

Berle, A. A., Jr., and G. C. Means (1932). *The Modern Corporation and Private Property*. New York: Macmillan.

Berry, B. J. L. (1961). "City Size Distributions and Economic Development," *Economic Development and Cultural Change* 9: 573–588.

Berry, B. J. L. (1963). *Commercial Structure and Commercial Blight*. University of Chicago, Department of Geography Research Paper No. 85.

Berry, B. J. L. (1967). *Geography of Market Centers and Retail Distribution*. Englewood Cliffs, N.J.: Prentice-Hall.

Berry, B. J. L., and H. G. Barnum (1962). "Aggregate Relations and Elemental Components of Central Place Systems," *Journal of Regional Science* 4: 35–68.

Berry, B. J. L., H. G. Barnum, and R. J. Tennant (1962). "Retail Location and Consumer Behavior," *Papers and Proceedings of the Regional Science Association* 9: 65–106.

Berry, B. J. L., and W. L. Garrison (1958a). "Functional Bases of the Central Place Hierarchy," *Economic Geography* 34: 145–154.

Berry, B. J. L., and W. L. Garrison (1958b). "A Note on Central Place Theory and the Range of a Good," *Economic Geography* 34: 304–311.

Berry, B. J. L., and W. L. Garrison (1958c). "Recent Developments of Central Place Theory," *Papers and Proceedings of the Regional Science Association* 4: 107–120.

Blaikie, P. M. (1971a). "Spatial Organization of Agriculture in Some North Indian Villages, Part 1," *Transactions of the Institute of British Geographers* 52: 1–40.

Blaikie, P. M. (1971b). "Spatial Organization of Agriculture in Some North Indian Villages, Part 2," *Transactions of the Institute of British Geographers* 53: 15–30.

Bluestone, B., and B. Harrison (1982). *The Deindustrialization of America: Plant Closings, Community Abandonment and the Dismantling of Basic Industry.* New York: Basic Books.

Bluestone, B., and B. Harrison (1986). *The Great American Jobs Machine: The Proliferation of Low Wage Employment in the U.S. Economy.* Washington, D.C.: Joint Economic Committee of the U.S. Congress.

Borchert, J. R. (1978). "Major Control Points in American Economic Geography," *Annals of the Association of American Geographers* 68: 214–232.

Borts, G. H., and J. L. Stein (1964). *Economic Growth in a Free Market.* New York: Columbia University Press.

Bottomore, T. (1983). *A Dictionary of Marxist Thought.* Oxford: Blackwell.

Braverman, H. (1974). *Labor and Monopoly Capital, or, The Degradation of Work in the Twentieth Century.* New York: Monthly Review Press.

Britton, J. N. H. (1969). "A Geographical Approach to the Examination of Industrial Linkages," *Canadian Geographer* 13: 185–198.

Britton, J. N. H. (1974). "Environmental Adaptation of Industrial Plants: Service Linkages, Locational Environment and Organization." In F. E. I. Hamilton (ed.), *Spatial Perspectives on Industrial Organization and Decision Making.* London: Wiley, pp. 363–390.

Brusco, S. (1982). "The Emilian Model, Productive Decentralization and Social Integration," *Cambridge Journal of Economics* 6, no. 2: 167–184.

Brusco, S. (1986). "Small Firms and Industrial Districts: The Experience of Italy," in D. Keeble and E. Wever, *New Firms and Regional Development in Europe.* London: Croom Helm, chap. 9, pp. 84–202.

Brush, J. E. (1953). "The Hierarchy of Central Places in Southwestern Wisconsin," *Geographical Review* 43: 380–402.

Bunge, W. (1966). "Theoretical Geography," *Lund Studies in Geography, Series C*, 1, 2d ed.

Carroll, G. R. (1982). "National City-Size Distributions: What Do We Know After 67 Years of Research? *Progress in Human Geography* 6: 1–43.

Carrothers, G. A. P. (1956). "A Historical Review of the Gravity and Potential Concepts of Human Interaction," *Journal of the American Institute of Planners* 22: 94–102.

Castells, M. (1977). *The Urban Question.* London: Edward Arnold.

Chandler, A. D., Jr. (1962). *Strategy and Structure: Chapters in the History of the Industrial Enterprise*. Cambridge, Mass.: MIT Press.

Chandler, A. D., Jr., and F. Redlich (1961). "Recent Developments in American Business Administration and Their Conceptualization," *Business History Review* 35: 1–27.

Chisholm, M. (1962). *Rural Settlement and Land Use*. London: Hutchinson.

Chisholm, M. (1966). *Geography and Economics*. New York: Praeger.

Christaller, W. (1966). *Central Places in Southern Germany*. C. W. Baskin, trans. Englewood Cliffs, N.J.: Prentice-Hall.

Christopherson (1989). "Flexibility in the US Service Economy and the Emerging Spatial Division of Labour," *Transactions of the Institute of British Geographers*, 14, no. 2, 131–143.

Clairmonte, F., and J. Cavanagh (1984). "TNCs and Services: The Final Frontier," *Trade and Development* 5: 215–273.

Clark, G. L. (1981). "The Employment Relation and the Spatial Division of Labor, A Hypothesis," *Annals Association of American Geographers*, 71, no. 3: 412–424.

Clark, G. L., M. S. Gertler, and J. E. M. Whiteman (1986). *Regional Dynamics: Studies in Adjustment Theory*. Boston: Allen & Unwin.

Coase, R. (1937). "The Nature of the Firm," *Economica* 4: 386–405.

Conference Board (1987). *Locating Corporate R&D Facilities*. New York: Conference Board.

Cosgrove, D. (1989). "Geography Is Everywhere: Culture and Symbolism in Human Landscapes," in D. Gregory and R. Walford (eds.), *Horizons in Human Geography*. London: Macmillan.

Craig, P. G. (1957). "Location Factors in the Development of Steel Centers," *Papers and Proceedings of the Regional Science Association* 3: 249–265.

Dacey, M. F. (1962). "Analysis of Central Place Patterns by Nearest Neighbor Analysis," *Lund Studies in Geography, Series B*, 24.

Daggett, S. (1955). *Principles of Inland Transportation*. New York: Harper & Row.

Daggett, S. (1968). "The System of Alfred Weber." In R. H. T. Smith, E. J. Taaffe, and L. J. King (eds.), *Readings in Economic Geography*. Skokie, Ill.: Rand McNally, pp. 58–64.

Danielsson, A. (1964). "The Locational Decision from the Point of View of the Individual Company," *Ekonomisk Tidskrift* 66: 47–87.

Darwent, D. F. (1969). "Growth Poles and Growth Centers in Regional Planning: A Review," *Environment and Planning* 1: 5–31.

Davies, W. K. D. (1967). "Centrality and the Central Place Hierarchy," *Urban Studies* 4: 61–79.

Davis, L. (1966). "The Capital Markets and Industrial Concentration," *Economic History Review* 19: 255–272.

Dicken, P. (1971). "Some Aspects of the Decision-Making Behavior of Business Organizations," *Economic Geography* 47: 426–437.

Dicken, P. (1976). "The Multiplant Enterprise and Geographical Space: Some Issues in the Study of External Control and Regional Development," *Regional Studies* 10: 401–412.

Dicken, P. (1986). *Global Shift: Industrial Change in a Turbulent World.* New York: Harper & Row.

Dicken, P., and P. E. Lloyd (1981). *Modern Western Society: A Geographical Perspective on Work and Well-Being.* London: Harper & Row.

Dunford, M. (1977). "The Restructuring of Industrial Space," *International Journal of Urban and Regional Research* 1: 510–520.

Dunford, M., and D. Perrons (1983). *The Arena of Capital.* London: Macmillan.

Dunn, E. S. (1954). *The Location of Agricultural Production.* Gainesville: University of Florida Press.

Durand, L. (1964). "The Major Milksheds of the Northeastern Quarter of the United States," *Economic Geography* 40: 9–33.

Edwards, R. C., M. Reich, and D. M. Gordon (1975). *Labor Market Segmentation.* Lexington, Mass.: D. C. Heath.

Eliot Hurst, M. E. (1972). *A Geography of Economic Behavior.* North Scituate, Mass.: Duxbury Press.

Emery, F. E., and E. L. Trist (1965). "The Causal Texture of Organizational Environments," *Human Relations* 18: 21–32.

Estall, R. C. (1972). "Some Observations on the Internal Mobility of Investment Capital," *Area* 4: 193–198.

Estall, R. C. (1985). "Stock Control in Manufacturing: The Just-in-Time System and Its Locational Implications," *Area* 17: 129–133.

Fine, B. (1984). *Marx's Capital,* (2d ed.). London: Macmillan.

Fothergill, S., and G. Gudgin (1982). *Urban and Regional Employment Change in the United Kingdom.* London: Heinemann.

Found, W. C. (1971). *A Theoretical Approach to Rural Land Use Patterns.* New York: St. Martin's Press.

Freeman, C. (1974). *The Economics of Industrial Innovation.* Harmondsworth, England: Penguin.

Freeman, C. (1984). "The Challenge of New Technologies." In *Interdependence and Co-operation in Tomorrow's World.* Paris: OECD.

Freeman, C. (1987). "The Challenge of New Technologies." In Organization for Economic Co-operation and Development, *Interdependence and Co-operation in Tomorrow's World.* Paris: OECD, chap. 5.

Freeman, C., et al. (1982). *Unemployment and Technical Innovation: A Study of Long Waves and Economic Development.* London: Pinter.

Friedmann, A. L. (1977). *Industry and Labour: Class Struggle at Work and in Monopoly Capitalism.* London: Macmillan.

Friedmann, J. (1966). *Regional Development Policy: A Case Study of Venezuela.* Cambridge, Mass.: MIT Press.

Friedmann, J. (1973). *Urbanization, Planning, and National Development.* Newbury Park, Calif.: Sage.

Friedmann, J. (1986). "The World City Hypothesis," *Development and Change* 17: 69–83.

Frobel, F. J. Heinrichs, and O. Kreye (1980). *The New International Division of Labor.* Cambridge: Cambridge University Press.

Fulton, M., and L. C. Hoch (1959). "Transportation Factors Affecting Location Decisions," *Economic Geography* 35: 51–59.

Galbraith, J. K. (1966). *The New Industrial State*. Boston: Houghton Mifflin.

Galbraith, J. K. (1974). *Economics and the Public Purpose*. Boston: Houghton Mifflin.

Garofoli, G. (1984). "Diffuse Industrialization and Small Firms: The Italian Pattern in the 1970's." In R. Hudson (ed.), *Small Firms and Regional Development*. Copenhagen: School of Economics and Business Administration.

Gertler, M. S. (1984). "Regional Capital Theory," *Progress in Human Geography* 8: 50–81.

Gertler, M. S. (1988). "The Limits to Flexibility: Comments on the Post-Fordist Vision of Production and Its Geography," *Transactions of the Institute of British Geographers,* 13: 419–432.

Gilmour, J. M. (1974). "External Economies of Scale, Inter-industrial Linkages and Decision Making in Manufacturing." In F. E. I. Hamilton (ed.), *Spatial Perspectives on Industrial Organization and Decision Making*. London: Wiley, pp. 363–393.

Gitlow, A. L. (1954). "Wages and the Allocation of Employment," *Southern Economic Journal* 21: 62–83.

Golledge, R. G., G. Rushton, and W. A. V. Clark (1966). "Some Spatial Characteristics of Iowa's Dispersed Farm Population and Their Implications for the Grouping of Central Place Functions," *Economic Geography* 42: 261–272.

Goodwin, W. (1965). "The Management Center in the United States," *Geographical Review* 55: 1–16.

Gottmann, J. (1961). *Megalopolis*. Cambridge, Mass: MIT Press.

Gould, P. R. (1963). "Man Against His Environment: A Game Theoretic Framework," *Annals of the Association of American Geographers* 53: 290–297.

Green, F., and B. Sutcliffe (1987). *The Profit System: The Economics of Capitalism*. London: Penguin.

Greenhut, M. L. (1956). *Plant Location in Theory and Practice*. Chapel Hill: University of North Carolina Press.

Gregory, D. (1982). "Alfred Weber's Location Theory," *Progress in Human Geography* 6: 115–128.

Gregory, D., and J. Urry (eds.) (1985). *Social Relations and Spatial Structures*. London: Macmillan.

Hagerstrand, T. (1960). "Aspects of the Spatial Structure of Social Communication and the Diffusion of Information," *Papers and Proceedings of the Regional Science Association* 16: 27–42.

Hagerstrand, T. (1967). "On Monte Carlo Simulation of Diffusion." In W. L. Garrison and D. F. Marble (eds.), "Quantitative Geography: Part 1. Economic and Cultural Topics," *Northwestern University Studies in Geography* 13: 1–32.

Haggett, P. (1965). *Locational Analysis in Human Geography*. New York: St. Martin's Press.

Haggett, P. (1979). *Geography: A Modern Synthesis* (3d ed.). New York: Harper & Row.

Haig, R. M. (1926). "Towards an Understanding of the Metropolis," *Quarterly Journal of Economics* 40: 421–433.

Hakim, C. (1987). "Trends in the Flexible Workforce," *Employment Gazette* 95, no. 11: 549–561.

Hall, M. (ed.) (1959). *Made in New York: Case Studies in Metropolitan Manufacturing.* Cambridge, Mass: Harvard University Press.

Hall, P. (1962). *The Industries of London.* London: Hutchinson.

Hall, P. (1985). "The Geography of the Fifth Kondratiev." In P. Hall and A. Markusen (eds.), *Silicon Landscapes.* Boston: Allen & Unwin, chap. 1.

Hamilton, F. E. I. (1967). "Models of Industrial Location." In R. J. Chorley and P. Haggett (eds.), *Models in Geography.* New York: Barnes & Noble, pp. 361–424.

Hansen, N. M. (1967). "Towards a New Approach in Regional Economic Policy," *Land Economics* 43: 377–383.

Harper, R. A. (1987). "A Functional Classification of Management Centers in the United States," *Urban Geography* 8: 540–549.

Harris, C. D. (1954). "The Market as a Factor in the Localization of Industry in the U.S.," *Annals of the Association of American Geographers* 44: 315–348.

Harvey, D. W. (1963). "Locational Change in the Kentish Hop Industry and the Analysis of Land Use Patterns," *Transactions of the Institute of British Geographers* 33: 123–144.

Harvey, D. W. (1969). *Explanation in Geography.* New York: St. Martin's Press.

Harvey, D. W. (1973). *Social Justice and the City.* London: Edward Arnold.

Harvey, D. W. (1982). *The Limits to Capital.* Oxford: Blackwell.

Harvey, D. W. (1983). "Geography." In T. Bottomore (ed.), *A Dictionary of Marxist Thought.* Oxford: Blackwell, pp. 189–191.

Harvey, D. W. (1988). "The Geographical and Geopolitical Consequences of the Transition from Fordism to Flexible Accumulation." In G. Sternlieb and J. W. Hughes (eds.), *America's New Market Geography.* New Brunswick N.J.: Center for Urban Policy Research.

Hay, A. M. (1973). *Transport for the Space Economy: A Geographical Study.* Seattle: University of Washington Press.

Hay, A. M. (1979). "Transport Geography," *Progress in Human Geography* 3: 267–272.

Haynes, K. E., and S. Fotheringham (1984). *Gravity and Spatial Interaction Models.* Newbury Park, Calif.: Sage.

Healey, M. J. (1984). "Spatial Growth and Spatial Rationalization in Multiplant Enterprises," *Geo Journal* 9: 133–144.

Heilbroner, R. L. (1972). *The Economic Problem* (3d ed.). Englewood Cliffs, N.J.: Prentice-Hall.

Helvig, M. (1964). *Chicago's External Truck Movements.* University of Chicago, Department of Geography Research Paper No. 90.

Hermansen, T. (1972). "Development Poles and Related Theories." In N. M. Hansen (ed.), *Regional Economic Development.* New York: Free Press, pp. 160–203.

Hewings, G. J. D. (1985). *Regional Input-Output Analysis.* Newbury Park, Calif.: Sage.

Higgins, B., and D. J. Savoie (eds.) (1988). *Regional Economic Development: Essays in Honor of François Perroux.* Boston: Unwin Hyman.

Hirschman, A. O. (1958). *The Strategy of Economic Development.* New Haven, Conn.: Yale University Press.

Hoare, A. G. (1985). "Industrial Linkage Studies." In M. Pacione (ed.), *Progress in Industrial Geography.* London: Croom Helm.

Holmes, J. (1986). "The Organization and Locational Structure of Production Subcontracting." In A. J. Scott and M. Storper (eds.), *Production, Work and Territory: The Geographical Anatomy of Industrial Capitalism.* London: Allen & Unwin.

Hoover, E. M. (1937). *Location Theory and the Shoe and Leather Industries.* Cambridge, Mass.: Harvard University Press.

Hoover, E. M. (1948). *The Location of Economic Activity.* New York: McGraw-Hill.

Horvath, R. J. (1969). "Von Thünen's Isolated State and the Area Around Addis Ababa, Ethiopa," *Annals of the Association of American Geographers* 59: 308–323.

Huff, D. L. (1960). "A Topographical Model of Consumer Space Preferences," *Papers and Proceedings of the Regional Science Association* 6: 159–173.

Hunt, E. K., and H. J. Sherman (1986). *Economics: An Introduction to Traditional and Radical Views* (5th ed.). New York: Harper & Row.

Hurd, R. M. (1924). *Principles of City Land Values.* New York: Record and Guide.

Hymer, S. (1972). "The Multinational Corporation and the Law of Uneven Development." In J. N. Bhagwati (ed.), *Economics and World Order.* London: Macmillan, pp. 113–140.

Isard, W. (1956). *Location and Space Economy.* Cambridge, Mass.: MIT Press.

Isard, W. (1960). *Methods of Regional Analysis.* Cambridge, Mass.: MIT Press.

Isard, W., and R. E. Kuenne (1953). "The Impact of Steel on the Greater New York–Philadelphia Industrial Region," *Review of Economics and Statistics* 35: 289–301.

Isard, W., and T. A. Reiner (1962). "Aspects of Decision Making Theory and Regional Science," *Papers and Proceedings of the Regional Science Association* 9: 25–34.

Janelle, D. G. (1969). "Spatial Reorganization: A Model and a Concept," *Annals of the Association of American Geographers* 59: 348–364.

Johnson, G., and K. Scholes (1988). *Exploring Corporate Strategy* (2d ed.). Hemel Hempstead, England: Prentice-Hall.

Johnston, R. J. (1983a). *Geography and Geographers: Anglo-American Human Geography Since 1945* (2d ed.). London: Edward Arnold.

Johnston, R. J. (1983b). *Philosophy and Human Geography: An Introduction to Contemporary Approaches.* London: Edward Arnold.

Kaplan, A. D. H. (1948). *Small Business: Its Place and Problems.* New York: McGraw-Hill.

Karaska, G. J., and D. F. Bramhall (1969). *Locational Analysis for Manufacturing: A Selection of Readings.* Cambridge, Mass.: MIT Press.

Katz, D., and R. L. Kahn (1966). *The Social Psychology of Organizations.* New York: Wiley.

Katzman, M. T. (1974). The von Thünen Paradigm, the Industrial-Urban Hypothesis, and the Spatial Structure of Agriculture," *American Journal of Agricultural Economics* 56: 683–697.

Kennelly, R. A. (1954). "The Location of the Mexican Steel Industry." In R. H. T. Smith, E. J. Taaffe, and L. J. King (eds.), *Readings in Economic Geography*. Skokie, Ill.: Rand McNally, pp. 126–157.

King, L. J. (1962). "A Quantitative Expression of the Pattern of Urban Settlements in Selected Areas of the U.S.," *Tijdschrift voor economische en sociale geografie* 53: 1–7.

Kirk, W. (1963). "Problems of Geography," *Geography* 48: 357–371.

Knos, D. S. (1962). *The Distribution of Land Values in Topeka, Kansas*. Lawrence: University of Kansas Press.

Kondratiev, N. (1925). "The Major Economic Cycles," *Voprosy Konjunctury* 1: 28–79. English translation in *Review of Economic Statistics* 18, November 1935, pp. 105–115.

Krumme, G. (1969). "Toward a Geography of Enterprise," *Economic Geography* 43: 30–40.

Krumme, G. (1981). "Making It Abroad: The Evolution of Volkswagen's North American Production Plans." In F. E. I. Hamilton and G. J. R. Linge (eds.), *Spatial Analysis, Industry and the Industrial Environment: Vol. 2. International Industrial Systems*. Chichester, England: Wiley.

Kuhn, A. (1966). *The Study of Society*. London: Associated Book Publishers.

Lash, S., and J. Urry (1987). *The End of Organized Capitalism*. Cambridge: Polity Press.

Lasuen, J. R. (1969). "On Growth Poles," *Urban Studies* 6: 137–161.

Leborgne, D., and A. Lipietz (1988). "New Technologies, New Modes of Regulation: Some Spatial Implications," *Environment and Planning D* 6, no. 3: 263–280.

Lindberg, O. (1953). "An Economic Geographic Study of the Swedish Paper Industry," *Geografiska Annaler* 35: 28–40.

Lionberger, H. F. (1960). *Adoption of New Ideas and Practices*. Ames: Iowa State University Press.

Lipietz, A. (1987). *Mirages and Miracles*. London: Verso.

Lipsey, R. G., P. O. Steiner, and D. D. Purvis (1987). *Economics* (8th ed.). New York: Harper & Row.

Lloyd, P. E. (1989). "Fragmenting Markets and the Dynamic Restructuring of Production," *Environment and Planning*, A, vol. 21, pp. 429–444.

Lloyd, P. E., and J. Shutt (1983). "Recession and Restructuring in the Northwest Region: The Implications of Recent Events." In D. Massey and R. Meegan (eds.), *Politics and Method*. London: Methuen, chap. 2, pp. 13–61.

Locklin, D. P. (1960). *The Economics of Transportation*. Homewood, Ill.: Irwin.

Lorsch, J. W., and S. A. Allen (1970). *Managing Diversity and Interdependence: An Organizational Study of Multidivisional Firms*. Cambridge, Mass.: Harvard University Press.

Lösch, A. (1954). *The Economics of Location*. New Haven, Conn.: Yale University Press. (Originally published 1939.)

McCarty, H. H., and J. B. Lindberg (1966). *A Preface to Economic Geography*. Englewood Cliffs, N.J.: Prentice-Hall.

McDermott, P. J., and M. J. Taylor (1982). *Industrial Organization and Location.* Cambridge: Cambridge University Press.

MacInnes, A. (1987), "The Question of Flexibility," *Research Papers,* Dept. of Social and Economic Research, University of Glasgow.

Mackay, J. R. (1958). "The Interactance Hypothesis and Boundaries in Canada," *Canadian Geographer* 11: 1–8.

McNee, R. B. (1960). "Toward a More Humanistic Economic Geography: The Geography of Enterprise," *Tijdschrift voor economische en sociale geografie* 51: 201–205.

McNee, R. B. (1974). "A Systems Approach to Understanding the Geographic Behavior of Organizations, Especially Large Corporations." In F. E. I. Hamilton (ed.), *Spatial Perspectives on Industrial Organization and Decision Making.* London: Wiley, pp. 47–75.

Madden, C. H. (1956). "Some Indicators of Stability in the Growth of Cities in the United States," *Economic Development and Cultural Change* 4: 236–252.

Malecki, E. J. (1979). "Locational Trends in R&D by Large U.S. Corporations, 1965–1977," *Economic Geography* 55: 309–323.

Malecki, E. J. (1980). "Dimensions of R&D Location in the United States," *Research Policy* 9, 2–22.

Mandel, E. (1980). *Long Waves of Capitalist Development: The Marxist-Interpretation.* Cambridge: Cambridge University Press.

Mansfield, E. (1968). *The Economics of Technical Change.* New York: Norton.

March, J. G., and H. A. Simon (1958). *Organizations.* New York: Wiley.

Marshall, J. U. (1969). *The Location of Service Towns.* University of Toronto, Department of Geography Research Paper No. 3.

Marshall, M. (1987). *Long Waves of Regional Development.* London: Macmillan.

Martin, R., and R. Rowthorne (eds.) (1986). *The Geography of Deindustrialization.* London: Macmillan.

Marx, K. (1859) *A Contribution to the Critique of Political Economy.* Berlin: Franz Duncker.

Massey, D. (1984). *Spatial Divisions of Labour: Social Structures and the Geography of Production.* London: Macmillan.

Massey, D. (1985). "New Directions in Space." In D. Gregory and J. Urry (eds.), *Social Relations and Spatial Structures.* London: Macmillan.

Massey, D. (1988). "A New Class of Geography." *Marxism Today,* May, pp. 12–17.

Massey, D., and J. Allen (eds.) (1984). *Geography Matters: A Reader.* Cambridge: Cambridge University Press.

Massey, D., and J. Allen (eds.) (1988). *Restructuring Britain: The Economy in Question.* London: Sage.

Massey, D., and R. Meegan (1978). "Industrial Restructuring Versus the Cities," *Urban Studies* 15: 273–288.

Massey, D., and R. Meegan (1982). *The Anatomy of Job Loss: The How, Why and Where of Employment Decline.* London: Methuen.

Massey, D., and R. Meegan (1985). *Politics and Method: Contrasting Studies in Economic Geography.* London: Methuen.

Mattingly, P. F. (1972). "Intensity of Agricultural Land Use Near Cities: A Case Study," *Professional Geographer* 24: 1, 7–10.

Meinig, D. W. (1962). "A Comparative Historical Geography of Two Railnets: Columbia Basin and South Australia," *Annals of the Association of American Geographers* 52: 394–413.

Meinig, D. W. (ed.) (1979). *The Interpretation of Ordinary Landscapes.* New York: Oxford University Press.

Mensch, G. (1979). *Technological Stalemate.* New York: Ballinger.

Miernyk, W. H. (1965). *The Elements of Input-Output Analysis.* New York: Random House.

Morgan, K. (1986). "Reindustrialization in Peripheral Britain: State Policy, the Space Economy and Industrial Innovation." In R. Martin and B. Rowthorne (eds.), *The Geography of Deindustrialization,* chap. 10, pp. 322–357. London: Macmillan.

Morgan, K., and A. Sayer (1985). "A Modern Industry in a Declining Region." In D. Massey and R. Meegan, *Politics and Method.* London: Methuen, chap. 6, pp. 144–168.

Morgan, K., and A. Sayer (1987). *Micro-circuits of Capital: The Electronics Industry and Regional Development.* Oxford: Blackwell.

Moroney, J. R., and J. M. Walker (1966). "A Regional Test of the Heckscher-Ohlin Hypothesis," *Journal of Political Economy* 74: 573–586.

Moses, L. N. (1958). "Location and the Theory of Production," *Quarterly Journal of Economics* 72: 259–272.

Mulligan, G. F. (1981). "Lösch's Single Good Equilibrium," *Annals of the Association of American Geographers* 71: 84–94.

Mulligan, G. F., and R. W. Reeves (1983). "The Theory of the Firm: Some Spatial Implications," *Urban Geography* 4: 156–172.

Murdie, R. A. (1965). "Cultural Differences in Consumer Travel," *Economic Geography* 41: 211–233.

Myrdal, G. (1957). *Rich Lands and Poor.* New York: Harper & Row.

Norcliffe, G. (1975). "A Theory of Manufacturing Places." In L. Collins and D. F. Walker (eds.), *Locational Dynamics of Manufacturing Activity.* London: Wiley, pp. 19–57.

Offe, C. (1985). *Disorganized Capitalism.* Cambridge: Polity Press.

Olsson, G. (1965). *Distance and Human Interaction: A Review and Bibliography.* Philadelphia: Regional Science Research Institute.

Park, R. E., and E. W. Burgess (1925). *The City.* Chicago: University of Chicago Press.

Parr, J. B. (1966). "Outmigration and the Depressed Area Problem," *Land Economics* 42: 149–159.

Parr, J. B. (1978). "Models of the Central Place System: A More General Approach," *Urban Studies* 15: 35–49.

Peet, J. R. (1969). "The Spatial Expansion of Commercial Agriculture in the Nineteenth Century," *Economic Geography* 45: 283–301.

Peet, R. (1987). *International Capitalism and Industrial Restructuring.* Boston: Allen & Unwin.

Perrons, D. (1981). "The Role of Ireland in the New International Division of Labour." *Regional Studies* 15, vol. 2, pp. 81–100.

Perroux, F. (1950). "Economic Space: Theory and Application," *Quarterly Journal of Economics* 64: 89–104.

Perroux, F. (1955). "Note sur la notion de 'pôle de croissance,'" *Économie Appliqué* 8. Translated version in D. L. McKee, R. D. Dean, and W. H. Leahy (eds.) (1970). *Regional Economics.* New York: Free Press, pp. 93–103.

Piore, M., and C. Sabel (1984). *The Second Industrial Divide.* New York: Basic Books.

Porter, M. E. (1985). *Competitive Advantage.* New York: Free Press.

Pratten, C. F. (1971). *Economies of Scale in Manufacturing Industry.* Cambridge: Cambridge University Press.

Pred, A. R. (1965). "Industrialization, Initial Advantage and American Metropolitan Growth," *Geographical Review* 55: 158–185.

Pred, A. R. (1966). *The Spatial Dynamics of U.S. Urban-Industrial Growth, 1800–1914.* Cambridge, Mass.: MIT Press.

Pred, A. R. (1971). "Large-City Interdependence and the Preelectronic Diffusion of Innovations in the U.S.," *Geographical Analysis* 3: 165–181.

Pred, A. R. (1973). "The Growth and Development of Systems of Cities in Advanced Economies." In A. R. Pred and G. E. Tornquist (eds.), "Systems of Cities and Information Flows," *Lund Studies in Geography, Series B* 38: 9–82.

Pred, A. R. (1974a). "Industry, Information and City-System Interdependencies." In F. E. I. Hamilton (ed.), *Spatial Perspectives on Industrial Organization and Decision Making.* London: Wiley, pp. 105–139.

Pred, A. R. (1974b). *Major Job-Providing Organizations and Systems of Cities.* Association of American Geographers, Commission on College Geography Resource Paper No. 27.

Preston, R. E. (1985). "Christaller's Neglected Contribution to the Study of the Evolution of Central Places," *Progress in Human Geography* 9: 177–193.

Rainnie, A. (1984). "Combined and Uneven Development in the Clothing Industry," *Capital and Class* 22: 141–156.

Rawstron, E. M. (1958). "Three Principles of Industrial Location," *Transactions of the Institute of British Geographers* 25: 135–142.

Ray, D. M. (1967). "Cultural Differences in Consumer Travel Behavior in Eastern Ontario," *Canadian Geographer* 11: 143–156.

Rees, J. (1974). "Decision Making, the Growth of the Firm and the Business Environment." In F. E. I. Hamilton (ed.), *Spatial Perspectives on Industrial Organization and Decision Making.* London: Wiley, pp. 189–211.

Roemer, J. (1987). *Free to Lose.* London: Radius.

Rubenstein, J. (1989). "The Impact of Japanese Investment in the United States." In C. M. Law (ed.), *Motor of Change.* London: Routledge, chap. 1.

Rumelt, R. P. (1974). *Strategy, Structure and Economic Performance.* Cambridge, Mass.: Harvard University Press.

Sabel, C. (1982). *Work and Politics: The Division of Labour in Industry*. Cambridge: Cambridge University Press.

Sampson, R. J., and M. T. Farris (1980). *Domestic Transportation: Practice, Theory and Policy* (4th ed.). Boston: Houghton Mifflin.

Sayer, A. (1986). "New Developments in Manufacturing: The Just-in-Time system," *Capital and Class* 30: 43–72.

Schlebecker, J. T. (1960). "The World Metropolis and the History of American Agriculture," *Journal of Economic History* 20: 187–208.

Schumpeter, J. A. (1939). *Business Cycles: A Theoretical Historical and Statistical Analysis of the Capitalist Process*. New York: McGraw-Hill.

Scott, A. J. (1982). "Locational Patterns and Dynamics of Industrial Activity in the Modern Metropolis," *Urban Studies* 19: 111–142.

Scott, A. J. (1983a). "Industrial Organization and the Logic of Intrametropolitan Location: I. Theoretical Considerations," *Economic Geography* 59: 233–249.

Scott, A. J. (1983b). "Industrial Organization and the Logic of Intrametropolitan Location: II. A Case Study of the Printed Circuit Industry in the Greater Los Angeles Region," *Economic Geography* 59: 343–367.

Scott, A. J. (1984). "Industrial Organization and the Logic of Intrametropolitan Location: III. A Case Study of the Women's Dress Industry in the Greater Los Angeles Region," *Economic Geography* 60: 3–27.

Scott, A. J. (1988). "Flexible Accumulation and Regional Development: The Rise of New Industrial Spaces in North America and Western Europe," *International Journal of Urban and Regional Research* 12: 171–186.

Scott, A. J. (1989). *New Industrial Spaces*. London: Pion Ltd.

Scott, A. J., and D. Angel (1987). "The US Semi-conductor Industry: A Locational Analysis," *Environment and Planning*, A, Vol. 19, pp. 875–912.

Scott, A. J., and P. Cooke (1988). "The New Geography and Sociology of Production," *Environment and Planning D*, 6: 241–244.

Scott, A. J., and M. Storper (eds.) (1986). *Production, Work Territory: A Geographical Anatomy of Industrial Capitalism*. Winchester, Mass.: Allen & Unwin.

Semple, R. K., and A. G. Phipps (1982). "The Spatial Evolution of Corporate Headquarters Within an Urban System," *Urban Geography* 3: 258–279.

Seyfried, W. R. (1963). "The Centrality of Urban Land Values," *Land Economics* 39: 275–285.

Sheard, P. (1983). "Auto Production Systems in Japan: Organizational and Locational Features," *Australian Geographical Studies* 21: 49–68.

Sheppard, E. (1982). "City Size Distributions and Spatial Economic Change," *International Regional Science Review* 7: 127–151.

Shutt, J., and R. Whittington (1986). "Fragmentation Strategies and the Rise of Small Units," *Regional Studies* 21, 1: 13–23.

Siebert, H. (1969). *Regional Economic Growth: Theory and Policy*. Scranton, Pa.: International Textbook.

Simon, H. A. (1960). *The New Science of Management Decision*. New York: Harper & Row.

Sinclair, R. (1967). "Von Thünen and Urban Sprawl," *Annals of the Association of American Geographers* 57: 72–87.

Smith, D. M. (1966). "A Theoretical Framework for Geographical Studies of Industrial Location," *Economic Geography* 42: 95–113.

Smith, D. M. (1981). *Industrial Location* (2d ed.). New York: Wiley.

Smith, D. M. (1987). "Neoclassical Location Theory." In W. F. Lever (ed.), *Industrial Change in the United Kingdom*. Harlow, England: Longman.

Smith, W. (1955). "The Location of Industry," *Transactions of the Institute of British Geographers* 21: 1–18.

Soja, E. W. (1989). *Post-modern Geographies: A Reassertion of Space in Critical Social Theory*. New York: Verso.

Stephens, J. D., and B. P. Holly (1980). "The Changing Patterns of Industrial Control in Metropolitan America." In S. D. Brunn and J. O. Wheeler (eds.), *The American Metropolitan System*. New York: Halstead.

Stevens, B. H. (1961). "An Application of Game Theory to a Problem in Location Strategy," *Papers and Proceedings of the Regional Science Association* 7: 143–157.

Stewart, C. T. (1958). "The Size and Spacing of Cities," *Geographical Review* 48: 222–245.

Stewart, J. Q. (1947). "Empirical Mathematical Rules Governing the Distribution and Equilibrium of Population," *Geographical Review* 37: 461–485.

Storper, M. (1985). "Oligopoly and the Product Cycle: Essentialism in Economic Geography," *Economic Geography* 61: 260–282.

Storper, M., and R. Walker (1983). "The Theory of Labor and the Theory of Location," *International Journal of Urban and Regional Research* 7: 1–41.

Taaffe, R. N. (1962). "Transportation and Regional Specialization," *Annals of the Association of American Geographers* 52: 80–98.

Taaffe, E. J., and H. Gauthier (1973). *Geography of Transportation*. Englewood Cliffs, N.J.: Prentice-Hall.

Taaffe, E. J., R. L. Morrill, and P. R. Gould (1963). "Transport Expansion in Underdeveloped Countries," *Geographical Review* 53: 503–529.

Tarrant, J. (1973). "Comments on the Lösch Central Place System," *Geographical Analysis* 5: 113–121.

Taylor, M. J. (1980). "Space and Time in Industrial Linkage," *Area* 12: 150–152.

Taylor, M. J. (1984). "Industrial Geography and the Business Organization." In M. J. Taylor (ed.), *The Geography of Australian Corporate Power*. Sydney: Croom Helm.

Taylor, M. J. (1986). "The Product Cycle Model: A Critique," *Environment and Planning* A18: 751–761.

Taylor, M. J., and N. J. Thrift (1982). "Industrial Linkage and the Segmented Economy: Some Theoretical Proposals," *Environment and Planning* A14: 1601–1613.

Taylor, M. J., and N. J. Thrift (1983). "Business Organization, Segmentation and Location," *Regional Studies* 17: 445–465.

Thompson, J. D. (1967). *Organizations in Action*. New York: McGraw-Hill.

Thompson, W. R. (1965). *A Preface to Urban Economics*. Baltimore: Johns Hopkins Press.

Thompson, W. R. (1968). "Internal and External Factors in the Development of Urban Economies." In H. S. Perloff and L. Wingo, Jr. (eds.), *Issues in Urban Economics*. Baltimore: Johns Hopkins Press, pp. 43–62.

Thrift, N. J. (1985). "Taking the Rest of the World Seriously: The State of British Urban and Regional Research at a Time of Crisis," *Environment and Planning* A17: 7–24.

Thünen J. H. von (1842). *Der Isolierte Staat*. In P. Hall (ed.), *Von Thünen's Isolated State*. London: Pergamon.

Tiebout, C. M. (1962). *The Community Economic Base Study*. New York: Committee for Economic Development.

Tornquist, G. (1968). "Flows of Information and the Location of Economic Activities," *Lund Studies in Geography, Series B* 30.

Ullman, E. L. (1953). "Human Geography and Area Research," *Annals of the Association of American Geographers* 43: 54–66.

Ullman, E. L. (1954). "Amenities as a Factor in Regional Growth," *Geographical Review* 44: 119–132.

Ullman, E. L. (1956). "The Role of Transportation and the Bases for Interaction." In W. L. Thomas (ed.), *Man's Role in Changing the Face of the Earth*. Chicago: University of Chicago Press, pp. 862–880.

Ullman, E. L. (1958). "Regional Development and the Geography of Concentration," *Papers and Proceedings of the Regional Science Association* 4: 179–198.

Vance, J. E., Jr. (1970). *The Merchant's World: The Geography of Wholesaling*. Englewood Cliffs, N.J.: Prentice-Hall.

Vernon, R. (1979). "The Product Cycle Hypothesis in a New International Environment," *Oxford Bulletin of Economics and Statistics* 41: 255–268.

Vining, R. (1955). "A Description of Certain Spatial Aspects of an Economic System," *Economic Development and Cultural Change* 3: 147–195.

Walker, R., and M. Storper (1981). "Capital and Industrial Location," *Progress in Human Geography* 5: 473–509.

Wallace, I. (1985). "Towards a Geography of Agribusiness," *Progress in Human Geography* 9: 491–514.

Warntz, W. (1965). *Macrogeography and Income Fronts*. Philadelphia: Regional Science Research Institute.

Watts, H. D. (1981). *The Branch Plant Economy*. London: Longman.

Webber, M. J. (1972). *The Impact of Uncertainty on Location*. Cambridge, Mass.: MIT. Press.

Webber, M. J. (1982). "Location of Manufacturing Activity in Cities," *Urban Geography* 3: 203–223.

Weber, A. (1909). *Theory of the Location of Industries*. Chicago: University of Chicago Press.

Werner, C. (1968). "The Law of Refraction in Transportation Geography: Its Multivariate Extensions," *Canadian Geographer* 12: 28–40.

Whitman, E. S., and W. J. Schmidt (1966). *Plant Relocation.* New York: American Management Association.

Williamson, O. E. (1975). *Markets and Hierarchies.* New York: Free Press.

Williamson, O. E. (1986). *The Economic Institutions of Capitalism: Firms, Markets, Relational Contracting.* New York: McGraw-Hill.

Winsberg, M. D. (1981). "Intensity of Agricultural Production and Distance from the City: The Case of the Southeastern United States," *Southeastern Geographer* 21: 54–63.

Wise, M. J. (1949). "On the Evolution of the Jewelery and Gun Quarters in Birmingham," *Transactions of the Institute of British Geographers* 15: 57–72.

Wolfe, R. I. (1962). "Transportation and Politics: The Example of Canada," *Annals of the Association of American Geographers* 52: 176–190.

Wood, S. (ed.) (1982). *The Degradation of Work: Skill, De-Skilling and the Labour Process.* London: Hutchinson.

Yeates, M. H. (1965). "Some Factors Affecting the Spatial Distribution of Chicago Land Values," *Economic Geography* 41: 55–70.

Zimmermann, E. W. (1951). *World Resources and Industry.* New York: Harper & Row.

Zipf, G. K. (1946). "The $P_1 P_2 / D$ Hypothesis and Intercity Movement of Persons," *American Sociological Review* 11: 677–686.

Zipf, G. K. (1949). *Human Behavior and the Principle of Least Effort.* Reading, Mass.: Addison-Wesley.

Index

centers of control in, 172, 316, 386–387
hierarchy in, 385–387, 389
peripheral regions in, 387–388
sector and, 385–386
Spatial impact, 7
Spatial margins to profitability, 99–100, 103
Spatial organization, 7, 67–70
Spatial pricing policies, 203–204
Spatial substitution, 176
Specialization, 149, 191–193, 212–215, 288–289, 302, 319
agglomeration and, 212–214
capital and, 193–196
in capitalism, 353–355
flexible. *See* Flexible specialization
Spread effects, 241–243, 314
Standard Industrial Classification (SIC), 259
Steady state of the central place system, 38
Steel industry, 92, 131, 138, 142, 229–234, 241
Stepped freight rates, 125, 128
Strategic decisions, 290–292
Strategic withdrawal, 297
Strategy
competitive, 292–316
defined, 291
game theory and, 209–211, 257–258
organizational structure and, 301–305
Structural approach, 11. *See also* Marxism
Subcontracting relationship, 333–338, 394
Substitution analysis, 93–97, 134, 135, 176–177, 202–207, 216–217
Suburbanization, 242
Supply
alternative sources of, 73
defined, 20
effect of transportation costs on, 132–134
in a market economy, 5–6
Supply curve, 5–6
Support activities, 271
Surface of population potential, 182
Surplus value, 348–352, 357, 360
Survival needs, 2
SWOT analysis (analysis of strengths, weaknesses, opportunities, and threats), 293

Synergy, 296
System, as term, 7

Tactics, 291
Tapering transportation costs, 121–125, 132–139
Task environment, 273
Tax holidays, 160
Technical knowledge, 169–174. *See also* Technology
availability of, over space, 173–174
from induced invention, 169–174
as an input factor, 169–170
mobility of, 172–173
variations in spatial distribution of, 170–172
Technical substitution, 152, 163, 177
Technology, 151. *See also* Technical knowledge
agricultural, 105–106
business segmentation and, 266–267
cumulative economic development and, 238–239, 243
defined, 169
economies of scale and, 199–201
environmental turbulence and, 284–288
in flexible specialization, 394–397
fuel, 102–104, 144–145, 195–196
impact of changes in, 102–104
information, 287–288
labor process debate and, 380
in production function, 152–156
as replacement for labor, 152, 163, 177
resource localization and, 102–104
specialization and, 193–196
transportation revolution and, 144–147, 177
waves of investment and, 369–375
Technostructure, 269n
Telecommunications, 287
Telephone traffic, 120–121
Temperature, agricultural production and, 105–106, 108–109
Terminal costs, 121–124, 128–130, 137
Textile industry, 167
Thatcherism, 393
Threshold, 237–238, 250
Threshold value
central place theory and, 25–38
multiple producers, multiple goods, 24–25
multiple producers, one good, 19–24
one producer, one good, 18–19